Lecture Notes in Computer Science 13230

More information about this series at https://link.springer.com/bookseries/558

Antonio Cerone · Marco Autili ·
Alessio Bucaioni · Cláudio Gomes ·
Pierluigi Graziani · Maurizio Palmieri ·
Marco Temperini · Gentiane Venture (Eds.)

Software Engineering and Formal Methods

SEFM 2021 Collocated Workshops

CIFMA, CoSim-CPS, OpenCERT, ASYDE
Virtual Event, December 6–10, 2021
Revised Selected Papers

Springer

Editors
Antonio Cerone (ID)
Nazarbayev University
Nur-Sultan, Kazakhstan

Marco Autili (ID)
University of L'Aquila
L'Aquila, Italy

Alessio Bucaioni (ID)
Mälardalen University
Västerås, Sweden

Cláudio Gomes (ID)
Aarhus University
Aarhus, Denmark

Pierluigi Graziani (ID)
University of Urbino
Urbino, Italy

Maurizio Palmieri (ID)
University of Pisa
Pisa, Italy

Marco Temperini (ID)
Sapienza University of Rome
Rome, Italy

Gentiane Venture (ID)
Tokyo University of Agriculture
and Technology
Tokyo, Japan

ISSN 0302-9743   ISSN 1611-3349 (electronic)
Lecture Notes in Computer Science
ISBN 978-3-031-12428-0   ISBN 978-3-031-12429-7 (eBook)
https://doi.org/10.1007/978-3-031-12429-7

Preface

The 19th International Conference on Software Engineering and Formal Methods (SEFM 2021) was held online during December 6–10, 2021. The main conference was held during December 8–10 and the collocated events were held during December 6–7. The decision to hold the conference and its collocated events in virtual mode was due to the difficult situation, in terms of health and mobility, caused by the COVID-19 pandemic. The organization of the virtual events was a joint effort of Carnegie Mellon University (USA), Nazarbayev University (Kazakhstan), and the University of York (UK). In particular, Nazarbayev University led and sponsored the organization of the collocated events and coordinated the editing of this proceedings volume.

This volume collects the proceedings of four collocated workshops:

- CIFMA 2021 - the 3rd International Workshop on Cognition: Interdisciplinary Foundations, Models and Applications;
- CoSim-CPS 2021 - the 5th Workshop on Formal Co-Simulation of Cyber-Physical Systems;
- OpenCERT 2021 - the 10th International Workshop on Open Community approaches to Education, Research and Technology; and
- ASYDE 2021 - the 3rd International Workshop on Automated and verifiable Software sYstem DEvelopment.

The workshop organizers ensured that all papers received at least three reviews. Another collocated event, the 10th International Symposium "From Data to Model and Back" (DataMod 2021), had the proceedings published as a separate LNCS volume. The variety of focused themes and application domains addressed by these workshops greatly enriched the SEFM program and demonstrated that software engineering and formal methods can be used together in a large variety of ways and attract the interest and usage of important, often interdisciplinary, scientific communities.

We would like to thank the workshop organizers, program chairs, keynote speakers, and authors for their effort in contributing to a rich and interesting program. We also thank the SEFM Program Committee chairs, Radu Calinescu and Corina Pasareanu, for taking care of the collocated event registration process, and Ioannis Stefanakos for managing the SEFM 2021 website.

March 2022   Antonio Cerone

Sponsor

NAZARBAYEV
UNIVERSITY

Contents

CoSim-CPS 2021 - 5th Workshop on Formal Co-Simulation of Cyber-Physical Systems

OpenCERT 2021 - 10th International Workshop on Open Community approaches to Education, Research and Technology

ASYDE 2021 - 3rd International Workshop on Automated and verifiable Software sYstem DEvelopment

CIFMA 2021 - 3rd International Workshop on Cognition: Interdisciplinary Foundations, Models and Applications

CIFMA 2021 Organizers' Message

The workshop on Cognition: Interdisciplinary Foundations, Models and Applications (CIFMA) aims first and foremost to bring together practitioners and researchers from academia, industry, and research institutions who are interested in the foundations and applications of cognition from the perspective of their areas of expertise and aim at a synergistic effort in integrating approaches from different areas. It also aims to nurture cooperation among researchers from different areas and establish concrete collaborations, and to present formal methods to cognitive scientists as a general modeling and analysis approach, whose effectiveness goes well beyond its application to computer science and software engineering.

The third edition of this workshop (CIFMA 2021), held on the December 6, 2021, included presentations by 11 speakers. The edition received 13 submissions, out of which five papers were accepted for both presentation and publication, five papers were accepted only for presentation with a conditional acceptance for publication, two papers were rejected, and one was withdrawn by the authors. All papers intially underwent a single-blind peer review process in which three reviewers were assigned to each paper. Conditionally accepted papers received a further round of review after presentation at the workshop: only four of them were accepted for publication and one was withdrawn. The workshop program also featured a keynote talk titled "The brain as a computer" by Martin Davis (New York University, USA).

We would like to thank the Program Committee members and the external reviewers for their enthusiasm and effort in actively participating in the review process. We are also grateful to the Workshop Chair, Antonio Cerone, for his management commitment. Finally, we would like to thank all workshop attendees for their active participation in discussions and for the feedback they provided to the authors.

March 2022

Pierluigi Graziani
Gentiane Venture

Organization

Program Committee Chairs

Pierluigi Graziani   University of Urbino, Italy
Gentiane Venture   Tokyo University of Agriculture
     and Technology, Japan

Steering Committee

Antonio Cerone   Nazarbayev University, Kazakhstan
Pierluigi Graziani   University of Urbino, Italy
Gentiane Venture   Tokyo University of Agriculture
     and Technology, Japan

Program Committee

Samuel Alexander   U.S. Securities and Exchange Commission,
     New York, USA
Oana Andrei   University of Glasgow, UK
John A. Barnden   University of Birmingham, UK
Francesco Bianchini   University of Bologna, Italy
Stefano Bonzio   University of Cagliari, Italy
José Creissac Campos   University of Minho, Portugal
Antonio Cerone   Nazarbayev University, Kazakhstan
Peter Chapman   Edinburgh Napier University, UK
Gianluca Curzi   University of Birmingham, UK
Luisa Damiano   IULM University, Italy
Anke Dittmar   University of Rostock, Germany
Alan Dix   Swansea University, UK
Pierluigi Graziani   University of Urbino, Italy
Yannis Haralambous   IMT Atlantique, France
Bipin Indurkhya   Jagiellonian University, Poland
Reinhard Kahle   NOVA University Lisbon, Portugal
Karl Lermer   ZHAW, Switzerland
Kathy L. Malone   Nazarbayev University, Kazakhstan
Paolo Masci   US National Institute of Aerospace (NIA), USA
Mieke Massink   Institute of Information Science
     and Technologies (CNR-ISTI), Italy

Paolo Milazzo   University of Pisa, Italy
Marco Nørskov   Aarhus University, Denmark
Eugenio Omodeo   University of Trieste, Italy
Antti Oulasvirta   Aalto University, Finland
Graham Pluck   Nazarbayev University, Kazakhstan
Giuseppe Primiero   University of Milan, Italy
Ka I Pun   Western Norway University of Applied
                             Sciences, Norway
Pedro Quaresma   University of Coimbra, Portugal
Giuseppe Sergioli   University of Cagliari, Italy
Sandro Sozzo   University of Leicester, UK
Mirko Tagliaferri   University of Urbino, Italy
Gentiane Venture   Tokyo University of Agriculture
                             and Technology, Japan

Additional Reviewers

Alessandro Aldini   University of Urbino, Italy
Stefano Calboli   University of Minho, Portugal
Stefano Nicoletti   University of Twente, The Netherlands

What Does It Mean to Inhibit an Action?

A Critical Discussion of Benjamin Libet's Veto in a Recent Study

Robert Reimer[✉]

Universität Leipzig, Leipzig, Germany
rreimer@posteo.de

Abstract. In the 1980s, physiologist Benjamin Libet conducted a series of experiments to test whether the will is free. Whilst he originally assumed that the will functions like an immaterial initiator of cerebral processes culminating in actions, he later began to think that it rather functions like an immaterial veto inhibiting unwanted actions by preventing unconsciously initiated cerebral processes from unfolding. Libet's veto was widely criticized for its Cartesian dualist and interactionist implications. However, in 2016, Schultze-Kraft et al. adopted Libet's idea of an action-inhibiting veto and conducted a new experiment. Its goal was to test until which moment agents can inhibit an action that they already intended to do. Despite insisting on the material nature of the veto, the researchers also described the activitiy of the veto in interactionist terms, namely as an act of the test subjects performed against their own cerebral processes. The purpose of this paper is to explain in which sense the veto in Schultze-Kraft's study is interactionist, too, and to provide a non-interactionist reinterpretation of the test subjects' action inhibitions.

Keywords: Benjamin Libet · René Descartes · Interactionism · Action inhibition

1 Introduction

In the 1980s, physiologist Benjamin Libet conducted a series of experiments to proof that the will is free. He asked his test subjects to flex their wrists at a moment of their own choice. Before acting, the test subjects should determine the point in time when they felt the urge to perform this movement. After a series of trials, Libet found out that the awareness of the urge was preceded by an unconscious, slow electrical change recordable on the scalp at the vertex by 300 ms or more [8]. This 'readiness potential' (RP) was already discovered by Kornhuber and Deecke [5] and was associated with the performed hand movement. Libet concluded that the action of the test subjects did not originate in the test subjects' urge to act but rather in the RP.

Disappointed by these results, Libet considered another option of how the freedom of the will could be saved. In some of his early experiments, test subjects reported that they were able to suppress the urge to act before any actual

© The Author(s), under exclusive license to Springer Nature Switzerland AG 2022
A. Cerone et al. (Eds.): SEFM 2021 Workshops, LNCS 13230, pp. 5–14, 2022.
https://doi.org/10.1007/978-3-031-12429-7_1

movement occurred [9]. Based on this observation, Libet argued that the will rather functions like a veto. Instead of being the initiator of the action, it is able to interfere with or control the unconscious processes triggered by the original RP by aborting or selecting some of them consciously before they culminate in bodily movement. This assumption was supported by the results of one of his later experiments in which the test subjects were explicitly instructed to veto the development of their action [7]. Assuming that these mental acts of veto are not themselves caused by any unconscious cerebral activities, Libet concluded that the will can indeed be free. He notes:

> Although the volitional process may be initiated by unconscious cerebral activities, conscious control of the actual motor performance of voluntary acts definitely remains possible. The findings should therefore be taken not as being antagonistic to free will but rather as affecting the view of how free will might operate. Processes associated with individual responsibility and free will would 'operate' not to initiate a voluntary act but to select and control volitional outcomes. [7]

Many criticized Libet's notion of an action inhibiting veto because it is based on an implausible *dualist interactionist paradigm*. Dualist interactionism is a theory of the mind according to which mind and brain are two distinct substances whose internal processes, in general, unfold independently from one another. However, the mind can still influence the processes of the brain by causing an effect in one of its parts.

Despite this criticism, Libet's idea of the veto was adapted in one way or another by scientists in some recent studies. Schultze-Kraft et al., for instance, designed an experimental setup in which test subjects should perform a simple foot movement while facing a green light. If the light turned red, however, they were supposed to inhibit their action by 'exerting a veto' against the already prepared and upcoming movement [12]. With their study, the researchers tried to determine the point in time after which people can no longer inhibit their action.

The purpose of this paper is to show that, despite explicitly rejecting Libet's dualist interpretation of the veto, Schultze-Kraft's study still falls prey to the interactionism that is underlying Libet's theory of veto. In section two, I will specify the general form of Libet's veto and discuss in which sense it is an expression of a dualist interactionist view. I will also say a few words on the historical background of Libet's idea and show that it can already be found in René Descartes' account on the body-soul interaction. In section three, I will pin down the interactionist character of the veto in Schultze-Kraft's interpretation and criticize it. In section four, I will then suggest a more plausible and non-interactionist interpretation of what actually happened when the test subjects inhibited their action in Schultze-Kraft's study.

2   Descartes, Libet, and the Veto

Any kind of interactionism presupposes the distinction between two distinct and mutually independent systems *in* a person. These are commonly mind and brain, or soul and body. But, as we will see later, they can also be the person herself and her brain. In each of these systems, many processes take place independently from the processes of the other system. However, one system can interfere with the processes of the other.[1]

Philosophers often call Descartes the 'founding father' of interactionism. In 'The Passions of the Soul', Descartes develops an account on action and perception that is consistent both with his own biological research and his religious faith. Against this background, Descartes posits the existence of two different substances constituting a person (an agent) - soul and body. The soul (the first system) is an immaterial and immortal substance that thinks and calculates. It is connected with the body through the pineal gland in the brain. The body (the second system) is a material and mortal substance, in turn. Various material processes such as digestion and limb movement take place in the body. But whilst the soul is absolutely free from the influences of the body, agents can feel and move their bodies *through* their souls by performing an act of will. The act of will (mediated by the soul and operating through the pineal gland) causes the nerves in the brain to 'vibrate' and then the muscles in the rest of the body to contract. This account on action is *interactionist* because it suggests that the way how the agent acts is by interfering with the processes of her body (system one) mediated through her soul (system two).

The standard function of the will, according to Descartes, is the initiation of cerebral processes. However, the agent can also abort an unfolding cerebral process by willing so. Descartes notes that the passions (fear, anger, lust, etc.) are sometimes in conflict with the agent's endeavors. The passions (as some of the body's processes) sometimes 'lead' the body to perform unwanted actions. The agent, however, can detect the unwanted passion, 'not give consent' to it, and

---

[1] I am thankful to two anonymous reviewers who pointed out the ambiguity of the terms 'interactionism' and 'interactivism'. In the modern cognitive sciences and in the philosophy of embodiment, these terms are used differently. Bickhard [1], for instance, understands interactivism as a theoretical system, according to which mental phenomena, such as perceptions, emerge from the interactions of the agent with her environment. Enactivists, such as Varela, Thompson, and Rosch [14], argue similarly and claim that cognition is essentially embodied and consists in the enactment of a world through structural coupling, that is, through the constant interaction between the agent's body and her environment. In none of these works, interaction is understood as mind-brain, soul-body, or person-brain interaction but as the interaction between the person's body and her environment. In this paper, however, I will use the terms 'interactionism' exclusively for the theory that agents act in virtue of their mind (or soul) interacting with their own body, or by interacting with their own brain (or body) directly.

withhold the upcoming movement with her willpower by stopping the movement of the respective limb [4].[2]

Libet's account on action in general resembles Descartes' account on action to a large extent, and Libet's interpretation of the vetoing will, more specifically, resembles Descartes' idea of the aborting or withholding will. However, whilst Descartes describes the person (the agent) as constituted of two distinct and independent substances - the material body and the immaterial soul -, Libet does not explicitly distinguish between two distinct and independent substances. But when it comes to action, Libet indeed conceives of the agent as her own system alienated from her body but with the ability to interfere with the brain's processes through an immaterial act of will.

This kind of dualist brain-person interactionism is implicit in many parts of Libet's work. To begin with, Libet assumes that whatever originated in the brain, cannot have been originated in the agent. In a reply to Libet's work, Velmans argues that Libet's veto could still be a free choice *of the agent* even if it has been initiated unconsciously in the brain and just became conscious [15]. Libet disagrees with him and remarks that, in this case, the agent would not be in *control* of the action. She would only be *aware* of an originally cerebrally initiated choice [8]. Velmans, similar to MacKay [7], wants to defend a position according to which the agent is *embodied* in the processes of his own brain so that it does not matter if her choice was caused by unconscious neural processes or not. Simply put, the agent is nothing but her body. Libet, however, does not even consider this option. For Libet, brain and agent are two mutually independent systems, with each system having its own independently unfolding processes. The agent is primarily just an observer of all of her brain's neural activities (including her urgers, desires, choices, intentions, etc.).

In spite of this observational stance, Libet grants the agent indeed some 'control of her (bodily) action', as he calls it. However, for Libet, being in 'control of the action' does not mean that the agent *takes over* some of the body's processes and *does* them.

There are many ways in which the word 'control' can be used. The most common way, however, implies some kind of interaction. For A to control B in an interactionist sense requires that A and B are distinct systems so that the processes happening in A and B take place independently from each other. However, A can control B by performing a 'counter-action' that interferes with B's movement so that B's movement stops or becomes aligned in the way A wants. If A does not perform such a counter-action to control B, B would just

---

[2] To be fair, in several parts of his book, Descartes also defends a rather hylomorphistic account on the relation between soul and body. According to this account, the soul is not another substance 'nesting' in the body, but rather that which gives the body its form. However, if the soul is the body's form, the soul can no longer be immortal because it's existence would be dependent on the existence of the mortal body. As C. Wohlers remarked in a preface to the German Meiner edition of Descartes' 'The Passions of the Soul', Descartes could never free himself from this dilemma and was, therefore, forced to defend a theory that oscillates between hylomorphism and dualist interactionism [4].

continue moving. I assume that people often use the word 'control' in this way to describe the control people exert on objects such as vehicles or animals. An agent can, for instance, control her car by hitting the breaks or by operating the stirring wheel (counter-action) to change the direction of the car's movement.

Libet uses the word 'control' in this sense to describe an interfering and regulating kind of control. For him, 'being in control of her actions' means that the agent can control the upcoming volitional process that was initiated by unconscious cerebral processes by vetoing it. This becomes clear in various formulations [7–9]. The veto is the 'counter-action' that the agent 'performs' to stop or abort the progress of the volitional process. Again, none of the things that happen in and with the body including the decision and the action itself is ever done by the agent. The body (including the brain) is a self-contained system that develops decisions and acts in accordance with these decisions. If the agent never exerted a veto (because she is happy with the unconsciously made decisions), the body would just continue moving and acting on its own. The veto, in turn, is not an act of the body but an act of the agent alone interfering with neural processes. Or in other words, the agent is 'in' her body like a driver in her car, according to Libet, and the only thing that she can do to determine the movement of the car is hitting the breaks.

The dualist nature of Libet's person-body interactionism becomes apparent in the passages when he discusses the nature of the veto. Interestingly, Libet leaves it open whether his account of the vetoing will is materialist or dualist [7]. However, his criticism against Velmans shows that the veto cannot have been initiated unconsciously. It must be an act that "[...] can appear without prior initiation by unconscious cerebral processes [...]" [8]. According to Wood [7] and Roskies [11], a conscious act that has no neural signature and is not mediated by any physical process, as Libet describes it, can also not be embedded in the stream of material processes. Instead, it must be a purely immaterial act unfolding parallelly to the brain's neural processes. Now, if, for Libet, the agent is just an observer of her own cerebral activities, and if the only means by which she can influence these activities (the veto) is immaterial, we can assume that Libet conceives of the agent herself as an immaterial entity attached to her own body.

3   The Veto in the Study of Schultze-Kraft et al.

Many have rejected the various dualist elements in Libet's works. Schultze-Kraft et al., for instance, insist that the veto, whatever it is, must be mediated by a cortical region and therefore be embedded in the other (material) activities of the brain [12]. They also refer to a famous older experiment conducted by Brass and Haggard in which such a determination was done [2].

Researchers such as MacKay and Nelson [7] also criticize the interactionist aspect of Libet's veto according to which acting consists in an aborting act of the agent against some of her own cerebral processes. According to them, this interactionism is not consistent with the phenomenological character of acting.

Nelson notes that "[...] conscious voluntary control is part of a conscious stream parallel to, but not interacting with, cerebral processes" [7]. MacKay agrees that decisions have various neural correlates, and that agents can determine the course of their action by controlling it. However, it is a category mistake to assume that they control their action by interacting with these correlates [7]. Schultze-Kraft et al., in turn, do not reject the interactionist paradigm in their paper. In fact, they use an analogy to describe the function of the veto that reminds us of a clear person-body interaction:

> One important question is whether a person can still exert a veto by inhibiting the movement after onset of the RP [...] The onset of the RP in this case would be akin to tipping the first stone in a row of dominoes. If there is no chance of intervening, the dominoes will gradually fall one-by-one until the last one is reached. This has been coined a ballistic stage of processing. A different possibility is that participants can still terminate the process, akin to taking out a domino at some later stage in the chain and thus preventing the process from completing. [12]

If we take the analogy seriously, the neural activities caused by the RP and culminating in the action (the row of dominoes) are processes that unfold independently from the agent herself (the domino master). The agent, however, might have a chance to stop this unfolding process by vetoing it (take out a domino). Otherwise, the process unfolds, and the body just starts to move (the dominoes continue falling).

But even if we do not take the analogy seriously, the interactionist paradigm underlies many formulations of the study paper, especially the verbs, that Schultze-Kraft et al. use throughout the paper to describe the function of the veto. Before I begin with a linguistic analysis of their formulations, however, let me describe the setup of the experiment.

Schultze-Kraft et al. conducted a stop-signal experiment that they describe as a 'duel' against a computer. Unlike the test subjects in Libet's studies, the participants in this experiment were not asked to abort their bodily movement spontaneously but as a response to a computer signal. They were placed in front of a green button and were allowed to press it with their foot whenever they felt like, as long as the button is still green. However, if the button turned red suddenly, they were no longer allowed to move. A computer tried to 'predict' the time of the test subjects' button press based on the collected EEG data of the test subjects. If it predicted an upcoming button press, the button turned red. In the case of an *early cancellation*, the test subjects intended to press the button; and after the button suddenly turned red, they managed to not move at all. In the case of a *late cancellation*, in turn, the test subjects intended to press the button, too; however, after the button turned red, they were not able to *not* move. They moved a bit. Some test subjects even touched the button. This latter scenario was called '*completed button press*'. According to the results of the study, there is a 'point of no return', after which agents are unable to not move, namely 200 ms before movement onset. However, that also means that, as the authors of the study note, "[...] a decision to move can be cancelled up until 200 ms before movement onset" [12].

Note that I chose neutral verb phrases to describe the target-phenomenon of the stop-signal experiment, namely 'being able not to move' and 'not being able not to move'. The formulations that the authors of the study use to describe the function of the veto resemble the formulations that Libet also used: 'Participants can/cannot *prevent themselves* from moving', 'participants can/cannot *veto/cancel/abort* prepared movements, upcoming movements, overt signs of movement, movement plans, etc.', 'test subjects also *exerted a veto*', and 'the stop process can/cannot *catch up* with the go process'. For Schultze-Kraft et al., all these acts of veto indeed have underlying neural mechanisms. However, the veto is still described as a (successful or unsuccessful) act of the agent directed *against* her own intentions and her upcoming or unfolding movements.

Consider the ditransitive verb 'prevent (from)' in "[...] an explanation is needed to clarify why people cannot prevent themselves from moving [...]" [12]. *Prima facie*, using the verb implies a distinction between two things - the one who prevents and the one who is prevented from doing something. The police can prevent the thief from running away by handcuffing him. I can also prevent myself from doing something, for instance, from sneezing loudly by squeezing my nose and mouth shut. In the case of holding my sneeze, I, as the agent, act on my own body. In both cases of prevention, there are two distinct and antagonistic processes involved. The counter-action (handcuffing, squeezing the nose), that the one who prevents performs, is a process distinct from the process that is supposed to be prevented (running away, sneezing loudly). If no counter-action would have been performed to prevent the other process, it would just have continued unfolding. This shows that acts of prevention fall in the category of interactionist control, as I described it in the previous chapter. But is this plausible? Did the test subjects prevent themselves from moving in the experiment, in a similar way in which I prevent myself from sneezing loudly?

Researchers have already determined what happens in the case of a successful action inhibition in stop-signal experiments on the sub-personal level. Logan et al., for instance, describe the phenomenon as a 'race between a (sub-personal) go-process and a (sub-personal) stop-process' [10]. When an agent is about to perform an action (intended or out of habit), brain signals are sent to the peripheral muscles (go-process). The subsequent perception of the stop signal (the red button), however, causes another parallel brain signal (stop-process). If the stop-process brain signals can 'catch up' with the go-process brain signals, they override the growth of activation of the go-process brain signals and *prevent* them from reaching a certain threshold. However, if the go-process brain signals reach that threshold earlier, the go-process brain signals 'escape overriding' and the limb of the person starts to move.

If this model is correct, whenever the test subjects in Schultze-Kraft's experiment managed to not move, one group of brain signals (the stop-process) overrode another group of brain signals (the go-process) and *prevented* them from culminating in overt movement. If, however, the test subjects failed to not move, the go-process brain signals were 'too fast' culminating in overt movement and 'escaping' the stop-process brain signals. So, on the sup-personal level, there were indeed two distinct and antagonistic neural processes involved in the experiment of Schultze-Kraft et al., namely the stop-process brain signals and go-process brain signals interfering with each other.

However, the interference between the stop-process and the go-process was an interference between two processes *only* happening on the sub-personal level. The go-process brain signals and the stop-process brain signals accompanied or underlay the action inhibition of the participants. That does not mean that they were the agent's action inhibition. It would be a mistake to conclude from these research results that the participants themselves overrode the go-process brain signals (accompanying their original action intention). The stop-process brain signals (accompanying the participants' action inhibition) overrode them. This remark resembles the criticism that MacKay and Nelson brought forth against Libet's interaction. Like them, I do not deny that the participants in the experiment inhibited their actions, and I also do not deny that their action inhibition was accompanied by various interfering neural processes, but I deny that the participants *themselves* interacted with their own neural processes.[3]

4   An Alternative Interpretation

Simply rejecting the interactionist interpretation suggested by Schultze-Kraft et al. is not enough. I also want to suggest an alternative non-interactionist reinterpretation of what the participants did when they inhibited their action.

To begin with, (failed) action inhibition is a common phenomenon. Assume that you want to throw a banana peel in your organic waste trash can. You press the pedal, the lid opens, and suddenly you realize that you forgot to put a new plastic bag in the trash can. You immediately stop moving. However, it is possible, since you were already in the act of throwing, that your hand still releases the peel accidentally. How would you describe what happened? Certainly, in the case of a successful action inhibition, you, upon realizing that there is no plastic bag in the trash can, simply changed your action intention from throwing the peel to holding still, and you successfully held still. You slowed

---

[3] I am not the first person referring to Schultze-Kraft's study. However, most researchers, such as Lavazza [6] and Uithol and Schurger [13] agree with the theoretical framework of the study and with the study results. Some researchers criticized the study but primarily for the experimental setup. Deecke and Soekader [3], for instance, remark that the action of the test subjects was not self-initiated in a proper sense because their movement intention was influenced by the presence of perceptual cues (green and red light) and also by the set time window. In the case of a real self-initiated action, the 'point of no return' might, therefore, be different. However, none of the researchers that I found questions or, at least, discusses the general assumption that agents can interact with their own neural processes (in order to inhibit a movement). Many of them even adopt the interactionist vocabulary of Schultze-Kraft et al.

down your current movement and finally stopped it. In the case of an unsuccessful action inhibition, you also changed your action intention from throwing the peel to holding still. However, your hand still released the peel. That is so, because the go-process brain signals, that started in the moment when you opened the lid, could not be overridden by the stop-process brain signals, that started in the moment when you saw the empty trash can. Accordingly, the go-process brain signals reached the threshold causing your muscles to move.

I assume that the test subjects in the Schultze-Kraft experiment experienced something similar. In the case of an early cancellation, they first intended to press the button and almost began to move, but, upon seeing the button turning red, their intention to move became an intention to hold still, and they held still. There is no interaction implied in this description. The test subjects simply replaced one intention early enough with another intention resulting in a non-movement of the foot. In the case of a late cancellation, the test subjects also intended to press the button, and, upon seeing the button turning red, their intention to move became an intention to hold still. However, the brain signals accompanying the new intention to hold still could not 'catch up' with the brain signals accompanying the intention to move. The latter brain signals reached a certain threshold and thereby caused a short and sudden twitch in the foot.

This twitch might have given the test subjects the feeling of being alienated from their body. But that does not mean that person-body-interactionism is correct, and that the test subjects tried but failed to veto or abort a neural process starting in their brain and culminating in the respective foot movement. It rather means that, as Velmans and MacKay argued, agents are *normally* embodied in their own body with all its neural processes. However, this feeling of embodiment can also be interrupted if neural processes accompanying an earlier action intention cause a muscle contraction that does not conform with the agent's current action intention. This non-conformity, in turn, is grounded in the delay of those neural processes, that accompany the agent's earlier action intention, and the slowness of her current intention's neural processes. And this is exactly what happened to the test subjects in the case of a late cancellation.

5 Conclusion

The purpose of this paper was to reveal the interactionist paradigm in a study conducted by Schultze-Kraft and to provide a non-interactionist reinterpretation of the test subjects' action inhibition. I did so by showing that we should not confuse what the test subjects did when they inhibited their action with the neural processes that accompany their action inhibition. Agents do not interact with their own body (or brain) when they inhibit an action. They rather change their intention or the course of their action. Unless the inhibition fails, they are fully embodied in their own moving limbs.

References

1. Bickhard, M.H.: The interactivist model. Synthese **166**(3), 547–591 (2009). https://doi.org/10.1007/s11229-008-9375-x
2. Brass, M., Haggard, P.: To do or not to do: the neural signature of self-control. J. Neurosci.: Official J. Soc. Neurosci. **27**(34), 9141–9145 (2007). https://doi.org/10. 1523/JNEUROSCI.0924-07.2007
3. Deecke, L., Soekadar, S.R.: Beyond the point of no return: last-minute changes in human motor performance. Proc. Natl. Acad. Sci. U.S.A. **113**(21), E2876 (2016). https://doi.org/10.1073/pnas.1604257113
4. Descartes, R.: Die Passionen der Seele. (French) [Les Passions de l'âme]. Meiner, Hamburg [1649] (2014)
5. Kornhuber, H.H., Deecke, L.: Hirnpotentialänderungen bei Willkürbewegungen und passiven Bewegungen des Menschen: Bereitschaftspotential und reafferente Potentiale. Pflügers Archiv für Gesamte Physiologie **284**, 1–17 (1965)
6. Lavazza, A.: Free will and neuroscience: from explaining freedom away to new ways of operationalizing and measuring it. Front. Hum. Neurosci. **10**(262) (2016). https://doi.org/10.3389/fnhum.2016.00262
7. Libet, B.: Unconscious cerebral initiative and the role of conscious will in voluntary action. Behav. Brain Sci. **8**(4), 529–566 (1985). https://doi.org/10.1017/ s0140525x00044903
8. Libet, B.W.: Do we have free will? J. Conscious. Stud. **6**(8–9), 47–57 (1999)
9. Libet, B., Gleason, C.A., Wright, E.W., Pearl, D.K.: Time of conscious intention to act in relation to onset of cerebral activities (readiness-potential): the unconscious initiation of a freely voluntary act. Brain **106**, 623–642 (1983). https://doi.org/10. 1093/brain/106.3.623
10. Logan, G.D., Van Zandt, T., Verbruggen, F., Wagenmakers, E.J.: On the ability to inhibit thought and action: general and special theories of an act of control. Psychol. Rev. **121**(1), 66–95 (2014). https://doi.org/10.1037/a0035230
11. Roskies, A.: Why Libet's studies don't pose a threat to free will. In: Sinott-Armstrong, W., Nadel, L. (eds.) Conscious Will and Responsibility. A Tribute to Benjamin Libet, pp. 11–22. University Press, Oxford (2011)
12. Schultze-Kraft, M., et al.: The point of no return in vetoing self-initiated movements. Proc. Natl. Acad. Sci. U.S.A. **113**(4), 1080–1085 (2016). https://doi.org/ 10.1073/pnas.1513569112
13. Uithol, S., Schurger, A.: Reckoning the moment of reckoning in spontaneous voluntary movement. Proc. Natl. Acad. Sci. U.S.A. **113**(4), 817–819 (2016). https:// doi.org/10.1073/pnas.1523226113
14. Varela, F., Thompson, E., Rosch, E.: The Embodied Mind Cognitive Science and Human Experience. MIT, Cambridge (2016)
15. Velmans, M.: Is human information processing conscious? Behav. Brain Sci. **14**(4), 651–669 (1991). https://doi.org/10.1017/S0140525X00071776

Regret from Cognition to Code

Alan Dix[1] and Genovefa Kefalidou[2]([⊠])

[1] Computational Foundry, Swansea University, Swansea, Wales, UK
alan@hcibook.com
[2] School of Computing and Mathematical Sciences, University of Leicester,
Leicester, UK
gk169@leicester.ac.uk
https://alandix.com/academic/papers/regret-2021/

Abstract. Regret seems like a very negative emotion, sometimes even debilitating. However, emotions usually have a purpose – in the case of regret to help us learn from past mistakes. In this paper we first present an informal cognitive account of the way regret is built from a wide range of both primitive and more sophisticated mental abilities. The story includes Skinner-level learning, imagination, emotion, and counterfactual reasoning. When it works well this system focuses attention on aspects of past events where a small difference in behaviour would have made a big difference in outcome – precisely the most important lessons to learn. The paper then takes elements of this cognitive account and creates a computational model, which can be applied in simple learning situations. We find that even this simplified model boosts machine learning reducing the number of required training samples by a factor of 3–10. This has theoretical implications in terms of understanding emotion and mechanisms that may cast light on related phenomena such as creativity and serendipity. It also has potential practical applications in improving machine leaning and maybe even alleviating dysfunctional regret.

Keywords: Regret · Cognitive model · Emotion · Machine learning

1 Introduction

Regret seems such a negative emotion, worrying about what might have been rather than about what could be. It seems maladaptive as it tends to focus on past temporal actions, perceived effects and imagined 'lost' possibilities. However looking at regret more deeply it turns out to not only be a well-adapted feeling, but one that demonstrates the rich interactions between different levels of cognition: rational thought, vivid imagination and basic animal conditioning. Particularly interesting is the role that quite complex assessments of probability plays in regret – the closer you were to averting a disaster but failed, the worse it seems! Furthermore, regret is associated with risk identification, risk-taking and prediction of potential outcomes. For example, making a life decision, which is quite risky (with high levels of uncertainty), can induce potential regret from not making this life decision, which can be greater than from having made it.

A. Cerone et al. (Eds.): SEFM 2021 Workshops, LNCS 13230, pp. 15–36, 2022.
https://doi.org/10.1007/978-3-031-12429-7_2

The origins of the models presented in this paper date back to an exploratory essay [15] that led to the cognitive model described in Sect. 2. This was followed by an early version of the computational model in Sect. 3, which suggested potential positive learning effects but was only reported in informal talks. The current work formulates these models more thoroughly and systematically explores and evaluates the wide range of parameters and options within the model allowing more rigorous and reliable analysis. Before proceeding to this, we will examine some of the psychological literature on regret.

1.1   The Psychology of Regret

Rationality and Agency. The ability to perform hypothetical comparisons (i.e. between an imagined state and a factual state) necessitates rational thinking and the capability to mentally represent these (e.g. *counterfactual thinking* [35] in defining *anticipated regret*), intention and risk-taking. Epstude and Roese [19] defined two different pathways as responsible for experiencing and perceiving regret. Information-based pathways directly affect intentions and consequently behaviour. On the other hand, content-neutral pathways can facilitate indirect effects from one's mind-set, positive and negative affect and motivational factors. Both pathways act as functional regulatory platforms for managing goals and 'controlling' behaviours within a socio-cognitive context. The foundational concept of the functional theory of counterfactual thinking is goal-setting and the comparison between current state and desired state.

At the same time, responsibility and agency appear to be critical in defining and experiencing regretful emotions. For example, Zeelenberg et al. [45], suggested that regret manifests -as an emotion- primarily to those that account themselves as responsible for a regretful action or interaction. A theory developed that reflects this is the Decision Justification Theory [14], according to which, decision-related regret is associated to comparative evaluation outcomes and 'self-blame'. Although self-blame can cause distress and negative emotions, experiencing regret has the potential to lead to better decision-making in the future. Developmental research has also found that children's decision-making can be improved if regret is experienced [30, 33].

Theories of Regret. The above frameworks have lead to the development of a number of theories of regret. These aimed to define or model regret incorporating insights from economic theories to socio-cognitive and interactional, including prospect theory [24] and regret theory [26]. According to Prospect theory, the losses and gains someone perceives are different depending on how these are formulated and on what types of affect generate. For example, if an investor is presented with two 'equal' options for investment opportunities, of which one is presented as associated to potential gains while the other one as associated to potential losses, the investor will tend to choose the former as an attempt to avoid the emotional impact that losses could cause (also known as "loss-aversion" theory). Two fundamental components of Prospect theory are the certainty (linked to probabilities) and the isolation effect (when outcomes are the same with the same probabilities, but where there are different routes or pathways to achieve these). In such cases, investors tend to follow 'the known' path or the one of 'least resistance' in an attempt to minimise their cognitive load. It is important to note that Prospect theory refers to pairs of choices; when the number of choice options increase, the complexity increases as well, due to additional interplaying factors. This is something that 'regret theory' [26] aimed to address. In regret theory, a key aspect is the so-called 'choiceless utility function', which represents the consequential state one experiences if no specific choice is made. In regret theory, where an individual has a choice to make, this occurs amongst multiple choice options, aiming to maximise the so-called "expected modified utility". Loomes and Sugden [26] posit that independently of whether someone experiences regret or not, they will attempt to maximise the "expected modified utility" when making decisions under uncertainty. In either case (i.e. choiceless utility function and expected modified utility), utility, in this context, is associated with the psychological and human-computer interaction (HCI) notions of pleasurability, desirability and user experience (UX) as a Gestalt impression.

Recapitulation and Reflection. Prospect and regret theories –as discussed above– focus on prospects and probabilities estimations. In contrast, Norm theory [23] includes forward-looking recapitulation of past events based on prior (or even currently experienced) 'norms'. Such 'norms' can be very personal and specific for different people suggesting individual differences in perception. Regret in such cases is dependent on memories and the capability to recall and process these (e.g. through mental simulation that can involve decision-making heuristics, biases and meta-cognition such as reflection).

The notion of mental simulation (and its role in perceiving regret) is further linked with the concept of 'mental models' whereby mental representations (or 'mappings') of the world facilitate a network of interconnected pieces of information that provide the ground-basis for reasoning and inferences generation [10–12]. Other 'retrospective' accounts of understanding regret includes the Reflection Evaluation Model that supports that reflection and evaluation mechanisms intertwine to promote comparative judgements that involve social, counterfactual and temporal aspects [28,29]. Reflections tend to be inherently experiential and in support of a quasi-realistic simulated state or scenario that could be considered as 'true' at a given moment. On the other hand, evaluations tend to be based on factual (and not fictional) past incidents that are then compared and assessed on the basis of goals fulfilment.

Regret and Human-Computer Interaction. Regret has an emerging role within the HCI community. Recent research suggests that regretful behaviours (i.e. in the form of interactions) can encourage 'remediation strategies' such as deleting unwanted or regretful messages, an action that can in turn become a source of confusion, uncertainty and information gaps. This is something prominent within certain demographic groups (e.g. teenagers), often associated with perceptions of trust and privacy. In all cases, feelings of regret are operationalised on the basis of psychological developmental theories for learning such as Pavlovian classical conditioning (stimulus-based associations that support learning) [6,13,31], Skinnerian operant (or instrumental) conditioning (consequential-based behavioural learning) [37,39], and Bandura's Social Learning Theory which posits that learning occurs by observation (Bobo doll experiment [5]). Indeed, psychological learning theories have been applied in the past in the design and evaluation of technologies, including robotic and automated systems (see e.g. Touretzky and Saksida's operant conditioning Skinnerbots [40] and [41]).

2   A Cognitive Model of Regret

This section presents a cognitive model of regret as an *adaptive learning mechanism* building on [15]. As regret is a complex emotion, the model builds incrementally from simpler leaning mechanisms each step of which has benefit in itself, as is necessary for plausible evolutionary development. The basic steps are: (a) an unpleasant effect is experienced; (b) potential actions that might have been causes are brought to mind; (c) a counter-factual assessment is made of how close the key actions were to averting or reducing the bad effect; (d) this modifies the emotional feeling of regret; (e) the image in memory of the past action is available simultaneously with the current (modified) emotion; (f) simple associative memories are then formed. As is evident, this includes very basic associative learning, with emotion, imagination and even counterfactual reasoning. We will now look at these building blocks in more detail and see how they come together to form the emotion we call regret.

2.1   Underlying Cognitive Systems that Enable Regret

Associative Learning. We begin with basic Pavlovian and Skinnerian behavioural conditioning (see Sect. 1.1) as it is present in all but the most basic animals. In simple associative learning, if you perform an action resulting in negatively perceived outcomes, then you learn not to repeat this action again. In effect, associative learning is helping to identify relationships between two or more stimuli (Pavlovian classical conditioning [6]). When associations are already learnt, if a condition repeats (e.g. in the form of stimuli-trigger), then feelings and experiences associated before to this stimuli re-emerge and give rise to aversion or compliance to perform a task or make a decision [34]. Figure 1 (left) illustrates this with the example of touching a sharp thorn and learning it is painful.

From Reactions to Foresight – Proto-Imagination. Basic associative learning requires near simultaneous presentation of action and consequence. For more complex learning some form of memory imagery is required - this is also

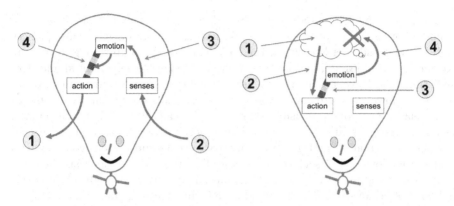

Fig. 1. (left) Simple associative memory: (1) touch thorn; (2) thorn pricks finger; (3) evaluation – Ow! it hurts! (4) learnt association – touching thorn is bad. (right) Momentary planning: (1) prospective imagination of planned action; (2) causes similar brain activity to actually doing it; (3) learnt association fires; (4) veto of planned action.

needed for momentary forward planning. Complex planning allows both re-active and pro-active behaviours for action and decision-making, both of which can lead to 'regretful' outcomes. More advanced conscious planning behaviours are part of meta-cognitive abilities related to socio-cultural and socio-cognitive activity [3] and are affected by attitudes, beliefs, motivations and goals, all of which contribute to imagination and forward-planning [22]. Forward planning (and/or its lack of such as in the case of impulsive behaviours) can both lead to regret depending on the context and knock-on effects' severity levels. Indeed, adaptive and dynamic self-regulatory mechanisms are necessary to support scaffolding and planning both within short-term and long-term contexts [2]. However, there is a lower and more primitive level of foresight when, for example, one half-imagines what is going to happen as one reaches for a door handle and hence surprised if the door is jammed. At multiple levels we have predictive abilities that enable us to prepare to act even before sensing the world [9]. However, this momentary foresight still needs a level of proto-imagination as illustrated in Fig. 1 (right), which leverages the way intention and action cause similar neural activity.

Dealing with Delays. As noted, simple associative learning is attenuated by any delay between action and consequence (for a detailed review of the delay literature see [25]); such effects can also be associated with feedback, reinforcement loops and time retention [20,38]. Indeed, effects of delays and impact on neural dynamics and learning have been explored before within the context of neural network simulations [4]. In reality incidents never happen absolutely simultaneously, but if brain activation decays slowly enough by the time a consequence occurs the areas associated with the last action may still be active enough to cause learning.

These learning effects reduce significantly once the delay is more than a few seconds. For simple creatures this means that learning of delayed consequences is all but impossible either due to the complexity of the context (e.g. too many extraneous variables and factors that interplay within the experiential and 'factual' timeline of events) or due to mere inability to recall and retrieve order of events. However, more cognitively complex animals do appear to be able to learn delayed consequences using some form of recall, even without full human memory. Figure 2 (left) shows how this can occur bringing past related events into one's imagination and hence making past events and present consequences available for low-level associative learning. This then forms the basis for more complex learning in the interplay of memory, imagination, procedural skills [17] and later of organising knowledge and structuring learning [43].

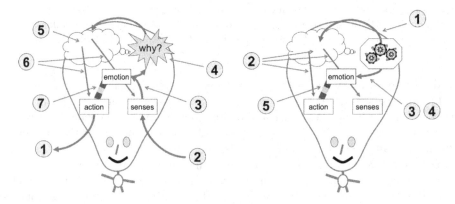

Fig. 2. (left) Dealing with delayed feedback: (1) touch unusual plant; (2) some hours later finger is sore; (3) evaluation – Ow that hurts! (4) desire to make sense; (5) recent salient events brought to mind – retrospective imagination; (6) causes simultaneous activation in relevant areas; (7) learnt association – don't touch that plant again! (right) Regret reinforcing critical learning: (1) logical deduction of what mattered determines what is brought to mind; (2) imagination causes simultaneous activation in relevant areas; (3) causes negative emotion "if only I hadn't" … regret; (4) counter factual deduction of how much it matters influences strength of emotion; (5) learnt association stronger or weaker depending on strength of emotion.

2.2   Regret as a Learning Mechanism

Attention in Memory and Counter-Factual Reasoning. A crucial element of association-making so far is the near simultaneity of imagined events. For example, recalling high school memories can evoke imagination processes whereby multiple separate events from that time intertwine and fuse in one go. For a sequence of events, because we focus on just a few highly salient events (e.g. touching a plant and feeling), we can replay potentially lengthy sequences in fast-forward and hence bring events close enough for learning to occur. However, if all past events are *equally* present, there cannot be effective learning. It is crucial that the *most appropriate* past events are remembered and that there is no confusion caused by information overload and information 'pollution'.

If through this more rational consideration of events we decide that particular actions or stimuli were not associated and relevant to the good or bad consequences, these 'drop out' of the story of the events so that the ones that are recalled most strongly –as we replay the events in our mind– are precisely those which were part of the causal chain leading to the effect. In particular, in the case of negative effects, it is the things that if we 'only had not done' that are recalled – regret. Because of this regret we are able to associate negative emotions with the right actions, those that we would wish to avoid in future. This is shown in steps 1–3 in Fig. 2 (right).

Note that this association is different from declarative knowledge that one might remember and think about logically later (e.g. in the form of self–reflection). Whilst the potential ill effects of a night in the town are something one can consider at leisure, in situations requiring 'fight or flight' responses the difference between rapid emotional reactions and more rational consideration is crucial [27].

Boosting Near Misses. Regret has another adaptive mechanism – the tuning of the strength of emotional response depends on the probability that our actions were the principal cause (*credit assignment* in AI terms). Some of the actions we perform may have contributed to a bad effect, but we'd have had to do something very unusual or perhaps some other additional actions to avoid the effect. However, other actions are ones where we feel they could very nearly have been different and changed things. Steps 4 and 5 in Fig. 2 illustrate this, as well as bringing relevant events to mind, the counter-factual *"what if"* reasoning amplifies the emotional response where a small change in behaviour could have made a large difference to outcomes.

For example, if you had bought a lottery ticket with the right number on it you would have won a million pounds, but you could easily have not got the right one – you don't feel too bad. However, imagine you almost chose to buy the lottery ticket with your birth date on, but decide not; later you find that the number came up – you are likely to feel more regret; the reason for that being that there was a 'personal' association with the 'successful' stimuli. Indeed research has suggested that regret aversion can partially be responsible for not exchanging lottery tickets even when there is a possibility of high material gain - once you purchase the ticket, you associate with it [42].

This lesser feeling of regret when things you did were less significant in the result and greater when what you did almost tipped the scales and made a difference is perfectly sensible. Higher emotional intensity can lead to higher levels of learning and stronger negative feelings attached to the action can affect actions and decisions to be made in future incidents. This applies to other complex emotions such as jealousy and envy [44] as well as regret.

Recapitulation. The final part of the learning armoury of regret is also the aspect that is most problematic in day-to-day life. Some events cause immediate regret, such as burning the toast, but are soon forgotten (perhaps due to the severity level of the incident or the impact value). However others, that have had especially large impact, become a repetitive rumination, which can be psychologically crippling, but, when not pathological, is also a learning mechanism. In machine learning (ML) it is common to use several copies of examples that are both rare and significant in order to improve learning. In a similar way, the repeated exposure to the imagined events means that their learning impact is increased. In a situation, such as a near miss from being eaten by a wild animal, this is clearly critical for survival as we do not want to have to repeatedly be exposed to such experiences in order to learn.

2.3   Cognitive Model Summary

In summary, regret is a subtle and well-adapted mechanism that enables us to learn effectively from the past recruiting deep (evolutionarily–old) mechanisms. In particular, although regret allows us to manage *"if only"* statements, the mechanisms do not deal with more complex modalities such as *"if only but I couldn't have known"*, or *"if only, but it will never happen again"*. In effect, certain contextual parameters are not considered when processing (or experiencing) regret either due to memory capacity limitations or due to personality traits and information load. We are able to reason explicitly about these *"if only but ..."* situations (although the former does seem to elude many); however, this explicit reasoning is not usually sufficient to mollify the negative emotional reactions of regret.

It is interesting to note that regret is often considered purely as a negative emotion. Indeed there no single word for the positive equivalent of regret "it worked but only because"? As we saw in Sect. 1.1, empirical studies in economics and psychology show that humans have a tendency to weigh negative results more strongly than positive ones, perhaps because not learning 'in the wild' to avoid bad things may kill you whereas missing good things simply means you have to try another time. As an adaptive mechanism regret shows that not only are negative effects stronger, but that we have additional mechanisms for negative emotion that may not exist for their positive counterparts.

3   A Computational Model of Regret

Aspects of the cognitive model have been built into a computational model. This was initially intended solely as a means of exploring and validating the cognitive model. For this, the core question was *"does regret aid learning"*, and this will be addressed by looking at two metrics:

– *asymptotic score* – does it do better in the long run
– *rate of learning* – does it get to the same score with less exposures

The former is the obvious quality metric, but the second is critical in real-life situations where experience is precious: for early humankind experiments could be fatal. The computational model is not expected to mimic (or use) real human data as human learning typically involves multiple simultaneous mechanisms. The intention instead is to explore the plausibility of the cognitive model, by evaluating the efficacy of the cognitively-inspired computational model.

While these initial aims are about cognitive understanding, we will see that the results also show promise as a technique to boost machine learning efficiency, especially in contexts when obtaining learning examples is costly. Computational experiments may not risk immediate death, but they still consume time and energy, contributing to global carbon emissions and ultimately cataclysmic climate change [16, 21].

3.1 Overall Architecture

Figure 3 shows the overall architecture of the model. The machine learning module interacts with an environment, which in our experiments is a simplified variant of the card game Pontoon. The machine learning module has two main parts a basic learning component and the regret module. The architecture is designed so that the regret module is loosely coupled and can be added to different forms of underlying learner and environment.

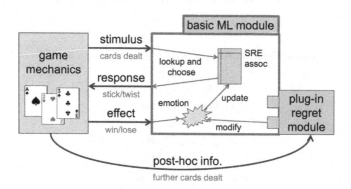

Fig. 3. Computational model architecture.

3.2 Example Game – Simple Pontoon

A simple version of Pontoon was chosen for these experiments as it is stochastic, offering challenge to the learner, but also simple enough to have perfect posterior knowledge, allowing very simple counter-factual evaluation.

Pontoon, or *vingt-un* is played with a normal pack of 52 cards. The player(s) and banker receive two cards each. Players may choose to 'twist' (receive an extra card) or 'stick' (keep the cards they have). The banker has a fixed rule for doing this: twist if the total is less than 17, stick otherwise. The aim is to have a hand with a higher total than the banker without going over 21 (called 'going bust'). The simplified version used in the experiments only has cards with values 1, 2 and 3 and the limit is 4 rather than 21 (so that a total of 5 or 6 is 'bust'). The player and banker initially are dealt one card each and can have only one extra card. Furthermore, the extra cards are 'dealt' even if the player or banker stick to enable perfect posterior knowledge, but the additional cards are only counted in the score if the player/banker had chosen to twist before 'seeing' their card. Table 1 shows the rules for when the player wins or looses. This is then translated into a score of 10 for a win or −10 for a loss, but this may be modified by the regret module.

Because the game is so simple it is possible to exhaustively calculate all possible games (81 in total) and the expected winnings for each play. The best possible play strategy leads to winning just over one third of the time. Scoring a win as 10 and loss as −10, this gives an expected best possible score of −3.09 and corresponding worst possible score of −7.53. (Note that in the long-term the banker always wins, hence even the optimal strategy yields a negative expected score.) The actual scores from learning can be compared with these to see how close to optimal the learner becomes or normalised as a percentage of optimal gain by transforming a raw score of x into $100 \times (x + 7.53)/3.44$, so that the worst possible strategy corresponds to a normalised score of 0 and the optimal strategy is 100%. This normalised value is used in subsequent discussion and graphs.

Table 1. Simple Pontoon Rules (cards 1–3 only). Rules apply in order..

Condition	Outcome
Player bust (>4)	Lose
Banker bust (>4)	Win
Player > banker	Win
Player = banker	Lose
Player < banker	Lose

3.3 The Environment

Although we use a specific game, this is presented to the machine learning component as a generic stimulus–response–effect environment (Fig. 4). When requested, the environment generates a new stimulus (`generateBefore` – the 'before' state); in the case of the Pontoon game, this is two cards (in the range

1–3), one each for the player and banker. It also provides a set of potential 'plays' (getPlays), possible actions that the player can choose take; in the case of Pontoon just 'stick' (keep the current card only) or 'twist' (have an additional card). The machine learning component chooses a play and the environment returns an 'after' state (generateAfter), which may depend on the chosen play, but also may involve stochastic elements; in the case of Pontoon a second card each for the player and banker. In some games, there is a clear arc of progress from start to finish, hence the before and after states are potentially of different types, but in other kinds of ongoing situations, these may be the same. Finally, an evaluation function (evaluate) gives a feedback score based on the before and after states and chosen play.

```
Before generateBefore();
Play[] getPlays( Before before );
public After generateAfter( Before before, Play play );
double evaluate( Before before, Play play, After after);
```

Fig. 4. Abstract interface of the game/environment component.

3.4   The Basic Learning Component

The abstract interface of the basic learning module (Fig. 5) has two principal methods. The first, getResponse, takes a stimulus and set of possible responses and returns a chosen response. The second, condition, takes an evaluation of the response (e.g. win or lose in Pontoon) and uses this to update its internal state.

The basic learning module used in experiments is a Skinner-like stimulus–response engine. This has an exhaustive table of all previous stimulus–response pairs and evaluation weight for each. For the simple learner the stimulus and response are completely opaque, simply used to look up or set relevant values; however a more complex learner, such as a neural network, might need a more detailed representation in order to generate generalised strategies.

When asked to provide a response, the Skinner module consults its table to find the evaluation of each matching stimulus–response pair. Previously unseen pairs are given a default weight. Variations in this default weight alter the extent of novelty seeking vs. risk aversion of the learner. The weights of the possible responses are used to generate a probability and the module then stochastically selects a response. The probability distribution of the response is parameterised by a power value: a power of 1 giving a linear probability (response with weight 2 twice as likely as weight 1); a power of 2 using squared weights; and a nominal power of 11 that acts as a 'winner takes all' where the response with the highest weight is always chosen.

The update function simply adds the effect to the current weight of the stimulus–response pair, with non-linear Sigmoid applied to keep it within a ±100 range.

```
Response getResponse( Stimulus stimulus, Response[] responseSet );
void     condition( Stimulus stimulus, Response response, double effect );
```

Fig. 5. Abstract interface of the learning component.

3.5   The Regret Module

The regret module is surprisingly simple, perhaps underlining how it builds on previous aspects of cognition. Figure 6 is the core function that implements regret. Without regret, the basic learning algorithm (`learner.condition`) would use the raw effect to modify the conditioning feedback. The counter-factual reason is embodied in the function `findBestResponse`. This uses posterior knowledge to predict what would have happened if any other play had been done and then returning the best possible result. In simple situations we may have perfect posterior knowledge, but in complex ones this may involve some form of uncertain or probabilistic inference. For example if you slip whilst rock climbing and are saved by the rope, you may factor in an element of *"but the rope might not have held"* even though you survived this time.

```
learn( stimulus, response, afterwards, effect):
    best = findBestResponse( stimulus, afterwards)
    regret = best - effect
    emotion = effect    // equal to effect without regret
    IF ( effect >= 0 ) {
    THEN   emotion = effect * POS_NO_REGRET_FACTOR - regret * POS_REGRET_FACTOR
    ELSE   emotion = effect * NEG_NO_REGRET_FACTOR - regret * NEG_REGRET_FACTOR
    learner.condition(stimulus,response,emotion)
```

Fig. 6. Pseudocode for regret thinker. The only adaptation to the basic learning function is to modify the strength of positive or negative feedback.

The regret thinker has several adjustable parameters: two each for positive and negative outcomes. The NO_REGRET factors are about modifying the effect when the outcome was as good as it could be. The REGRET factors about modifying things when there was a better option (`regret > 0`). These are separate factors because human emotional responses tend to be different to otherwise 'equal' positive and negative situations. For example, we might expect a 'no regret' situation to potential reduce the negative feelings for a negative outcome *"well I did the best I could"*, but boost the positive feelings for a positive outcome.

The earliest code only had 'negative' regret, that is only the second arm of the 'if statement when the initial effect was less than zero (a loss), as this corresponded to the day-to-day meaning of the term. However, the code looked 'messy' and hence, we experimented with adding the alternative and found that this boosted learning. Essentially this is a *'grass is greener'* emotion, for example if eating and enjoying a meal in a restaurant you might see someone else eating a different meal that looks very tasty and then feel less happy about your own meal!

4 Experimental Results

The underlying learner and regret engine have a significant number of parameters. We ran experiments over a wide range of configurations to avoid chance conclusions. We present typical examples, and summary views; the full data is available online at https://alandix.com/academic/papers/regret-2021/.

4.1 Obtaining Learning Saturation and Reducing Stochastic Noise

Both the basic Skinner-like learning and regret-enhanced learning make rapid initial gains in average game scores with the majority of learning gains over the first 1000 exposures. Learning slows but continues at a slower pace, so we have run all experiments to 10,000 iterations to examine asymptotic saturated learning. This approach to saturation is evident in the example run in Fig. 7.

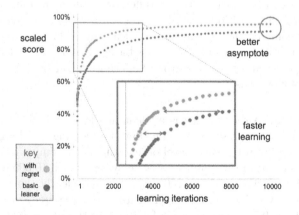

Fig. 7. Learning scaled to min/max possible scores showing better asymptotic values and faster learning. (linear weight, mild risk aversion, negative regret only).

Each run is stochastic, both in terms of the cards played and also the response of the low-level learning algorithm. Running over large numbers of iterations means that much of the randomness averages out as learning progresses, but substantial variation remains. The standard deviation of scores in the Pontoon game is approximately 1, 0.3 and 0.1 at 100, 1,000 and 10,000 iterations respectively. We therefore ran each configuration 10,000 times and created average learning traces with distribution statistics. This replication and averaging therefore means that the standard deviations of reported means are about 0.01, 0.003 and 0.001 at 100, 1,000 and 10,000 iterations respectively, or approx. 5%, 1.5% and 0.5% for normalised scores such as Fig. 7. Note that the variation is still quite large early in learning until about 100 iterations, and this is evident in some of the results (e.g. Fig. 8). Detailed studies of early learning would need more replications, but we will confine ourselves to further along the learning process when a greater level of learning has been achieved. Future studies will explore in more detail learning performance within different 'time-windows' (e.g. early learning vs. mid-term vs. later learning).

4.2   Observed Behaviour

Figure 7 shows a typical run, in this case parameterised for linear weighting of the Skinner learner, mild risk aversion and negative regret only. The graph highlights two aspects. First is the *improvement in learning*: the scores at the end of this particular rule are 90.8% for the basic learner rising to 95.3% when regret is added. Both are still slowly rising with further learning even at 10,000 iterations, so it is possible that the basic learning will eventually reach similar levels, but clearly only after vast numbers of learning steps. Second is *faster learning*: the regret learner obtains the same level of learning after fewer iterations. We look at both in more detail below.

While the exact numbers vary for different parameter configurations, the overall pattern is similar. For example, in the winner takes all, high risk averse, positive regret configuration, the asymptotic learning is closer to 100%, but there is still substantial improvement (97.3% for basic learning vs. 98.4% for regret). There is also consistent speedup of around 2.5 times faster (regret reaches 97.3% after only 3600 iterations).

Faster Learning. Although this is evident in Fig. 7, it is hard to assess the precise gain. Figure 8 represents the same data by plotting how long it takes the regret learner to obtain the same level of learning as the simple learner without regret. The lower part of the graph is quite noisy (even with 10,000 replications!), but the data is stable after about 100 iterations and clearly shows that adding regret substantially reduces the number of iterations required. In this case, the difference is about 0.5 on the log scale for much of the range, corresponding to a speed-up ratio (as measured by number of iterations) of just over 3. This is typical over the range of configurations, with speed-up ratios between 2.5 and 10 times in the central part of the range. Note too that the speed up increases towards the higher values in Fig. 8. This is possibly because the saturation value is higher for regret (better asymptotic learning), so that an extended version of the graph would see the lower curve level off with the simple learner never reaching the same levels of learning.

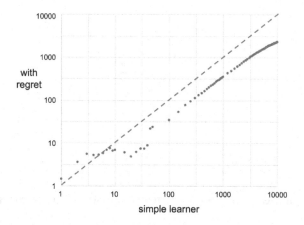

Fig. 8. Speed up, log-scale – number of regret exposures to reach same quality of learning as simple learner. Note higher variation for smaller numbers of trials. (linear weight, mild risk aversion, negative regret only).

Better Learning. As noted, the example run in Fig. 7 appears to show an improvement in asymptotic learning. This is observed across all configurations of parameters tested. Figure 9 compares learning of the basic learner compared to those with regret for a variety of parameter configurations for both the Skinner-like learner and regret module. The graph on the left shows performance after 1000 learning iterations and the right shows 10,000 learning iterations. Each dot shows the normalised average score over all 10,000 replications of the same parameter configuration. The x and y axes show the percentage of maximum score obtained by the simple learner and the same leaner with regret added. Note that the axes show different ranges on the left and right graphs as for both simple and regret learners the performance continues to improve over this range. The six vertical lines of dots on each graph are because there are six configurations of the simple learner and for each of these six alternative configurations of regret were added.

The dashed line on each graph denotes equal learning. As is evident in every configuration of the simple learner, regret improves performance. Some of the regret configurations improve it more than others, but all make substantial improvements. The dotted line shows the half-way point between the simple learner performance and perfect performance. As is evident, adding regret is close to or exceeds this mark, especially at higher levels of learning. That is, in many cases, regret halves the gap between the basic learning performance and perfect learning.

If one examines the cases in more detail, there are patterns amongst the various regret parameterisations. Positive regret on its own is not as effective as negative regret or both combined. However, we will not explore these in detail here as these differences may relate to the specific example game, where there was only win or lose, hence no way for an alternative action to have been better

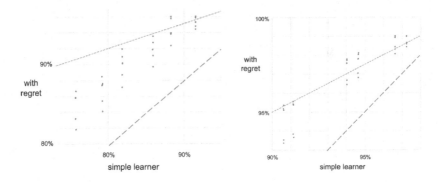

Fig. 9. Comparative scores after (left) 1000 iterations (right) 10,000 iterations.

than a win. One point that is promising is that –if anything– the proportionate improvements (in the sense of how much they make the outcome nearer to optimal) are better with the more effective Skinner-learner configurations. Although we would be cautious in generalising this, it does suggest the boosting effect of regret is not limited to poor learners.

4.3   Computational Model Discussion

Our initial reason for creating a computational model of regret was to validate and explore properties of the cognitive model. However, we have seen that it also shows promise as a way to enhance other machine learning algorithms.

Regret is often used as a *metric* within machine learning for both practical algorithms and as part of theoretical analysis (e.g. [7,36,46]). In addition, some algorithms such as Counterfactual Regret Minimization [47] and Deep Counterfactual Regret Minimization [8] work explicitly to minimise regret. However, the ability to 'bolt on' regret to existing algorithms does not appear to have been exploited previously.

Variants of regret minimisation have also been used to create explainable models of human moral decision making [1]. In this and other uses of regret minimisation, the focus is on the *average* disparity between actual and best behaviour for a decision strategy/algorithm as a whole, whereas the emotion of human regret, as captured in the computational model here, is *instance based*. From machine learning we know that in many cases point-based changes can lead to global optimisation (e.g. gradient descent, back-propagation), so the human-inspired instance assessment and learning is likely to end up with global 'regret' (in the ML sense) optimisation. In addition, some of the pathological issues with human regret may be about an overly strong emphasis on the instance; so there may be therapeutic lessons too.

Over recent years there has been growing interest in what has been termed 'human-like computation' [18,32], not least in order to emulate the single-shot learning of higher-cognition compared to the vast number of exposures needed by sub-symbolic learning. Regret sits somewhere between the two achieving *fewer shot* learning and offering insights into the way higher- and lower-level cognition can work together. This is important as forms of hybrid learning are likely to be essential for next generation AI and Cognitive Computing.

The computational model has also yielded some promising insights into regret itself, not least the importance of what we have called 'positive regret', that is the '*grass is greener*' effect after making even a decision with positive consequences. This can of course be problematic if one does not fully appreciate the positive things that happen, but it does improve learning, especially to help escape local maxima.

Finally, although not a new insight, the experimental results emphasise the methodological importance of (a) exploring the space of free-parameters and (b) running sufficient replications.

On the first of these, it is common in the machine learning literature to see papers that quote many fine-tuning parameters, such as network sizes or relaxation constants without any explanation of why the values were chosen or whether they are critical to the results. In early explorations of the regret model it appeared that positive regret was actually substantially more important than negative regret in terms of faster learning, however when a wider parameter-space was explored this effect was not found to be consistent, and restricted to 'winner takes all' low-level learning where local maxima are harder to escape. It would have been easy to publish these early results, which would have not only been misleading, but not exposed the underlying properties of positive regret.

On the second methodological point, many machine learning methods include stochastic or pseudo-random elements. The results are therefore also likely to have variability. The models used in this paper have very small numbers of learned weights (a few dozen) compared to many millions or billions in deep neural networks (DNNs). However, even so, many replications were needed in order to obtain statistically reliable results. The sizes of many DNNs makes this level of replication all but impossible, creating significant challenges in assessing sensitivity and reliability for safety critical applications and in interpreting results theoretically.

4.4   How the Computational Model Feeds Back to the Cognitive Model

The mere process of formulating a basic cognitive model of regret that informs a computational model has provided us with the opportunity to explore complex affect-driven processes in a simple game-based context. A number of insights have been presented and discussed above as outcomes of this process i.e. findings from modelling computationally a –traditionally– negative emotion. However, we also found that this process is cyclic and bi-directional instead of 'one-way'; indeed, the derived computational model (and its current results) have provided us with additional insights that feed back to the initial cognitive model.

For example, the computational model has highlighted the manifestation of positive regret alongside negative regret. This is something that has been –only recently– acknowledged in popular literature and certainly it was not accounted for in traditional 'Skinnerian' models of learning. This provides further support to more recent psychological affirmations in regret theories. Furthermore, the computational model has indicated an 'interaction' between positive/negative regret with parameters setting, namely, 'positive' effects of regret on learning were altered as parameters altered in the learning cycle of the computational model. The way that 'parameter' elements are acknowledged in psychological theories is through the role of context and complexity of the environment; context and its complexity –albeit acknowledged in past literature– are not explicitly mapped onto cognitive–computational models of regret; the present research is demonstrating the need for considering the inclusion of contextual elements in creating test-beds for regret models to firstly optimise them and secondly to formulate a more ecologically valid modeling environment.

There is also another interesting way to view parameters and context in regret i.e. in regards to the temporal aspect. The natural context whereby we –as humans– experience regret is dynamic, with a number of different ecological parameters interplaying and/or co-participating into the formation and trigger of feelings of regret. In effect, parameters' choice could facsimile both the dynamic and temporal features of our natural ecology. We do acknowledge that this may be a simplistic way to view and 'observe' phenomena of regret, however, we argue that 'parameterisation' could act as a potential simulation of more realistic scenarios whereby models of regret (cognitive and computational) can be tested and –perhaps more importantly– verified.

5   Conclusions and Future Work

The models presented offer both theoretical insight into regret as a human emotion and practical potential as a way to enhance machine learning. Future work will explore both of these directions and also the way that understanding of regret and associated cognitive functions can be used as part of interactive applications that can help users to enhance the positive aspects of 'regretful' decision-making and control the negative ones.

Future regret modelling can include mapping associative emotions to regret to create an ecological context whereby a richer (perhaps more enhanced and realistic) model of regret can be generated. In particular, we would like to explore the recapitulation aspect of regret that is not included in the current computational model and is critical in pathological regret. Given the dynamic nature of emotions (positive or negative), a next step would also include to explore more dynamic (or interactive) models of emotions that can be calibrated on-the-fly through different technological means but also through parameterisation (i.e. parameterisation as 'simulation' of real world and as a feedback mechanism). In that way platforms of 'in-the-wild' modelling that supports different layers of 'human-in-the-loop' interactions can be further designed. Indeed, there is an ongoing research on interactive Machine Learning approaches, utilised to provide 'verification' mechanisms to the data quality fed to the computational model and to its generated outputs.

Furthermore, the explorations and incorporation of positive emotion dynamics (even when modelling negative emotional responses) would be another direction to advance modelling approaches and techniques for experiential phenomena, acknowledging, in that way, the complexity and contextuality of human emotions. A way to do so would be to model context explicitly within the computational model (e.g. as another 'plug-in module that connects with the 'plug-in regret module' and design in –at least– a set of indicative individual differences present from within the literature in perceiving regret. To do so we wish to employ formal mechanisms to analyse and represent the model (e.g. employing Task Analyses and mental models theory).

In conclusion, next steps will include:

Internal validity of the computational model – constructing different and more complex simulated environments through parameterisation as 'simulation'.

Extending the computational model – incorporating recapitulation and incomplete posterior knowledge (simple predictive reasoning/heuristics).

Calibrating the computational model – design and run interactive sessions and 'human-in-the-loop' approaches to cross-validate parameterisation/output.

External validity of cognitive model – design experiments to test elements of the model.

Extending the cognitive model (temporal aspects of emotion) – recapitulation has an inherent temporal dimension, so far cognitive and computational models involve instant emotion only, but we know emotional state has a complex temporal dynamic.

Applying the model (in the wild) – looking at ways in which the lessons from the models can be embedded in, say, CBT-style and rehabilitation apps and interventions, for example, helping people to consider and imagine vividly *"if only but..."* scenarios to help them overcome trauma (Sect 2.3).

Broadening scope – build on existing work on creativity and serendipity, which also involve rich interactions between similar cognitive elements (response to uncertain situations, prediction, imagination, emotion) but also apply in other contexts e.g. problem-solving (e.g. optimisation) as opposed to games and decision-making.

Employing formal analytical methods and theoretical constructs – use of e.g. Task Analyses and theory of mental models to re-construct and recapitulate in a 'step-wise' format the stages/tasks involved in the model running.

References

1. Agrawal, M., Peterson, J.C., Griffiths, T.L.: Scaling up psychology via scientific regret minimization. Proc. Natl. Acad. Sci. **117**(16), 8825–8835 (2020). https://doi.org/10.1073/pnas.1915841117
2. Azevedo, R., Hadwin, A.F.: Scaffolding self-regulated learning and metacognition-implications for the design of computer-based scaffolds. Instr. Sci. **33**(5/6), 367–379 (2005)

3. Baker-Sennett, J., Matusov, E., Rogoff, B.: Planning as developmental process. Adv. Child Dev. Behav. **24**, 253–281 (1993)
4. Baldi, P., Atiya, A.F.: How delays affect neural dynamics and learning. IEEE Trans. Neural Netw. **5**(4), 612–621 (1994)
5. Bandura, A., Walters, R.H.: Social Learning Theory, vol. 1. Prentice Hall, Englewood Cliffs (1977)
6. Bitterman, M.: Classical conditioning since Pavlov. Rev. Gen. Psychol. **10**(4), 365–376 (2006)
7. Blum, A., Mansour, Y.: Learning, regret minimization, and equilibria. In: Nisan, N., Roughgarden, T., Tardos, E., Vazirani, V.V. (eds.) Algorithmic Game Theory, pp. 79–102. Cambridge University Press, Cambridge (2007). https://doi.org/10.1017/CBO9780511800481.006
8. Brown, N., Lerer, A., Gross, S., Sandholm, T.: Deep counterfactual regret minimization. In: International Conference on Machine Learning, pp. 793–802. PMLR (2019). https://proceedings.mlr.press/v97/brown19b.html
9. Butz, M.V., Achimova, A., Bilkey, D., Knott, A.: Event-predictive cognition: a root for conceptual human thought (2021). https://doi.org/10.1111/tops.12522
10. Byrne, R.M.: Cognitive processes in counterfactual thinking about what might have been. In: Medin, D. (ed.) The Psychology of Learning and Motivation: Advances in Research and Theory, pp. 105–154. Academic Press, San Diego (1997)
11. Byrne, R.M.: Mental models and counterfactual thoughts about what might have been. Trends Cogn. Sci. **6**(10), 426–431 (2002)
12. Byrne, R.M.: The Rational Imagination: How People Create Alternatives to Reality. MIT Press, Cambridge (2007)
13. Clark, R.E.: The classical origins of Pavlov's conditioning. Integr. Physiol. Behav. Sci. **39**(4), 279–294 (2004)
14. Connolly, T., Zeelenberg, M.: Regret in decision making. Curr. Dir. Psychol. Sci. **11**(6), 212–216 (2002)
15. Dix, A.: The adaptive significance of regret (2005). https://alandix.com/academic/essays/regret.pdf, unpublished essay
16. Dykes, J.: The carbon footprint of AI and cloud computing. Geographical (2020). https://geographical.co.uk/nature/energy/item/3876-the-carbon-footprint-of-ai-and-cloud-computing
17. Egan, K.: Memory, imagination, and learning: connected by the story. Phi Delta Kappan **70**(6), 455–459 (1989)
18. EPSRC: Human-like computing: Report of a workshop held on 17 & 18 February 2016, Bristol, UK (2016)
19. Epstude, K., Roese, N.J.: The functional theory of counterfactual thinking. Pers. Soc. Psychol. Rev. **12**(2), 168–192 (2008)
20. Fehrer, E.: Effects of amount of reinforcement and of pre-and postreinforcement delays on learning and extinction. J. Exp. Psychol. **52**(3), 167 (1956)
21. Freitag, C., Berners-Lee, M., Widdicks, K., Knowles, B., Blair, G.S., Friday, A.: The real climate and transformative impact of ICT: a critique of estimates, trends, and regulations. Patterns **2**(9), 1–18 (2021). https://doi.org/10.1016/j.patter.2021.100340
22. Friedman, S.L., Scholnick, E.K.: The Developmental Psychology of Planning: Why, How, and When Do We Plan? Psychology Press, London (2014)
23. Kahneman, D., Miller, D.T.: Norm theory: comparing reality to its alternatives. Psychol. Rev. **93**(2), 136–153 (1986). https://doi.org/10.1037/0033-295X.93.2.136

24. Kahneman, D., Tversky, A.: Prospect theory: an analysis of decision under risk. Econometrica **47**(2), 263–291 (1979). https://doi.org/10.2307/1914185. http://www.jstor.org/stable/1914185

25. Lattal, K.A.: Delayed reinforcement of operant behavior. J. Exp. Anal. Behav. **93**(1), 129–139 (2010). https://doi.org/10.1901/jeab.2010.93-129

26. Loomes, G., Sugden, R.: Regret theory: an alternative theory of rational choice under uncertainty. Econ. J. **92**(368), 805–824 (1982). https://doi.org/10.2307/2232669

27. MacIntyre, P., Gregersen, T.: Emotions that facilitate language learning: the positive-broadening power of the imagination. Stud. Second Lang. Learn. Teach. **2**(2), 193–213 (2012). https://pressto.amu.edu.pl/index.php/ssllt/article/view/5009/5229

28. Markman, K.D., McMullen, M.N.: A reflection and evaluation model of comparative thinking. Pers. Soc. Psychol. Rev. **7**(3), 244–267 (2003)

29. Markman, K.D., McMullen, M.N., Elizaga, R.A.: Counterfactual thinking, persistence, and performance: a test of the reflection and evaluation model. J. Exp. Soc. Psychol. **44**(2), 421–428 (2008)

30. McCormack, T., Feeney, A., Beck, S.R.: Regret and decision-making: a developmental perspective. Curr. Dir. Psychol. Sci. **29**(4), 346–350 (2020). https://doi.org/10.1177/0963721420917688

31. McSweeney, F.K., Bierley, C.: Recent developments in classical conditioning. J. Consum. Res. **11**(2), 619–631 (1984)

32. Muggleton, S., Chater, N.: Human-Like Machine Intelligence. University Press, Oxford (2021)

33. O'Connor, E., McCormack, T., Feeney, A.: Do children who experience regret make better decisions? A developmental study of the behavioral consequences of regret. Child Dev. **85**(5), 1995–2010 (2014). https://doi.org/10.1111/cdev.12253

34. Pearce, J.M., Bouton, M.E.: Theories of associative learning in animals. Annu. Rev. Psychol. **52**(1), 111–139 (2001). https://doi.org/10.1146/annurev.psych.52.1.111

35. Sanna, L.J., Stocker, S.L., Clarke, J.A.: Rumination, imagination, and personality: specters of the past and future in the present. In: Chang, E.C., Sanna, L.J. (eds.) Virtue, Vice, and Personality: The Complexity of Behavior, pp. 105–124. American Psychological Association (2003). https://doi.org/10.1037/10614-007

36. Shalev-Shwartz, S., et al.: Online learning and online convex optimization. Found. Trends Mach. Learn. **4**(2), 107–194 (2011). https://doi.org/10.1561/2200000018

37. Skinner, B.: Operant conditioning. In: The Encyclopedia of Education, vol. 7, pp. 29–33

38. Smith, T.A., Kimball, D.R.: Learning from feedback: spacing and the delay-retention effect. J. Exp. Psychol. Learn. Mem. Cogn. **36**(1), 80 (2010)

39. Staddon, J.E., Cerutti, D.T.: Operant conditioning. Annu. Rev. Psychol. **54**(1), 115–144 (2003)

40. Touretzky, D.S., Saksida, L.M.: Operant conditioning in Skinnerbots. Adapt. Behav. **5**(3–4), 219–247 (1997)

41. Touretzky, D., Saksida, L.: Skinnerbots. In: Maes, P., Mataric, M., Meyer, J.A., Pollack, J., Wilson, S.W. (eds.) From Animals to Animats 4: Proceedings of the Fourth International Conference on Simulation of Adaptive Behavior, pp. 285–294. MIT Press, Cambridge (1996)

42. Van de Ven, N., Zeelenberg, M.: Regret aversion and the reluctance to exchange lottery tickets. J. Econ. Psychol. **32**(1), 194–200 (2011)

43. Weick, K.E.: The role of imagination in the organizing of knowledge. Eur. J. Inf. Syst. **15**(5), 446–452 (2006). https://doi.org/10.1057/palgrave.ejis.3000634
44. Zeelenberg, M., Pieters, R.: Consequences of regret aversion in real life: the case of the Dutch postcode lottery. Organ. Behav. Hum. Decis. Process. **93**(2), 155–168 (2004)
45. Zeelenberg, M., Van Dijk, W.W., Manstead, A.S.R., der Pligt, J.: The experience of regret and disappointment. Cogn. Emot. **12**(2), 221–230 (1998)
46. Zhang, Z., Ji, X.: Regret minimization for reinforcement learning by evaluating the optimal bias function. In: Advances in Neural Information Processing Systems, vol. 32 (2019). https://proceedings.neurips.cc/paper/2019/file/9e984c108157cea74c894b5cf34efc44-Paper.pdf
47. Zinkevich, M., Johanson, M., Bowling, M., Piccione, C.: Regret minimization in games with incomplete information. In: Proceedings of the 20th International Conference on Neural Information Processing Systems (NIPS 2007), pp. 1729–1736. Curran Associates Inc., Red Hook (2007)

In Silico Simulations and Analysis of Human Phonological Working Memory Maintenance and Learning Mechanisms with Behavior and Reasoning Description Language (BRDL)

Antonio Cerone[1]([✉]) [iD], Diana Murzagaliyeva[1], Nuray Nabiyeva[1], Ben Tyler[1], and Graham Pluck[1,2] [iD]

[1] Department of Computer Science, School of Engineering and Digital Sciences, Nazarbayev University, Nur-Sultan, Kazakhstan
{antonio.cerone,diana.murzagaliyeva,nuray.nabiyeva,btyler}@nu.edu.kz
[2] Faculty of Psychology, Chulalongkorn University, Bangkok, Thailand

Abstract. Human memory systems are commonly divided into different types of store, the most basic distinction being between short-term memory (STM) and long-term memory (LTM). Phonological STM, as proposed in the working memory model is closely linked to semantic LTM. Nevertheless, the mechanisms of maintenance with STM, and transfer of information with LTM are poorly understood. Candidate mechanisms within phonological STM are rehearsal (either articulatory or elaborative), and refreshing. There is also evidence of long-term learning within STM.

In this paper we use the Behavior and Reasoning Descriptive Language (BRDL) to model human memory contents as well as the perceptions that allow humans to input information into STM. By using the Maude rewrite system to provide semantics to BRDL and dynamics to BRDL models, we can explore various cognitive theories about phonological STM maintenance and transfer of information for long-term retention, such as articulatory rehearsal, elaborative rehearsal, and refreshing. This approach has been implemented in a tool that allows cognitive scientists to carry out in silico the simulation of learning processes as well as the replication of experiments conducted with human beings in order to contrast alternative cognitive theories.

Keywords: Short-term memory · Working memory · Semantic memory · Behaviour and Reasoning Description Language (BRDL) · Formal methods

Work partly funded by Project SEDS2020004 "Analysis of cognitive properties of interactive systems using model checking", Nazarbayev University, Kazakhstan (Award number: 240919FD3916).

A. Cerone et al. (Eds.): SEFM 2021 Workshops, LNCS 13230, pp. 37–52, 2022.
https://doi.org/10.1007/978-3-031-12429-7_3

1   Introduction

Most theories of human learning processes make a fundamental distinction between short-term memory (STM) and long-term memory (LTM). This distinction is supported by a wealth of evidence from neuropsychology, experimental psychology, and analysis of the different computational needs of short-term and long-term forms of information processing [30]. Although multiple models exist of the interactions between STM and LTM, predominant is the Multicomponent Working Memory Model proposed by Baddeley and Hitch [7].

In our paper, the expression working memory should thus be interpreted as referring broadly to the model proposed by Baddeley and colleagues, unless stated otherwise. In the most recent form of the Multicomponent Working Memory Model, the most basic feature distinguishing short-term processing is that it is essentially fluid, while LTM is essentially crystalized [3,4]. The concepts of fluid and crystalized processing originated in studies of intelligence within differential psychology [14].

Crystalized knowledge is viewed as stored information, and is thus highly dependent on experiences, culture and education. In contrast, fluid processing describes processes which work 'online' and are conducted to deal with novel, current demands and are basically independent of cultural influences and education. That working memory is fluid is almost tautological, as its very definition includes, in addition to short-term storage of information, it "supports human thought processes by providing an interface between perception, long-term memory and action" [4, p. 829]. Indeed, some argue that working memory is so involved with high-level, domain-general cognitive processing that it is virtually equivalent to general intelligence [21], and concordant with that, is a substantial predictor of performance in formal education [35].

The most recent version of the Multicomponent Working Memory Model proposes a singular top-down, executive, limited capacity processor, and two STM stores. Visuospatial storage in STM, also known as the visuospatial sketchpad, interacts with visual semantics stored in LTM, while an acoustic-based STM, known as the phonological loop interacts with semantic LTM [3,4].

In our paper we focus on the latter of these two, the phonological loop, and how information is maintained (i.e., how it acts as an STM store) and how information may be parsed with semantic LTM. In our own computational modelling, we refer to this phonological store as pSTM, but we consider it more or less equivalent to the phonological loop in the Multicomponent Working Memory Model of Baddeley and Hitch [7] or the basic acoustic-based STM store in the Atkinson and Shiffrin's model [1].

Evidence that pSTM is closely linked to semantic LTM is known from various sources. One being that pSTM in the context of working memory ability is linked to normal vocabulary learning in children, and brain damage that impairs pSTM in adults effectively prevents new vocabulary learning, but not episodic LTM learning [6]. This suggests that pSTM is a crucial stage contributing to crystalized semantic LTM. In support of the opposite connections, from LTM to pSTM, the often quoted capacity of pSTM, seven plus or minus two items, is frequently disproven by prose recall [23] and chunking of verbal material [31]. Both of which suggest that semantic-lexical LTM can actively support pSTM, allowing much more information to be held ready for immediate use.

In fact, a recent conceptualization of the position of working memory in the greater memory system, proposes that while still basically separate, working memory operates between LTM and action/output systems [5].

In Sect. 2 we review experimental studies on the pSTM processes (articulatory rehearsal, elaborative rehearsal, and refreshing) that enable information to persist in the short-term storage and, possibly, move to LTM (Sects. 2.1 and 2.2), and we introduce fast and slow learning mechanisms (Sect. 2.3). In Sect. 3 we briefly recall the Behavior and Reasoning Descriptive Language (BRDL), which was introduced in the first author's previous work [15], and we describe how to use it to model maintenance and learning mechanisms (Sect. 3.1). In Sect. 4 we first review modelling approaches to the in silico simulation of working memory, and then describe our own approach and tool, which are based on our previous work [16–18], in which Real-Time Maude [34] is used to provide semantics to BRDL and dynamics to BRDL models. In particular, Sect. 4.1 describes the tool implementation and Sect. 4.2 illustrates the use of the tool for the in silico replication of the 1969 Collins and Quillian's experiment [20] on time retrieval from LTM and the 2017 Souza and Oberauer's experiment [37] comparing fast and slow learning mechanisms. Finally, in Sect. 5 we draw conclusions and discuss future work.

2 Maintenance in pSTM and Transfer to LTM

Despite wide-spread support for the concept of the Multicomponent Working Memory Model, the mechanism by which material is maintained in STM (e.g., the phonological loop), is poorly understood. Furthermore, based on experimental evidence, it is argued that whatever the mechanisms are, they also seem to contribute to information retention in LTM [26]. Thus, working memory maintenance and LTM trace formation or strengthening are intertwined. Two main mechanisms have been proposed for how information is maintained in pSTM, given that it is supposedly of short duration, with memory traces decaying in perhaps less than 3 seconds [13]. These are rehearsal and refreshing. Both have also been proposed as mechanisms of how information transfers from STM to LTM.

2.1 Rehearsal

As the phonological loop component of the Working Memory Model developed from the earlier Modal Model of memory [1,2], it inherited from that the concept of rehearsal. This suggests that information decays rapidly within phonological STM, unless it undergoes some form of repetition, hence the name, phonological loop. This appears to take the form of sub-vocal articulation. As it is proposed as a trace enhancing mechanism, it is often called articulatory rehearsal. Nevertheless, such rehearsal is also proposed as a principal mechanism by which information moves from STM to LTM storage [2]. These are thus quite closely related concepts, as information that is articulatory rehearsed (presumably increasing the probability of transfer to LTM) also remains available in STM for longer.

This has led to the suggestion that the length of time that information spends in phonological STM determines its likelihood of transfer to LTM [1]. However, experimental support for this, at least as caused by articulatory rehearsal, has proven to be quite limited [38]. One suggestion is that the intentional act of initiating articulatory rehearsal and the associated processes, such as preparing the articulatory code, are the mechanism that causes some transfer to LTM, with subsequent rehearsal loops maintaining information in STM, but not inducing further transfer [29]. Recent evidence suggests that the amount of time that information spends in phonological STM does indeed greatly influence whether or not it will become available for later recall from LTM, but this may be due to reasons other than greater opportunity for articulatory rehearsal, such as elaboration [37].

As articulatory rehearsal appears to not explain much LTM learning, an alternative version, elaborative rehearsal, has been proposed [28]. The expression is something of a misnomer, as it refers to processing of the meaning of information with phonological STM. However it is interpreted, it is clear that accessing meaning of material and making connections of meaning is a potent mechanism of LTM learning. This is demonstrated in the classical experiments used to form the Levels of Processing approach to human memory [22]. It is now widely accepted that elaborative semantic rehearsal of information that is held in phonological STM substantially increases the chance that the material will be available from LTM when tested later, particularly if this is for linguistic material [8].

2.2   Refreshing

An alternative mechanism for delaying trace decay in phonological STM is refreshing, which has become particularly popular within cognitive psychology over the past two decades. It is defined in general terms as "a domain-general maintenance mechanism that relies on attention to keep mental representations active" [11, p. 19]. This is core to the Time-Based Resource-Sharing (TBRS) model of working memory, which suggests that attention is a limited resource which can maintain representations, but they will decay if attention is moved to other tasks, unless it momentarily returns to refresh the traces [9]. This emphasizes the difference between the older Modal Model of STM and the more current working memory models, in which some form of resource limited executive control is required, such as to sporadically refresh memory traces. This executive resource produces a bottleneck as it is argued to work sequentially, and thus must be switched frequently to maintain not only STM traces but also various task goals, and other task-relevant information.

Experimental evidence supports the existence of attentional refreshing as an independent mechanism of maintenance of verbal material within STM. Furthermore, that it operates in addition to articulatory rehearsal, and that the two may have an additive effect [12]. Several studies have suggested that in addition to extending traces in STM, refreshing leads to better delayed recall, indicating it promotes learning in LTM. For example, Loaiza and McCabe [27] used a word learning task and found that articulatory rehearsal had no impact on later recall from LTM, but opportunities for refreshing did. Nevertheless, a recent experimental study that compared elaborative rehearsal and refreshing side-by-side, found evidence that word learning in LTM is improved by the former, but not by the latter [36]. The experimental evidence for LTM trace formation being enhanced by refreshing in phonological STM is therefore unclear, as is whether rehearsal can fulfil that function.

2.3 Fast and Slow Learning Mechanisms

It has been suggested that learning may take place within pSTM. This is implied by the Hebb repetition effect [32] in which lists that are surreptitiously repeated in immediate recall tasks are better recalled than novel lists, suggesting an incidental STM learning mechanism. Burgess and Hitch [10], in their neural network model of pSTM, implement a 'fast mechanism' responsible for storage of items with pSTM for active use, and a 'slow mechanism' (e.g. long-term potentiation) responsible for long-term learning that would form the LTM traces.

3 Behavioral and Reasoning Descriptive Language (BRDL)

BRDL [15] is a modelling language to describe human reasoning and human automatic and deliberate behavior. A BRDL model consists of a set of mental representations of facts, inference rules and behavior patterns, classified as one of the following types:

fact representation which is part of our knowledge, such as 'An animal can move';

inference rule which is acquired through our lives and applied deliberately; for example, when we are driving a car, we know that if we are approaching a zebra crossing and we see pedestrians ready to walk across the road, then we must give way to the pedestrians;

deliberate basic activity which is driven by a goal; for example the activity of grasping an object is driven by the goal of moving it to a specific place;

automatic basic activity which occurs as a reaction to perception from the environment in combination with some mental state; for example the activity of pushing the car brake may occur as a reaction to a red light while I am in a driving mental state.

The first three types of representation refer to information stored in the declarative part of LTM. The use of this information by working memory processing is driven by specific goals. Namely, we may retrieve a fact as the answer to a question [16] or because we need to use it in the achievement of a goal [17]. The retrieval of facts is modeled as an internal process of LTM. It operates through a pattern matching between elements of some information stored in pSTM (i.e., a question or some information stored by the goal-driven task execution) and elements of fact representations in LTM, possibly in combination with the identification of a semantic connection between the two matched elements of the two information items. For example, if pSTM contains the question 'Can a dog bark?', a full pattern matching with the fact 'A dog can bark' is directly identified in LTM (both 'dog' and 'bark' match). Instead, if the question is 'Can a dog move?', two partial pattern matches with the facts 'A dog is an animal' (match on 'dog') and 'An animal can move' (match on 'move') are identified in combination with the semantic connection between 'dog' and 'animal', which expresses generalization.

An alternative possibility is that one of the information items involved in the matching is actually stored in pSTM as a consequence of some working memory activity. In this case, the semantic connection may be identified between a fact in LTM and a fact in pSTM or even between two facts in pSTM. Representations of automatic activities are stored in the procedural part of LTM and are not driven by goals. For example, the behavior of an experienced car driver is mostly automatic: the driver is aware of the high-level tasks that are carried out, but is not aware of low-level activities such as changing gear, using the indicator and reacting to the presence of a traffic light or a zebra crossing. These low-level activities are performed automatically as a direct response to specific perceptions whose selection is controlled by the information stored in STM, with no involvement of actual reasoning activities. Thus, in this approach, attention is modeled as an STM/working memory process [17].

3.1   Modelling Maintenance and Learning Mechanisms

Inspired by Burgess and Hitch's 'fast' and 'slow' short-term learning processes [10], we model articulatory rehearsal by associating each information item (or chunk) stored in pSTM with a decay time and lifetime. The former is the remaining time after which the information item would be removed from STM for natural decay, in the absence of any form of maintenance or reinforcement. The latter is the entire lifetime of the item from the moment it is stored to the moment it is removed.

The 'fast' short-term learning process is controlled by the decay time, while the parallel 'slow' learning mechanism is represented by an increase in the information lifetime. Such an increase can be set in a way that can accommodate a specific hypothesis or theory by using appropriate equations to define operator maintenance-effect. For example, we can model a small, constant increase at each rehearsal loop or we may implement the suggestion by Naveh-Benjamin and Jonides [29] that the first rehearsal is the most important, because it involves producing the articulatory plan, with subsequent loops of that plan within pSTM adding little to the transfer to LTM.

Refreshing of an item in pSTM may be activated by questions which involve the identification of semantic connections between facts in pSTM and facts in LTM. For example, if we have just read the fact that 'a dog is an animal', which we were unaware of, this fact is stored in pSTM but is not present in LTM. In pSTM, the fact is associated with an initial decay time that equals the lifetime. If we are then asked to answer the question 'Can a dog move?', we may actually use the new fact to access the LTM fact that 'an animal can move' and find out that also 'a dog can move'. In our model this usage of the newly read fact in pSTM increases its lifetime and resets its decay time to the value of the lifetime. Such a process promotes learning by expanding the lifetime of facts in pSTM that are used to access information in LTM. We also consider a lifetime threshold for the transfer of information from pSTM to LTM, so that the repetition of the refreshing process will eventually lead to a transfer of the fact from pSTM to LTM, where it is finally stored permanently, thus completing the knowledge acquisition process.

4 A Tool for Simulating Learning Processes and Performing In Silico Experiments

Although the experimental evidence described in Sect. 2 has been greatly informative on elucidating the overall human memory system, other approaches are available. One of these is producing various forms of in silico simulations, which can then be tested in different circumstances, using hypotheses derived from the experimental cognitive literature. For example, Oberauer and Lewandowsky [33] have used a neural network approach to model their Time-Based Resource-Sharing (TBRS) model approach to working memory, showing that it can produce many of the phenomena associated with human performance on experimental tasks designed to assess working memory. In fact, a basic version of the TBRS neural network implementation has been used to compare articulatory rehearsal and refreshing mechanisms directly in pSTM, and reported that the former performed substantially below the level of the latter as an STM trace maintenance mechanism [25].

In this section we describe how to carry out in silico experiments by using our own formal-modelling approach, which we have implemented using the real-time extension of Maude [34]. Our approach supports the modeling of experiments in terms of sequences of perceptions present in the environment with which the human memory model interacts. Each perception is associated with a starting time and a duration. When the starting time is reached the perception may be transferred to pSTM, depending on the attentional mechanism controlled by the content of pSTM.

We use in silico experiments to compare alternative hypotheses and theories that describe the transfer of information from pSTM to LTM through rehearsal and refreshing. One way to carry out such a comparison is to determine and test alternative quantitative implementations of conceptual hypotheses or theories, as we proposed for the parallel learning mechanisms within pSTM proposed by Burgess and Hitch [10] and for the Naveh-Benjamin and Jonides hypothesis [29] about the initial articulatory plan for a rehearsal loop being the most important factor contributing to retention within LTM. Another approach is the direct comparison of alternative estimates from cognitive psychology or neuroscience. This is the case for pSTM decay time, and for inclusion of the time to read a sentence, i.e., the initial conversion of the orthographical format into the phonological format, estimated at about 100 msecs by Kolers [24].

Furthermore, the results of in silico experiments may also be compared with real datasets to evince which model best mimics reality. In addition to a manual comparison, we can use the methodology defined in our previous work [19] in order to formally validate hypothesis, automatically converting a dataset into a formal representation that can be composed in parallel with our human memory model. Finally, from a computer science perspective, the human memory model of a user can be combined with the model of the computer system or application in order to analyze properties of the interaction, such as usability, learnability and safety.

4.1   Tool Implementation

The purpose of our tool is to perform in silico experiments that allow researchers to compare and analyse hypothesis and theories related to human memory. Users of the tool are able to automatically generate a formal model of the human component and analyse the overall system by performing simulation and checking properties.

The tool can be downloaded from a GitHub repository[1]. It is equipped with a simple and concise interface which was implemented using the Python GUI framework, an MySQL database, and the Maude rewrite system. The tool interface allows the researcher to model human memory in a BRDL-like fashion and the resultant model is automatically translated into a formal model expressed using Real-Time Maude. The tool allows for the adjustment of memory parameters before running experiments. The results of the in silico experiments are then visualised by the tool by appropriately changing the content of the human memory model.

The tool supports project and version control to make the user's experience friendly, reliable and maintainable. The user can create a project with a number of versions to compare the results of different in silico experiments, using various combinations of the parameters of the memory. The project page (Fig. 1) represents a simplified simulation of human memory.

This main project page allows the researcher to define a model of human memory by entering the contents in the various memory components. Episodic memory and sensory memory are not implemented in the current version of the tool. However, sensory memory is implicitly given by the presence of timed perceptions in the environment. The timing may be actually used to characterise the duration of the perception in sensory memory. BRDL fact representations, inference rules and deliberate basic activities are stored in semantic memory, while automatic basic activities are stored in procedural memory. Normally, the short-term memory is initially empty, but several parameters can be set by the

---

[1] https://github.com/nuraynab/interactive-system-modelling.

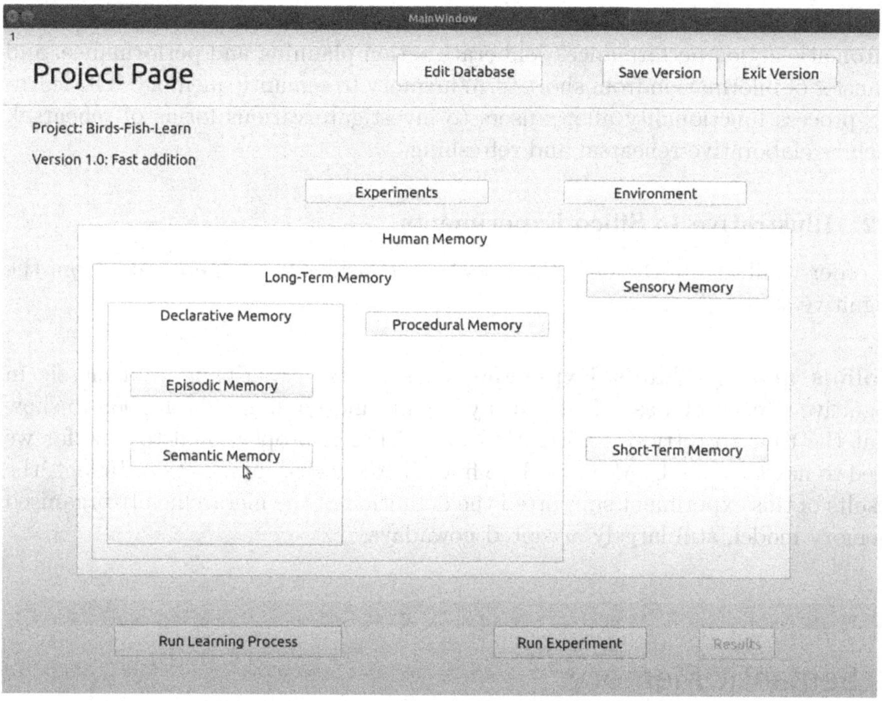

Fig. 1. Project main page.

experimenter: STM capacity and cognitive load (a load due to other tasks, which are not explicitly modeled) as well as information initial lifetime and its increase due to the information persistence in pSTM.

Experiments are timed sequences of facts and/or questions, which in a real experimental setting with human subjects may appear on a screen for a given duration at given time intervals. These timings are set by the user before running the experiment. As can be seen from the buttons in Fig. 1, the tool can run two kinds of experiments, that is, two functionalities: a 'proper' experiment or a learning process.

A 'proper' experiment may be a timed sequence of questions that are generated in the environment, are perceived and then enable a memory process that retrieves information from the semantic memory in order to produce the answer in STM and then transfer it to the environment. Another form of 'proper' experiment may be a timed sequence of generic perceptions that the 'virtual' subject has to rehearse. The experiment functionality can be used to study information retrieval from semantic memory and maintenance rehearsal in pSTM.

A learning process normally occurs over a long period of time, which can seldom be incorporated within any real experiment with human subjects. Here the 'experiment' is a timed sequence of facts and questions that are generated in the environment, are perceived, and then enable a number of memory processes, including retrieval of information from the semantic memory, inferences, automatic action performance, deliberate action planning and performance, and transfer of information from short-term memory to semantic memory. The learning process functionality allows users to investigate various forms of rehearsal, such as elaborative rehearsal and refreshing.

4.2   Illustrative In Silico Experiments

In order to illustrate the use of our tool, we consider two experiments from the cognitive science literature.

Collins and Quillian's Experiment. The first experiment is a classic in cognitive science. It was carried out by Collins and Quillian [20] in 1969 to show that the time to retrieve information from LTM is proportional to how far we need to navigate the LTM network to find the requested information. In fact, the results of this experiment supported the definition of the hierarchically organised memory model, still largely accepted nowadays.

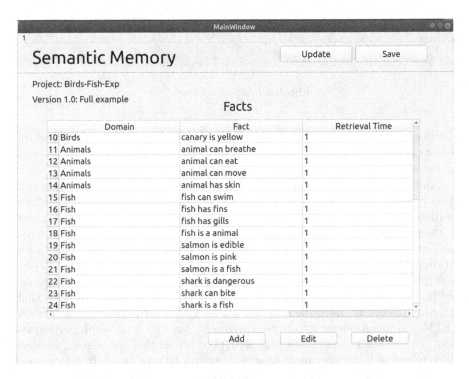

Fig. 2. Initial setting in the in silico replication of Collins and Quillian's experiment.

In this experiment, the semantic memory model consists of three domains "Animals", "Birds", and "Fish". Collins and Quillian's hypothesis was that instead of saving the fact “Shark can swim” in the model it is more memory efficient to retrieve this data from the category relation, in particular from the facts that shark is a fish and a fish can swim. However, in this way, the time to retrieve the fact from the hierarchical model will be longer than if it had stored and retrieved it directly from semantic memory. Therefore, they concluded that the deeper the information in semantic memory the longer it will take time to retrieve it. This works under the assumption that it takes an equal amount of time to get the fact from a single node regardless of its level. Collins and Quillian’s experiments produced results in accordance with their hypothesis.

To replicate this experiment in silico, we create a semantic memory with 24 facts about animals, birds and fish, using, for sake of simplicity, the same retrieval time of one unit. The initial setting of the experiment is shown in Fig. 2. The experiment component includes 34 questions, each available for 2 seconds. After running the experiments, 28 questions resulted answered, as shown in Fig. 3. Not all the questions were answered due to pSTM limited capacity and information decay time.

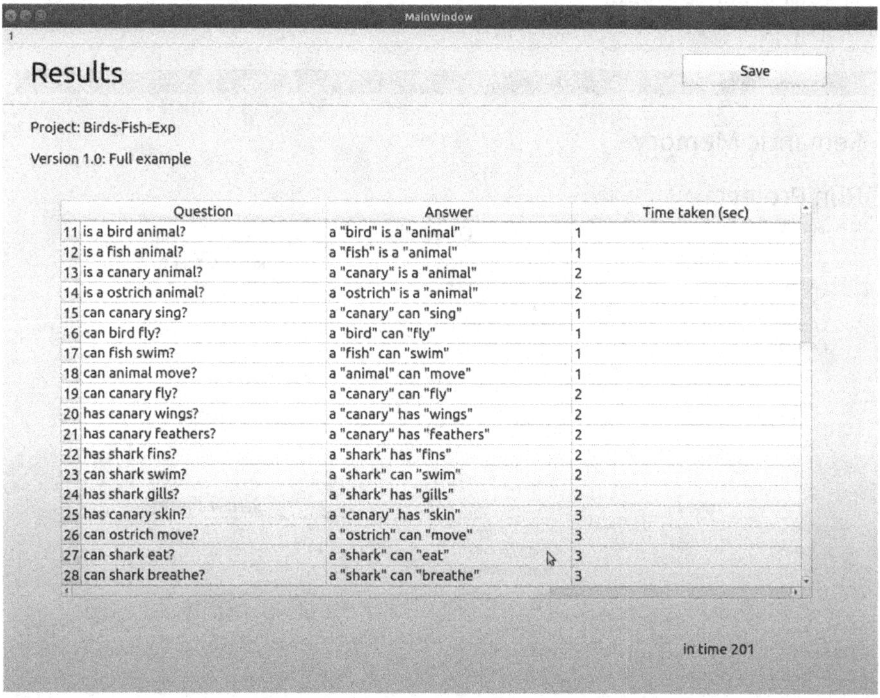
MainWindow
Results
Save
Project: Birds-Fish-Exp
Version 1.0: Full example
Question
Answer
Time taken (sec)
11 is a bird animal? | a "bird" is a "animal" | 1
12 is a fish animal? | a "fish" is a "animal" | 1
13 is a canary animal? | a "canary" is a "animal" | 2
14 is a ostrich animal? | a "ostrich" is a "animal" | 2
15 can canary sing? | a "canary" can "sing" | 1
16 can bird fly? | a "bird" can "fly" | 1
17 can fish swim? | a "fish" can "swim" | 1
18 can animal move? | a "animal" can "move" | 1
19 can canary fly? | a "canary" can "fly" | 2
20 has canary wings? | a "canary" has "wings" | 2
21 has canary feathers? | a "canary" has "feathers" | 2
22 has shark fins? | a "shark" has "fins" | 2
23 can shark swim? | a "shark" can "swim" | 2
24 has shark gills? | a "shark" has "gills" | 2
25 has canary skin? | a "canary" has "skin" | 3
26 can ostrich move? | a "ostrich" can "move" | 3
27 can shark eat? | a "shark" can "eat" | 3
28 can shark breathe? | a "shark" can "breathe" | 3
in time 201

Fig. 3. Results for the in silico replication of Collins and Quillian’s experiment.

Answers that could fully match the questions could be retrieved directly, thus requiring only one time units. For example, fact 18 “A fish is an animal” in Fig. 2 fully matches question 12 “Is a fish an animal?” in Fig. 3.

Instead, question 28 "Can a shark breathe?" in Fig. 3 is not fully matched by any fact representation. Answering the questions requires claiming the fact hierarchy in semantic memory (refer to the fact numbers in Fig. 2):

1. first by matching "shark" from the question with fact 24 "A shark is a fish";
2. then by matching "fish" from fact 24 with fact 18 "A fish is an animal";
3. finally by matching "animal" from fact 18 and "breathe" from the question with fact 11 "An animal can breathe".

For this reason answer 28 in Fig. 3 is retrieved in 3 time units.

Souza and Oberauer's Experiment. The second experiment was carried out by Souza and Oberauer [37] in 2017 to show that it is the total time duration that information spends in pSTM that influences whether or not it becomes represented in LTM. In their experiments, they used slower versus faster presentation of information. Slower presentation allows information to stay in STM for longer periods. In fact, slower presented information was found to be more likely to transfer to LTM.

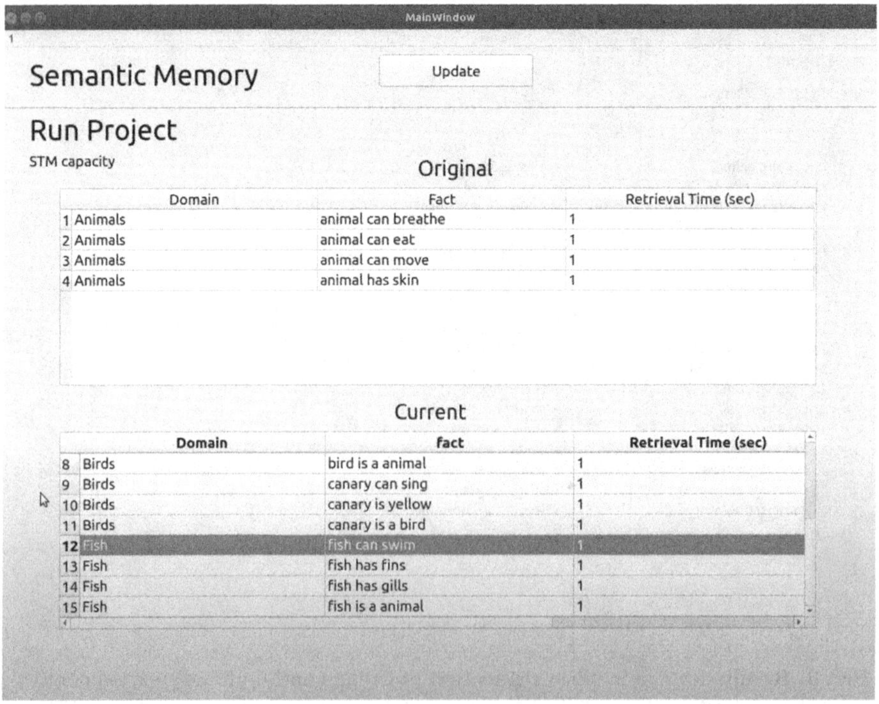

Fig. 4. Original and Final (Current) content of semantic memory for the in silico replication of Souza and Oberauer's experiment.

For this experiment, which is the in silico emulation of a learning process occurring over a long time, the semantic memory only contains the four facts in the table labelled "Original" in both Fig. 4 and Fig. 5. The same human memory model underwent two experiments. Also the facts and questions in the experiment setup were the same. However, the persistence time of the perceptions were different.

Semantic Memory   Update

Run Project

STM capacity   Original

	Domain	Fact		Retrieval Time (sec)
1	Animals	animal can breathe	1	
2	Animals	animal can eat	1	
3	Animals	animal can move	1	
4	Animals	animal has skin	1	

Current

	Domain	fact		Retrieval Time (sec)
15	Birds	ostrich is tall	1	
16	Birds	ostrich is a bird	1	
17	Fish	salmon is edible	1	
18	Fish	salmon is pink	1	
19	Fish	salmon is a fish	1	
20	Fish	shark can bite	1	
21	Fish	shark is dangerous	1	
22	Fish	shark is a fish	1	

Fig. 5. Original and Final (Current) content of semantic memory for the in silico replication of Souza and Oberauer's experiment.

To simulate fast learning facts and questions were presented every 2 time units, whereas to simulate slow learning they were presented every 10 time units. As shown in the tables labelled "Current" in Fig. 4 and 5, semantic memory contains 15 facts in the case of fast learning (Fig. 4) and 22 facts in the case of slow learning (Fig. 5), which is consistent with Souza and Oberauer's experimental results [37] and Burgess and Hitch's neural network model [10] where slow learning is more effective than fast learning.

5   Conclusion and Future Work

In this paper we presented an approach and tool for the simulations and analysis of memory processes and learning mechanisms underlying pSTM maintenance and transfer of information to LTM. We built up on our previous work by using BRDL [15] to model the human memory content, and its Real-Time Maude implementation [16–18] to provide dynamics to BRDL models. With respect to our previous work we extended the Maude implementation with the timing infrastructure to support the modelling of fast and slow learning mechanisms. We tested our approach by comparing the outcome of our in silico simulation with Souza and Oberauer's experimental results [37].

Moreover, the tool addresses cognitive scientists, who are obviously not familiar with formal methods and would have difficulty in using Maude directly as the modelling language. By using the tool interface they can replicate in silico their experiments with human subjects and compare their experimental results with various human memory models. This approach can provide a form of empirical validation for a number of cognitive models.

In our future work, we are planning to generalise the scope of the tool by combining the human memory component with the model of an interacting computer/physical system. Such an overall model could be formally verified using Real-time Maude model-checking features. Furthermore, we plan to have a web-based version of this generalised tool and equip it with features for supporting remote collaboration among research teams.

References

1. Atkinson, R.C., Shiffrin, R.M.: Human memory: a proposed system and its control processes. In: Spense, K.W. (ed.) The Psychology of Learning and Motivation: Advances in Research and Theory II, pp. 89–195. Academic Press (1968)
2. Atkinson, R.C., Shiffrin, R.M.: The control of short-term memory. Sci. Am. **225**(2), 82–90 (1971)
3. Baddeley, A.: The episodic buffer: a new component of working memory? Trends Cogn. Sci. **4**(11), 417–423 (2000)
4. Baddeley, A.: Working memory: looking back and looking forward. Nat. Rev. Neurosci. **4**, 829–839 (2003)
5. Baddeley, A.: Working memory: theories, models, and controversies. Ann. Rev. Psychol. **63**(1–29) (2012)
6. Baddeley, A., Gathercole, S., Papagno, C.: The phonological loop as a language learning device. Psychol. Rev. **105**(1), 158–173 (1998)
7. Baddeley, A.D., Hitch, G.: Working memory. In: Bower, G.H. (ed.) Psychology of Learning and Motivation: Advances in Research and Theory, pp. 47–89. Academic Press (1974)
8. Baddeley, A.D., Hitch, G.J.: Is the levels of processing effect language-limited? J. Mem. Lang. **92**, 1–13 (2017)
9. Barrouillet, P., Camos, V.: The time-based resource-sharing model of working memory. In: Logie, R., Camos, V., Cowan, N. (eds.) Working Memory: State of the Science. Oxford University Press (2007)
10. Burgess, N., Hitch, G.J.: Memory for serial order: a network model of the phonological loop and its timing. Psychol. Rev. **106**(3), 551–581 (1999)
11. Camos, V., Johnson, M., Loaiza, V., Portrat, S., Souza, A., Vergauwe, E.: What is attentional refreshing in working memory? Ann. N. Y. Acad. Sci. **1424**(1), 19–32 (2018)

12. Camos, V., Lagner, P., Barrouillet, P.: Two maintenance mechanisms of verbal information in working memory. J. Mem. Lang. **61**, 457–469 (2009)
13. Campoy, G.: Evidence for decay in verbal short-term memory: a commentary on Berman, Jonides, and Lewis (2009). J. Exp. Psychol. Learn. Mem. Cogn. **38**(4), 1129–1136 (2012)
14. Cattell, R.B.: Intelligence: Its Structure, Growth and Action. Advances in Psychology, vol. 35. Elsevier (1987)
15. Cerone, A.: Behaviour and Reasoning Description Language (BRDL). In: Camara, J., Steffen, M. (eds.) SEFM 2019. LNCS, vol. 12226, pp. 137–153. Springer, Cham (2020). https://doi.org/10.1007/978-3-030-57506-9_11
16. Cerone, A., Murzagaliyeva, D.: Information retrieval from semantic memory: BRDL-based knowledge representation and Maude-based computer emulation. In: Cleophas, L., Massink, M. (eds.) SEFM 2020. LNCS, vol. 12524, pp. 159–175. Springer, Cham (2021). https://doi.org/10.1007/978-3-030-67220-1_13
17. Cerone, A., Ölveczky, P.C.: Modelling human reasoning in practical behavioural contexts using Real-Time Maude. In: Sekerinski, E., et al. (eds.) FM 2019. LNCS, vol. 12232, pp. 424–442. Springer, Cham (2020). https://doi.org/10.1007/978-3-030-54994-7_32
18. Cerone, A., Pluck, G.: A formal model for emulating the generation of human knowledge in semantic memory. In: Bowles, J., Broccia, G., Nanni, M. (eds.) DataMod 2020. LNCS, vol. 12611, pp. 104–122. Springer, Cham (2021). https://doi.org/10.1007/978-3-030-70650-0_7
19. Cerone, A., Zhexenbayeva, A.: Using formal methods to validate research hypotheses: the Duolingo case study. In: Mazzara, M., Ober, I., Salaün, G. (eds.) STAF 2018. LNCS, vol. 11176, pp. 163–170. Springer, Cham (2018). https://doi.org/10.1007/978-3-030-04771-9_13
20. Collins, A.M., Quillian, M.R.: Retrieval time from semantic memory. J. Verbal Learn. Verbal Behav. **8**, 240–247 (1969)
21. Colom, R., Rebollo, I., Palacios, A., Juan-Espinosa, M., Kyllonen, P.C.: Working memory is (almost) perfectly predicted by g. Intelligence **32**(3), 277–296 (2004)
22. Craik, F.I., Lockhart, R.S.: Levels of processing: a framework for memory research. J. Verbal Learn. Verbal Behav. **11**(6), 671–684 (1972)
23. Jefferies, E., Ralph, M.A.L., Baddeley, A.D.: Automatic and controlled processing in sentence recall: the role of long-term and working memory. J. Mem. Lang. **51**, 623–643 (2004)
24. Kolers, A.P.: A pattern-analyzing basis of recognition. In: Cermak, L.S., Craik, F.I. (eds.) Levels of Processing in Human Memory, pp. 363–384. Psychology Press (2014)
25. Lewandowsky, S., Oberauer, K.: Rehearsal in serial recall: an unworkable solution to the nonexistent problem of decay. Psychol. Rev. **122**(4), 674–699 (2015)
26. Loaiza, V.M., Lavilla, E.T.: Elaborative strategies contribute to the long-term benefits of time in working memory. J. Mem. Lang. **117**, 104205 (2021)
27. Loaiza, V.M., McCabe, D.P.: The influence of aging on attentional refreshing and articulatory rehearsal during working memory on later episodic memory performance. Neuropsychol. Dev. Cogn. Sect. B Aging Neuropsychol. Cogn. **20**(4), 471–493 (2013)
28. Montague, W.E.: Elaborative strategies in verbal learning and memory. Psychol. Learn. Motiv. **6**, 225–302 (1972)
29. Naveh-Benjamin, M., Jonides, J.: Maintenance rehearsal: a two-component analysis. J. Exp. Psychol. Learn. Mem. Cogn. **10**, 369–385 (1984)

30. Norris, D.: Short-term memory and long-term memory are still different. Psychol. Bull. **143**, 992–1009 (2017)

31. Norris, D., Kalm, K., Hall, J.: Chunking and redintegration in verbal short-term memory. J. Exp. Psychol. Learn. Mem. Cogn. **46**(5), 872–893 (2019)

32. Oberauer, K., Jones, T., Lewandowsky, S.: The Hebb repetition effect in simple and complex memory span. Mem. Cogn. **43**, 852–865 (2015)

33. Oberauer, K., Lewandowsky, S.: Modeling working memory: a computational implementation of the time-based resource-sharing theory. Psychon. Bull. Rev. **18**(1), 10–45 (2011)

34. Ölveczky, P.C.: Real-time Maude and its applications. In: Escobar, S. (ed.) WRLA 2014. LNCS, vol. 8663, pp. 42–79. Springer, Cham (2014). https://doi.org/10.1007/978-3-319-12904-4_3

35. Pluck, G., Villagomez-Pacheco, D., Karolys, M.I., Montano-Cordova, M.E., Almeida-Meza, P.: Response suppression, strategy application, and working memory in the prediction of academic performance and classroom misbehavior: a neuropsychological approach. Trends Neurosci. Educ. **17**, 100121 (2019)

36. Bartsch, L.M., Singmann, H., Oberauer, K.: The effects of refreshing and elaboration on working memory performance, and their contributions to long-term memory formation. Mem. Cogn. **46**(5), 796–808 (2018). https://doi.org/10.3758/s13421-018-0805-9

37. Souza, A.S., Oberauer, K.: Time to process information in working memory improves episodic memory. J. Mem. Lang. **96**, 155–167 (2017)

38. Watkins, M.J., Peynircioglu, Z.F.: A perspective on rehearsal. Psychol. Learn. Motiv. **16**, 153–190 (1982)

Fostering Safe Behaviors via Metaphor-Based Nudging Technologies

Francesca Ervas[1]($\boxtimes$)(iD), Artur Gunia[2](iD), Giuseppe Lorini[1](iD),
Georgi Stojanov[3](iD), and Bipin Indurkhya[2](iD)

[1] University of Cagliari, Cagliari, Italy
ervas@unica.it
[2] Jagiellonian University, Kraków, Poland
artur.gunia@uj.edu.pl
[3] The American University of Paris, Paris, France

Abstract. Our research objective is to study the role of metaphor on the effectiveness of technologies that are designed to nudge people towards more healthy or socially appropriate behaviors. Towards this goal, we focus on the problem of motivating and encouraging appropriate social behaviors in the context of the ongoing COVID-19 pandemic, such as maintaining mandated social distance, wearing masks, and washing hands. Over the last two years, many countries have developed different approaches to promoting and enforcing the mandated behaviors. Here, we explore metaphor-based solutions to this problem by studying the following research questions: (1) How is it possible for artificial agents to recognize inappropriate behavior (mobile systems, robots)? (2) How to design metaphor-based interfaces of artificial agents that effectively influence relevant human decisions and choices in the event of improper behavior? Our approach is implemented in three steps: (1) Identifying inappropriate behaviors in the context of maintaining social distance. (2) Designing a persuasive metaphor-based interface to nudge people towards appropriate behaviors. (3) Designing a user study by deploying technologies that incorporate the interface. This research is interdisciplinary and concerns cognitive linguistics, IT, human-computer interactions, cognitive science, media ethics, and philosophy of law.

Keywords: Metaphor · Nudging technologies · Healthy behaviors

1 Introduction

In recent years, a number of persuasive technologies have been developed that are designed to nudge people towards more healthy or socially appropriate behaviors. As new technologies are introduced, such as mobile phones, computers, GPS systems, and self-trackers, they make possible a range of new behaviors, and new norms have to be established for these behaviors to determine what is appropriate and what is not in any given situation [9,18]. This might be of crucial importance in the context of sanitary emergencies, where appropriate collective behaviors, such as social distance, wearing masks, washing hands, could be nudged to reach a solution. Past research has shown that metaphors can be very persuasive in nudging people towards certain behaviors. Our goal is to explore this possibility by studying use cases, developing metaphor-based prototypes, and conducting preliminary user studies. Based on the results of these studies, we will propose some guidelines for designing metaphor-based interfaces to persuade people towards healthy and safe behaviors.

From the COVID-19 pandemic outbreak onward, advanced mobile and internet technologies have taken a giant leap in making our online activities take on a new dimension, including business meetings through video conferencing, e-learning, and deliveries of food ordered online. This has allowed new ways to channeling these technologies for correcting or nudging human behavior. The role of metaphors in persuading people to follow certain behavior patterns has been known for some time [46,47], but it is still not fully understood why some metaphors can be more persuasive than others in the context of sanitary emergencies. Our goal in this research is to study and design use cases, focusing on the problem of motivating and encouraging appropriate and safe social distance in the context of the ongoing COVID-19 pandemic.

In the last several months, many countries have developed different approaches to promoting and enforcing social distance. We explore the role of an interface based on the fire metaphor for COVID-19 to address this problem: COVID-19 is compared to a spreading fire, with people being "matches" that need to be far from each other to stop the fire from spreading. Metaphor has indeed been recently studied and discussed as a persuasive conceptual tool to let laypeople understand the need for social cooperation in preventing COVID-19 spread (see [34] for a critical analysis of different metaphors for COVID-19 prevention). In particular, our research is guided by the following research questions:

1. Can we design artificial agents to recognize inappropriate and unsafe behaviors?
2. How to design a metaphor-based interface of artificial agents that relevantly and effectively influence human decisions and choices in the event of improper behavior?

In what follows we provide a framework to address each question, based on which we will explore some challenges for testing nudging prototypes based on the fire metaphor.

2   How Artificial Agents Recognize Appropriate Social Distance

We adopt a conceptual perspective on the influence of artificial agents on human behavior in the context of sanitary emergencies. We focus on those artificial agents (mobile or web applications and/or social robots) that can recognize such behaviors: for example, an architecture of sensors that can collect data from the environment to recognize some relevant events for the mandated behavior. We also assume that these artificial agents have a nudging interface. First, we argue that "social distance" should be better understood by the artificial agent (but also by laypeople) in terms of "spatial distance" but allowing social closeness [1]. Second, we adopt the idea that decision-making techniques based on psychology and economics can be used to influence and correct human behavior, especially for maintaining social distance.

2.1 Social Distance as Spatial Distance and Social Closeness

During the COVID-19 pandemic, social distancing has been one of the crucial measures taken by many countries to slow down the spread of the virus. As per WHO recommendation on physical distancing[1], people were suggested to maintain a distance of at least 1 m from each other to reduce the possibility of transmission. Though spatial distance can be quite easily measured and maintained by an artificial agent, it is a difficult task for humans [48]. Because of this, explicit instructions and reminders (written and visual normative rules) were developed to enforce spatial distance in humans. However, despite the abundance of both linguistic and visual instructions and reminders, this restriction was (and still is) most often not properly respected. As has been pointed out, social distancing "pushes against human beings' fundamental need for connection with one another" [49], especially in difficult and stressful times. A lack of social closeness can also lead to serious mental health problems: affective and social support can help people during pandemics, but loneliness can bring anxiety and depression [1]. Thus, though spatial distancing is a safe behavior to prevent the transmission of COVID-19, social distancing leads to unsafe behavior, and so social closeness should be fostered to prevent unsafe mental health behaviors. In other words, effective technology should promote both spatial distance and social closeness.

The latest Information Technologies (ITs) can not only facilitate certain "spatial" tasks, such as spatial distancing (sometimes also by providing entertainment), but they can also influence our everyday social behavior. Previous research has studied how computer games can trigger emotions, including moral emotions (guilt, pride, compassion, gratitude, contempt, indignation, see e.g. [42]). Authors of these studies suggest that moral emotions are triggered when players perform especially immoral behavior in the game (primarily guilt) [19]. On this basis, they conclude that even while playing anti-social games, some pro-social effects may appear [16], such as civic engagement, establishing new social interactions, easier contacting of closed and shy people, and the need for making new contacts in the real world [15]. More recent research shows how certain behaviors in the virtual world influence behaviors in subsequent social interac-

---

[1] The WHO recommendation on physical distancing is available at https://www.who.int/westernpacific/emergencies/covid-19/information/physical-distancing (last accessed 2022/03/30).

tions. In particular, cognitive technologies can change the body schema/image and influence emotions, thereby affecting immersion and cyborgization [17].

Systems based on the idea of quantified self (or self-tracking) are a special group of mobile and wearable cognitive technologies that particularly affect our behavior. Using IT tools, it is possible to measure different aspects of our behavior, including spatial distance from other people. The 'quantified self' is the idea of measuring human behavior, biological signals, mood, or geographical location in order to optimize our life in various aspects, e.g., in order to increase our emotional and social intelligence, optimal sleep, prolong life, maintain health, track our preferences, habits, social and material practices. Optimization in these systems is associated with detecting various correlations, anomalies, and frequent occurrences in our behavior. Embedded decision-support systems may dictate what should be done to achieve optimization in a given field [41]. It is also connected with monitoring, controlling, and applying pressure to improve ourselves [26].

These tools can be regarded as elements of cognitive enhancement for supporting self-motivation, where the goal is to achieve the goals one sets for oneself. This motivational enhancement works by providing a better perception of one's needs, and then optimizing and controlling one's capabilities to fulfill these needs. Another goal is to stimulate and supplement cognitive processes and emotions for improving performance of the motivated tasks [17]. For example, apps for mobile phones for measuring physical activity, such as MapMyFintessor or Runkeeper, can monitor physical and physiological activity. These solutions are based on optimizing basic physiological processes. The second group of applications aims at self-improvement, based on scheduling, self-control, and (finally) self-awareness.

2.2 Nudging the Appropriate Behavior

Persuasive technology can positively influence social agency and modulate the way people connect to the environment, making sense of their relationships with the world and the others. Many people think that they have high multi-tasking skills, but our attention, as well as our abilities for epistemic vigilance, are limited and biased in specific ways both by the environment and by social relationships (see e.g. [38]). In the Nudge Theory proposed by [43], indirect reinforcement and suggestions are proposed as ways to influence the behavior and decision-making of groups or individuals [3,25,50]. The term "nudge" suggests that, instead of coercing a person, one may (only) "push" him/her (in a gentle manner), by taking advantage of certain human cognitive biases. Nudge is the modification of what Thaler and Sunstein call architecture, that is, the structure of the physical world (for example the arrangement of sweets in a canteen), to make it easier to determine the socially desired action.

Thaler & Sunstein [43] focus on "pushing" someone in some direction by mainly exploiting his/her cognitive biases. However, defining "nudging" as manipulation of choices, independently from "regulation", implies widening its meaning to include any attempt to influence someone else's behavior, even when merely providing information (see [23] for a critical perspective). Moreover, nudging effects are short-lived and their effectiveness depends on the correct identification of

1. the mechanisms through which information influences behavior;
2. the motivations for specific (in)appropriate behavior; and
3. the specific context in which the target behavior occurs [4].

More importantly for our research, language (and visual format) of the conveyed messages matters for nudging [33], thus different (metaphorical) framings of the messages can bring about the success or the failure of the nudging itself.

3   A Metaphor-Based Interface to Promote Social Distance

In our view, metaphorical representations, providing people with useful models of the (social) world [44,45] can be the basis of persuasive strategies used to build conceptual models of how artificial agents influence human behavior. Metaphor has been described as a cognitive process that leads people to grasp an unknown (often abstract) conceptual domain in terms of a better known (and often more concrete) conceptual domain [5,6,22]. Metaphor is thus a useful conceptual tool not only to design interfaces in human-computer interaction [8], but also to represent messages from artificial agents that can be easily understood by laypeople [21]. The analysis carried out in this way will allow us to develop an architecture of representation of messages from artificial agents that would reach people in their everyday life, thereby effectively influencing the mandated behavior in the case of social distance.

3.1   Metaphorical Framing as a Reasoning Device

Metaphor has been considered as a reasoning device that implicitly leads the audience to make some inferences from a source conceptual domain to a target conceptual domain. In the process, as has been noted widely [7,10,36], metaphor is never "neutral", because it provides a figurative frame that makes the audience ignore some properties of the source and select other properties, which become prominent. Thus, metaphor as a reasoning device has an "ignorance-preserving trait" [2,11]: it gently "pushes" the audience to select the relevant properties of the source to understand the target, while other properties of the source remain ignored or underrated.

In a well-known series of experiments, Thibodeau & Boroditsky [44,45] showed that when metaphorically framing a target, i.e., an important societal problem (e.g., crime) via different metaphors (virus vs. beast), participants consistently adopted different behaviors (enacting respectively social reforms vs. harsher enforcement laws). The authors argued that metaphor is a framing strategy that presents people with an implicit evaluation of the target (e.g. the societal problem), thus influencing the way they interpret and reason about it. In a similar framework, Scherer and colleagues described a disease (flu) literally vs. metaphorically (as a beast, riot, army, or weed) and showed that participants are more prone to get vaccinated when flu is described metaphorically rather than literally. The authors concluded that "describing the flu virus metaphorically in decision aids or information campaigns could be a simple, cost-effective way to increase vaccinations against the flu" [31].

However, other follow-up studies show that the metaphorical framing effect on reasoning and decision-making on the same societal issues was below the significance threshold. Steen and colleagues [40] found effects neither of the metaphorical frame for crime nor of the metaphorical textual support on reasoning. They concluded that a simple text exposure to the issue as an important social problem and the novelty of a metaphor might play a major role in enhancing the communicative effects of the text. In the same vein, the more recent study by Panzeri and colleagues [28] shows that describing COVID-19 as a war entailed no metaphorical framing effect on reasoning and decision making in the participants, but rather that the acceptance of metaphorical-consistent behavior was modulated by the participants' previous political views. In other words, reasoning about COVID-19 in terms of war just confirmed and reinforced right-wing oriented participants' previous beliefs.

Interestingly, an experimental study [30] on the use of metaphor for the solution of everyday dilemmas showed that the framing effect also depends on reasoning conditions. The authors proposed *the metaphor processing termination hypothesis*, suggesting that when a metaphor is unnecessary, not consistent with the reasoner's understanding process, or increases ambiguity, the metaphorical framing effect decreases: "reasoners were less likely to make decisions consistent with the metaphor, were less likely to rate the metaphor as apt, and were less likely to choose metaphor-consistent responses on a subsequent verbal analogy test" [30]. More recent empirical studies [13,14] have shown that conventional metaphors can lead people to revise the premises of the reasoning processes to hold their (already believed) conclusions, while creative metaphors can help them in finding alternative solutions, but this might crucially depend on *the affective coherence* of the metaphorical source with the target. Thus, the metaphorical framing might influence people's reasoning, also entailing a shift in their implicit attitudes toward the target, but this depends on the social context addressed by the metaphor, the reasoning aptness of the metaphor, and its affective coherence in light of a specific conclusion.

3.2   The Fire Metaphor for Spatial Distance and Social Closeness

In the conceptual metaphor theory [22], metaphors can modify people's behavior because they are not just "linguistic" manifestations, but rather constitute cognitive models for conceptualization. However, as we are mostly unaware of metaphorical framing, we might not be able to identify its effects on behaviors either. This is striking in the case of the latest crisis in our society during the COVID-19 pandemic where the metaphor of COVID-19 as war was widespread in political discourse, aiming to influence people's views about it. In the field of health communication, the WAR metaphor has been largely applied in discourse to describe illness and therapy management, especially in oncology [12, 36], thereby making it easier to conceptualize a phenomenon that is difficult to express literally in people's lives.

As the WAR metaphor is highly conventional and frequent in health communication, it is easier to understand than new and creative metaphors. However, scholars also highlighted some negative entailments of the metaphor [36, 37]: people reported feelings of anger or sadness in perceiving themselves as losers in a war that was not in their control. More recently, the WAR metaphor has been applied to the COVID-19 pandemic, thus moving from the individual health context to the collective health of entire communities threatened by COVID-19, where new key features of the WAR metaphor emerged [24]. Indeed, the war against COVID-19 expressed the urgency for masks and social distance as weapons against COVID-19, but was highly criticized in public debates for its negative consequences, especially from an affective point of view. Politicians used motivating language against COVID-19, targeting emotions to call for collective action [32]: by frightening people and instilling in them a fear of the unknown, social distancing has been imposed as physical isolation without connection with social spaces. Thus, as remarked by Schnell & Ervas [32], "new metaphors have been proposed in a variety of discourse genres (from social campaigns to political cartoons) to challenge the shortcomings of the WAR metaphor for COVID-19, and to find alternative and more suitable metaphors to talk about the social crisis engendered by the pandemic" (see the #ReframeCovid Initiative[2], for a collection of COVID-19 metaphors, in both verbal and visual shapes).

As previously pointed out [34, 35], the fire metaphor for COVID-19 could be more effective in communication and reasoning about health emergencies. In particular, the metaphorical frame of fire can highlight "different aspects of the pandemic, including contagion and different public health measures aimed at reducing it" [34]. Indeed, beyond evoking vivid and rich images, fire metaphors can convey danger and urgency, but also help in explaining the different phases of the pandemic, how transmission happens, and the role of individuals within that, how the pandemic connects with health inequalities, and other problems. Most importantly for our research, while the war metaphor cannot explain measures for reducing contagion, in the fire metaphors people are "trees" and "fuel", thus exploiting "the forest fire scenario to convey the effectiveness of quarantines and social distancing" [34]. These might be the basic elements to build a metaphor-based interface to promote social distancing as "spatial distancing", which can suggest how to avoid the spreading of the fire.

What is missing in Semino's analysis is that fire is also used as a metaphorical frame to entail affective "warmth" and "enlightenment". Indeed, other studies [32] show that COVID-19 as enlightenment entails a call to change and find a new direction in life: quarantine and spatial distance were also times and places

---

[2] The #ReframeCovid Initiative and the collection of COVID-19 metaphors is available at https://sites.google.com/view/reframecovid/home (last accessed 2022/03/30).

for thinking of and reflecting on our (pre-pandemic) life, making sense of the new situations and finding creative ways to socially connect to others. From this perspective, the fire metaphor for COVID-19 entails an opportunity (rather than a tragedy) to find creative ways of sharing our lives with others, rather than social isolation. These might rather be the basic elements of a metaphor-based interface to promote social closeness, besides "spatial distancing".

4   Design Challenges for Visual Metaphor-Based Prototypes

The fire metaphorical frame for COVID-19 has been proposed not only in verbal messages to promote social distance but also in the visual mode (images and video) to facilitate people's understanding, and to encourage them to think about the right thing to do to avoid contagion. An example of visual metaphor in a short video is provided by the graphic designer Juan Delcan[3], who depicted COVID-19 as a fire spreading via matches, representing people, who need the "right" distance to stop the spread. As in any metaphor, also this example can be criticized for missing analogies with the real-world situation, but in our perspective, it can be an effective way to gather people's attention on the (social) problem and nudge their behavior in the desired direction. The visual metaphor attracts people's attention precisely because it "creates" a (social) problem that urgently needs a solution.

As previously argued [20,29], a visual metaphor challenges our familiar conceptualization of the (social) world. A visual metaphor is "based on a disruption of existing familiar conceptualizations of objects and/or actions" [11]. Nudging technologies could therefore exploit the unfamiliar or changed "architecture of the world" presented via visual metaphor. To nudge the desired behavior and promote social distance, the change in architecture can be proposed via

1. homospatiality (i.e., the physical co-impossibility of the two discrete entities occupying the same space);
2. suspension of functionality (regarding objects or spaces);
3. unexpected affordances.

The visual version of the fire metaphor can thus be realized in different ways in the interface: in the example provided by Delcan, it is realized by suspending the functionality of the match, which suggests suspending the functionality of a person as a carrier for COVID-19 contagion.

In previous research [27], a system to generate visual metaphors based on algorithmic perceptual similarity has been proposed. In this research, a visual version of the fire metaphor can be exploited to create and test a persuasive interface for social distancing. The project first adopts some technical methods for identifying inappropriate behavior in spatial distancing, covering the use of

---

[3] The video is available at https://www.youtube.com/watch?v=8Hi9-5F2zW4 (last accessed 2022/03/30).

relevant sensors in mobile devices and social robots. The project thus aims to use open-access data from GPS transmitters, Bluetooth communication channels, and gyroscopes. A recognition and explanation system would then be designed using advanced machine- and deep-learning methods.

5 Conclusion

We explored the idea of metaphor-based interfaces for technologies designed to nudge social distance. Developing such tools requires guidelines, also for the privacy of data processed by artificial agents. The recognition of (appropriate) human behavior is associated with the collection of large amounts of data about the user and the user's environment, which can lead to the following problems:

1. legal issues related to methods of data management;
2. surveillance issues (for instance, the user of these technologies may have the feeling of being observed, which will reduce the effectiveness of such a solution).

In future research, it will also be important to consider, from an ethical point of view, if the metaphor can entail negative connotations, especially in different cultures.

References

1. Abel, T., McQueen, D.: The COVID-19 pandemic calls for spatial distancing and social closeness: not for social distancing! Int. J. Public Health **65**(3), 231–231 (2020). https://doi.org/10.1007/s00038-020-01366-7
2. Arfini, S., Casadio, C., Magnani, L.: Ignorance-preserving mental models thought experiments as abductive metaphors. Found. Sci. 1–19 (2018)
3. Berger, J.: The Catalyst: How to Change Anyone's Mind. Simon & Schuster, New York (2020)
4. Bicchieri, C., Dimant, E.: Nudging with care: the risks and benefits of social information. Public Choice **191**, 443–464 (2019). https://doi.org/10.1007/s11127-019-00684-6
5. Black, M.: Metaphor. Proc. Aristotelian Soc. **55**, 273–294 (1954)
6. Bowdle, B., Gentner, D.: The career of metaphor. Psychol. Rev. **112**(1), 193–216 (2005)
7. Burgers, C., Konijn, E.A., Steen, G.: Figurative framing: shaping public discourse through metaphor, hyperbole, and irony. Commun. Theory **26**, 1–21 (2016)
8. Colburn, T.R., Shute, G.M.: Metaphor in computer science. J. Appl. Log. **6**(4), 526–533 (2008)
9. Colley, A., et al.: The geography of Pokémon GO: beneficial and problematic effects on places and movement. In: Proceedings of the 2017 CHI Conference on Human Factors in Computing Systems, pp. 1179–1192 (2017)
10. Entman, R.: Framing: toward clarification of a fractured paradigm. J. Commun. **43**, 51–58 (1993)
11. Ervas, F.: Metaphor, ignorance and the sentiment of (ir)rationality. Synthese **198**(7), 6789–6813 (2019). https://doi.org/10.1007/s11229-019-02489-y

12. Ervas, F., Montibeller, M., Rossi, M.G., Salis, P.: Expertise and metaphors in health communication. Medicina Storia **9–10**, 91–108 (2016)
13. Ervas, F., Ledda, A., Ojha, A., Pierro, G.A., Indurkhya, B.: Creative argumentation: when and why people commit the metaphoric fallacy. Front. Psychol. **9**, 1815 (2018)
14. Ervas, F., Rossi, M.G., Ojha, A., Indurkhya, B.: The double framing effect of emotive metaphors in argumentation. Front. Psychol. **12**, 628460 (2021)
15. Ferguson, C.J.: Blazing angels or resident evil? Can violent video games be a force for good? Rev. Gen. Psychol. **14**(2), 68 (2010)
16. Grizzard, M., Tamborini, R., Lewis, R.J., Wang, L., Prabhu, S.: Being bad in a video game can make us morally sensitive. Cyberpsychol. Behav. Soc. Netw. **17**(8), 499–504 (2014)
17. Gunia, A.: Wzmocnienie poznawcze w kontekście transhumanistycznym: teoria, praktyka oraz konsekwencje wpływu technologii kognitywnych na człowieka (Doctoral dissertation) (2019)
18. Gunia, A., Indurkhya, B.A.: Prototype to study cognitive and aesthetic aspects of mixed reality technologies. In: 3rd IEEE International Conference on Cybernetics (CYBCONF), pp. 1–6. IEEE Press, New York (2017)
19. Hartmann, T., Toz, E., Brandon, M.: Just a game? Unjustified virtual violence produces guilt in empathetic players. Media Psychol. **13**(4), 339–363 (2010)
20. Indurkhya, B., Ojha, A.: An empirical study on the role of perceptual similarity in visual metaphors and creativity. Metaphor. Symb. **28**(4), 233–253 (2013)
21. Klingen, L.: The Intentionality of Interface Metaphors. Unpacking Assumptions, Thinking Design (2018)
22. Lakoff, G., Johnson, M.: Metaphors We Live By. Chicago University Press, Chicago (1980)
23. Lorini, G., Moroni, S.: Ruling without rules: not only nudges. Regulation beyond normativity. Global Jurist **20**(3) (2020)
24. Marron, J., Dizon, D.S., Symington, B., Thompson, M.A., Rosenberg, A.R.: Waging war on war metaphors in cancer and COVID-19. JCO Oncol. Pract. **16**(10), 624–627 (2020)
25. Matsumura, N., Fruchter, R., Leifer, L.: Shikakeology: designing triggers for behavior change. AI & Soc. **30**, 419–429 (2015)
26. Neff, G., Nafus, D.: Self-tracking. MIT Press, Cambridge (2016)
27. Ojha, A., Indurkhya, B.: On the role of perceptual similarity in producing visual metaphors. In: Barnden, J., Gargett, A. (eds.) Producing Figurative Expression: Theoretical, Experimental and Practical Perspectives, pp. 105–126. John Benjamins, Amsterdam (2020)
28. Panzeri, A., Rossi, F.S., Cerutti, P.: Psychological differences among healthcare workers of a rehabilitation institute during the COVID-19 pandemic: a two-step study. Front. Psychol. **12**, 578 (2021)
29. Pérez-Sobrino, P.: Multimodal metaphor and metonymy in advertising: a corpus-based account. Metaphor. Symb. **31**(2), 73–90 (2016)
30. Robins, S., Mayer, R.E.: The metaphor framing effect: metaphorical reasoning about text-based dilemmas. Discourse Process. **30**(1), 57–86 (2000)
31. Scherer, A.M., Scherer, L.D., Fagerlin, A.: Getting ahead of illness: using metaphors to influence medical decision making. Med. Decis. Making **35**(1), 37–45 (2015)
32. Schnell, Z., Ervas F.: Intercultural discussion of conceptual universals in discourse. Joint online methodology to bring on social change through novel conceptualizations of Covid-19. Palgrave Macmillan, London (2021)

33. Schultz, P.W., Nolan, J.M., Cialdini, R.B., Goldstein, N.J., Griskevicius, V.: The constructive, destructive, and reconstructive power of social norms. Psychol. Sci. **18**(5), 429–434 (2007)
34. Semino, E.: "Not soldiers but fire-fighters" – metaphors and Covid-19. Health Commun. **36**(1), 50–58 (2020)
35. Semino, E.: COVID-19: a forest fire rather than a wave? Metode Sci. Stud. J. **11**, 5 (2021)
36. Semino, E., Demjén, Z., Hardie, A., Payne, S., Rayson, P.: Metaphor, Cancer and the End of Life. A Corpus-Based Study. Routledge, London (2018)
37. Sontag, S.: Illness as Metaphor. Farrar Straus & Giroux, New York (1978)
38. Sperber, D., et al.: Epistemic vigilance. Mind Lang. **25**(4), 359–393 (2010)
39. Steen, G.J.: The paradox of metaphor: why we need a three-dimensional model for metaphor. Metaphor Symb. **23**, 213–241 (2008)
40. Steen, G.J., Reijnierse, G., Burgers, C.: When do natural language metaphors influence reasoning? A follow-up study to Thibodeau and Boroditsky (2013). PLoS ONE **9**(12), e113536 (2014)
41. Swan, M.: Sensor mania! the internet of things, wearable computing, objective metrics, and the quantified self 2.0. J. Sens. Actuator Netw. **1**(3), 217–253 (2012)
42. Tangney, J.P., Stuewig, J., Mashek, D.J.: Moral emotions and moral behavior. Annu. Rev. Psychol. **58**, 345–372 (2007)
43. Thaler, R.H., Sunstein, C.R.: Nudge: Improving Decisions About Health, Wealth, and Happiness. Penguin, Yale University Press (2009)
44. Thibodeau, P.H., Boroditsky, L.: Metaphors we think with: the role of metaphor in reasoning. PLoS ONE **6**(2), e16782 (2011)
45. Thibodeau, P.H., Boroditsky, L.: Natural language metaphors covertly influence reasoning. PLoS ONE **8**(1), e52961 (2013)
46. Thibodeau, P.H., Hendricks, R.K., Boroditsky, L.: How linguistic metaphor scaffolds reasoning. Trends Cogn. Sci. **21**(11), 852–863 (2017)
47. Thibodeau, P.H., Matlock, T., Flusberg, S.J.: The role of metaphor in communication and thought. Lang. Linguist. Compass. e12327 (2019)
48. Yamamoto, N.: Distance perception. In: Caplan, B., De Luca, J.J., Kreustzer, J. (eds.) Encyclopedia of Clinical Neuropsychology, pp. 1–5. Springer, Berlin (2017)
49. Zaki, J.: Integrating empathy and interpersonal emotion regulation. Annu. Rev. Psychol. **71**, 517–540 (2020)
50. Lorini, G., Moroni, S.: How to make norms with drawings: an investigation of normativity beyond the realm of words. Semiotica **233**, 55–76 (2020). https://doi.org/10.1515/sem-2018-0062

Developing the Semantic Web
via the Resolution of Meaning Ambiguities

Simone Pinna$^{(\boxtimes)}$, Francesca Ervas , and Marco Giunti

University of Cagliari, Cagliari, Italy
`simone.pinna@unica.it`

Abstract. The paper presents an interdisciplinary project in cognitive linguistics, computer science, and mathematical logic, aimed at the development of a theoretical framework and correspondent logical tools for the treatment, in the Semantic Web context, of some typical linguistic phenomena in natural languages, such as lexical ambiguities and figures of speech. In particular, we focus on some specific features of metaphor that need to be addressed in order to enhance the overall quality of knowledge representation in the Semantic Web. To this extent, we briefly present PROL (Parametric Relational Ontology Language) as a novel ontological approach to the representation of the whole semantic content of n-ary relations usually expressed in natural language. Lastly, we show how specific instances of metaphorical expressions can be represented and dealt with via PROL.

Keywords: Semantic Web · Meaning ambiguities · Contextual knowledge

1 Introduction

The main goal of the Semantic Web is the organization of web contents as a network of linked data through a machine-readable language [2]. However, the languages used for this purpose (RDF, RDFS, OWL, etc.) have several expressive limitations and many semantic ambiguities need to be solved in order to reach the main goal. For instance, the translation of n-ary relations in RDF-based languages is inherently difficult, as they can directly express only binary relations. The proposals of the W3C Working Group for dealing with these issues [37] have several weaknesses that may be solved by devising more general and effective ontological patterns [18]. Semantic ambiguities in natural language are indeed a problem that needs to be solved in order to better classify the knowledge available on the Web and to enable an effective use, by either people or artificial agents, of the cultural and scientific contents [22]. Our approach highlights the importance of having formal tools able to supply a semantically

Supported by Fondazione Banco di Sardegna (FdS 2019, research grant n. F72F20000420007).

faithful representation of a knowledge base. To this end, the semantic richness of a knowledge base expressed in natural language should be preserved in all its aspects, including those which are most difficult to deal with from a formal standpoint, such as semantic ambiguities. Indeed, a more context-sensitive and cognitive-oriented approach can help to discern not only the different linguistic phenomena included under the term "semantic ambiguities", but also the cognitive mechanisms underlying the understanding of those phenomena. Semantic ambiguities are omnipresent in natural language, in a continuum that ranges from literal to non-literal cases. Homonymy and polysemy are indeed cases in which a word has (completely and partially) different literal meanings [38]. Figurative language uses, as metonymy and metaphor, are instead cases in which a word has a non-literal meaning and a literal one [4, 29].

A unitary framework was proposed for both polysemy and conventional metaphor [43], but still, novel metaphors cannot be included in the framework [10], as well as other figurative language uses, such as ironic metaphors, hyperbole, litotes [34]. The role of context in the disambiguation of semantic ambiguities was also discussed, ranging from the pre-semantic context required by homonymy understanding to the semantic and post-semantic context required, respectively, by conventional vs. novel metaphor comprehension [33]. Still, there is a heated debate on how context shapes the understanding of semantic ambiguities [5, 35] that pervasively occur in everyday language and in different genres of discourse [36].

This paper focuses on the disambiguation of meaning ambiguities in the Semantic Web framework. We assume that meaning ambiguities are widespread and omnipresent in the natural languages of Web users, ranging from literal to non-literal ambiguities. The problem of understanding the intended meaning is thus crucial not only for human-human interaction but also for human-machine interaction. An easy solution would be to represent in our formal language all the intended meaning of a text, allowing ambiguities to appear only in the corresponding natural language expressions. However, this solution would bypass the problem of machine understanding of natural language. In the solution proposed in this paper we presuppose only a light processing of natural language, namely the recognition of statements representing facts, that are then translated into the formal language with the support of a very simple ontology expressed in a RDF compatible ontological language (PROL, see below), and an algorithm that supplies measures of semantic proximity between concepts. These elements should then allow the machine to identify semantically ambiguous expressions in the text like metaphorical ones.

The disambiguation of meaning ambiguities, especially for non-literal cases, depends on the use of context: the finer the knowledge coming from the context, the easier the disambiguation. Knowledge representation is a central issue for Artificial Intelligence and the Semantic Web. In particular, the ontological languages used for the Semantic Web have strong limitations, crucially the fact that they are able to directly express only binary relations. This complicates the representation of the context in which a statement appears, allowing for many kinds of semantic ambiguities. A correct representation of n-ary relations, can indeed provide the amount of context needed to solve several meaning ambiguities.

We, therefore, hypothesize that an effective representation of n-ary relations in RDF-based languages, such as RDFS or OWL, can enhance the match between the semantic content of their ontological representations and the intended meaning of these representations, by providing a sufficient amount of contextual information in order to solve problems of semantic ambiguities. Devising more general and effective ontological patterns, which provide for an adequate translation of n-ary relations in RDF, is crucial to handle contextual knowledge in order to properly understand users' intended meaning. To this extent, in Sect. 2 we briefly present PROL (Parametric Relational Ontology Language) as an alternative to the ontological patterns based on the reification model recommended by the W3C Working Group [37]. In Sect. 3, we consider metaphor and some specific features of this linguistic phenomenon that need to be addressed in order to enhance the overall quality of knowledge representation in the Semantic Web. In Sect. 4, we show how specific instances of metaphorical expressions can be represented and dealt with in PROL.

2   PROL and the Representation of Relations as Concepts

RDF is the declarative language that, together with its ontological extension RDFS (RDF Schema) and the more powerful OWL (an ontological language based on description logics and compatible with RDF), provides for formal representation of a knowledge base as a directed labeled graph. However, a strong limitation of RDF (as well as RDFS or OWL) is that its syntax can only express facts that involve binary relations. In a RDF graph, indeed, any subgraph expressing a specific fact consists of a subject-predicate-object triple. Unary relations and relations with $n \geq 3$ places can be formalized only indirectly, by a suitable translation into binary ones. While unary relations can be easily formalized in RDF by using RDF classes and the special property rdf:type, the lack of a standard pattern for the representation of n-ary relations (namely, relations with arity $n \geq 3$) leads to possible ambiguities in the interpretation of the corresponding graphs. Moreover, the logical concept of a n-ary relation implies that the relation holds, or does not, for n individuals in a given order, that is to say, it holds for ordered n-tuples.[1] This logical feature of n-ary relations is either completely lost

---

[1] We do not consider here relations with no fixed arity, or multigrade predicates [30]. In the conceptual graph framework, this issue has been tackled through the notion of variadic conceptual graphs [23, §2.1.4]. We pointed out elsewhere that the apparent variability of the arity of some relations can be dealt with by the concept of a sub-relation, which is a generalization to n-ary relations of the notion of a subproperty defined in RDFS [18, §2] Subrelations can be introduced in PROL by a straight-forward extension of its vocabulary. Another issue concerns the order of relations, which is not always relevant, such as in a binary symmetric relation. In our view, any relation holds, or does not, for a fixed number of individuals in a given order, so that it does not even make sense to ask whether a relation holds for some individuals

in the reification patterns proposed by the W3C Working Group [37] for representing such relations, or it is preserved but only to the cost of highly increasing their technical complexity and introducing a quite unnatural interpretation of the reified relation [18].

To tackle these problems (and other related ones) Giunti et al. [18] proposed PROL (Parametric Relational Ontology Language). PROL is a simple ontological language, compatible with RDF, which is designed to express an arbitrary n-ary fact (with $n \geq 1$) as a *parametric pattern*, namely, as a binary relation parameterized with respect to $n-2$ arguments (i.e., all the arguments except the first two).[2] The vocabulary of PROL includes just 6 terms (2 RDF classes and 4 RDF properties) defined by a simple RDFS ontology. Two terms (prol:type, prol:next) serve to represent any n-ary fact as a parametric pattern. The remaining four terms (prol:Relation, prol:Domain, prol:hasPlaces, prol:represents) serve to express the ontology that (a) defines the n-ary relations involved in the facts to be represented, as well as the corresponding parametric binary relations, and (b) allows for the correct detection and interpretation of the representing parametric patterns.[3]

Here we cannot give the full formal details of PROL (see [18]). However, we provide an illustration of its main features by the following paradigmatic example. Let us take the following fact: Irene gives her Teddy Bear to Laura. It is an instance of the ternary relation *()gives her()to()*, namely: (Irene) gives her (Teddy Bear) to (Laura). In PROL, this is represented by means of a binary relation that is parameterized with respect to the third individual of the triple, Laura, and holds for the first two individuals, Irene and Teddy Bear.

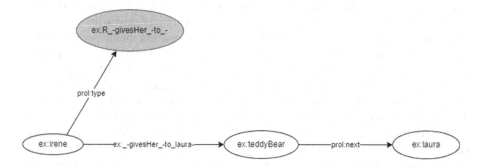

Fig. 1. PROL graph corresponding to the fact: Irene gives her Teddy Bear to Laura.

if they are not listed in some order [18, p. 709]. Thus, order is a necessary property of any relational fact, but this is not to say that order is relevant for any relation. In any case, the issue of the symmetry of a n-ary relation with respect to all, or even only some of its places, can also be treated through the concept of a subrelation.

[2] In this paper we do not consider the use of conceptual graphs for the representation of n-ary relations [11], for this approach is not directly implementable in some RDF compatible language, even if this is a viable possibility (see [1]).

[3] A possible extension of PROL may include a fuzzy treatment of n-ary relations in order to formally represent the use of fuzzy concepts in natural language (see [17].

In the corresponding graph (see Fig. 1) the grey ellipse, representing the relation, is linked via the term prol:type to the first individual of the triple (Irene). The arrow that links Irene to the second individual (Teddy Bear) includes a label representing the *parametric property* and it specifies the *parametric path* (individuated via the term prol:next) to be followed in order to fill up the remaining places of the relation. The choice of the right parametric path is obviously not determined just by the label on its first arrow, but by the definition in PROL (through its specific term prol:represents) of the parametric property "ex:_-givesHer_-to_Laura-" that represents the *n*-ary relation "ex:R:_-givesHer_-to_-" as a binary one. This definition specifies the parameter, namely "ex:Laura", which indicates the next node to be found in the corresponding parametric path. Once we have such definitions of the parametric properties we need, we can express the facts of a knowledge base as parametric patterns through the terms prol:type and prol:next (for formal details see [18, §5.1]). The parametric property, indeed, ensures the right path along the individuals connected via prol:next, in order to keep distinct the fact that Irene gives her Teddy Bear to Laura from, for example, the other fact that Irene gives her Teddy Bear to Marta. The second fact, which is a different instantiation of the same relation, leads to a different parametric path with respect to the first.

It is worth noting that, differently from standard RDF-based languages (such as RDFS or OWL), in PROL any n-ary relation ($n \geq 1$) is thought as a concept, namely, the intensional class of all ordered *n*-tuples that are instances of the relation. In a standard RDF graph labeled arrows represent binary relations or, that is the same, RDF properties, while the graph nodes are either individuals or classes. This means that the only concepts represented as nodes are classes. On the contrary, a knowledge base expressed in PROL will not include only classes and individuals as separate nodes, but relations of any arity ($n \geq 2$) as well. The result is that the corresponding graph will contain much more semantic information with respect to a knowledge base expressed as standard RDF triples, because any node representing a unary relation (which is the PROL equivalent of a RDF class) will be surrounded by nodes representing all the n-ary ($n \geq 1$) relations in which the individual instances of the unary relation also take part. As said, in a PROL graph unary relations can be identified with RDF classes. A knowledge base represented in PROL, then, will include as nodes only individuals and *n*-ary relations of any arity $n \geq 1$. Any node corresponding to a relation will be connected through prol:type to the first node of any parametric path that represents a *n*-tuple instantiating the relation (where it is intended that 1-tuples are identical to individuals).

3   Sources of Meaning Ambiguity in Metaphor Comprehension

In this paper, we focus on metaphorical expressions, as they are interesting cases of meaning ambiguities that have been proposed as a prototypical example of linguistic phenomena in the continuum ranging from literal to non-literal cases

[8, 9, 42, 44]. Metaphor is usually considered as a cognitive mechanism that leads the interpreter along a path of inferences from a *source* conceptual domain to a *target* conceptual domain. In the process, some properties of the source concept are selected (while other properties are ignored) to understand the target domain [12]. Metaphor is thus a device to fill in the conceptual distance between different conceptual domains, and to improve our knowledge of the target. Thus, far from being just a linguistic phenomenon, metaphor has also a conceptual and pedagogical function, that makes it a crucially important issue to be handled for the development of the Semantic Web.[4]

The conceptual distance between the source and the target can vary and can be covered by already "frozen" conceptual structures in a linguistic community and already lexicalized entries in the vocabulary of a language. This is the case of *conventional* metaphors, which have a status similar to polysemous terms and whose metaphorical meaning goes unnoticed by most native speakers [14, 16, 19]. *Novel* metaphors are rather new and creative uses of language that cannot be found in the vocabularies of languages, that create unprecedented connections between distant (or previously unthought as connected) conceptual domains. Of course, unless a metaphor is literalized [32], conventional metaphors can be revitalized, by creating new connections with some properties, while novel metaphors can "die" for overuse in a community and thus become conventional.

In their life, metaphors can thus vary in the continuum of literal to non-literal cases, depending on their use in the linguistic communities and the context where they appear. The context indeed provides useful information to select the relevant properties to be attributed to the target, especially in the case of novel metaphor, for which we do not have previous (linguistic) knowledge to rely on. The experimental literature has indeed shown that the processing of novel metaphors is rather different from that of conventional metaphors [3, 39]. Novel metaphor comprehension can involve perceptual properties coming from mental imagery [10, 26]. The information in a metaphorical expression such as "A woman is a Venice glass" would be too narrow to understand novel metaphors' typical imagistic effect [28]. Additional semantic information coming from the context and/or background encyclopedic knowledge (ex. Venice glasses are colorful) can thus help in novel metaphor comprehension (an advantage known as "context availability effect", see [15, 20]).

Not only the production of completely new metaphors (or new emergent properties, such as "being colorful"), but also the (i) linguistic structure of the metaphor and (ii) its directionality are challenges to be handled for the development of the Semantic Web. As to the linguistic structure of the metaphor, most of the previous literature on metaphor comprehension, especially the experimental literature, focused on *nominal* metaphors, especially of the form "A is

---

[4] Our choice of PROL, a RDF compatible ontological language, as the basis for our treatment of metaphor is motivated by the goal of contributing to the development of the Semantic Web, but at this early stage of the research project we cannot claim any superiority or advantage *per se* of our proposal with respect to other computational approaches to metaphor (for an overview, see [41]).

B" (ex. "The actor is a dog"). Less attention has been paid to *verbal* metaphors (ex. "Leo grasped an idea"), whose target is not explicitly mentioned and whose metaphoricity depends on the meaning of the verb (in relation to its object, in this case).

As to the *directionality* of a metaphor, concerning the direction of the attribution of properties from the source to the target conceptual domain, it depends on its linguistic structure, more precisely on the order of the terms in the relation (ex. "actor" and "dog"). Thus, the metaphor "The actor is a dog" cannot be reversed in "The dog is an actor": the latter would count as a different metaphor, as having a different source and thus different relevant properties. Of course, in specific contexts, where for instance we utter "The dog is an actor" referring to a dog who is actually an actor, the intended meaning of the latter may also be not metaphorical at all.

According to [40], the metaphorical directionality can be explained in terms of salience imbalance: the meaning of a metaphor depends on a matching process between high-salient properties of the source with low-salient properties of the target. Ortony [31] then «extended Tversky's model by defining the salience of a feature relative to the particular object of which it is an attribute: the same features may have different salience in two different objects» [27, p.95]. This is especially true in the case of conventional metaphors, while novel metaphors would be more prone to be interpreted as "bidirectional" [21], precisely because of their completely new and creative use. As pointed out [6,7,13,24,25], the source and the target conceptual domains interact, creating a more complex meaning or conceptual space, when compared to the individual concepts involved in the metaphor.

4   Solving Meaning Ambiguities in PROL

Consider a knowledge base that includes the following metaphorical expression: "Quell'attore è un cane", whose translation into English is "That actor is a dog". In Italian, this expression has a negative meaning and indicates that the subject in question is a very bad actor. Suppose that the same knowledge base also includes the expression: "That dog is an actor", where someone indicates an actual dog which is also an actual actor (for example, Lukas, the main character of the 2020 movie Lassie comes home). Here, we are not prone to attribute a metaphorical meaning to the statement. The attribution of a literal meaning in this case, indeed, is very likely. However, if the dog concerned is not an actual actor, then we will be forced to attribute to this utterance a metaphorical sense. Now, assume that our knowledge base has a sufficient amount of information in order to represent the concept of actor and dog via the relations in which individual actors and dogs take part. If we express the knowledge base in PROL and look at the formal properties of the corresponding graphs, would we be able to distinguish between metaphorical and literal meanings of natural language expressions? Figure 2 sketches a possible answer to this question. The two larger grey nodes correspond to the unary relations, or classes, *()is an actor* and *()is a dog*. Each class is surrounded by a semantic cloud formed by the relations that the individuals of that class are most likely to participate in. More precisely, the proximity of a relation to a class will be directly proportional to the number of individuals members of that class participating in the relation. For example, the two-place relation *()acted in a movie with()* will have more connections with the class of actors than the two-place relation *()takes()to the vet*, hence, it will be included in the semantic cloud surrounding the "actor" class.

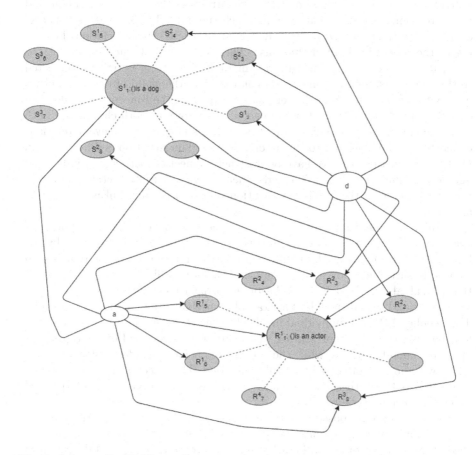

Fig. 2. Schema of a PROL graph representing the semantic clouds linked to the unary relations "dog" and "actor", and the connections between these two clouds and two individuals (in white), d (a dog) and a (an actor). Superscripts indicate the ariety of each relation, subscripts enumerate all the relations included in the same semantic cloud.

In Fig. 2 classes are connected to the respective relations via dashed lines, while the only two visible nodes representing individuals, namely "a" (the actor) and "d" (the dog), are connected via arrows to the relations or classes in which they take part. The dashed lines between the two classes *()is a dog, ()is an actor* and their connected relations represent paths whose intermediate nodes are all individuals. Indeed, any path in the graph that directly links any two concepts (i.e., classes or relations) is made of two intersecting parametric paths, p_1 and p_2, where p_1 represents an instance of one of the two concepts, and p_2 represents an instance of the other concept.

The graph sketched by Fig. 2 represents the context of two different utterances (1) "That actor is a dog" and (2) "That dog is an actor". In case (1) an individual actor (a), which is connected via arrows to the class ()is an actor and to the relations linked to that class in which the individual a takes part, is also connected to the class ()is a dog, but it is disconnected to the relations belonging to the semantic cloud of this class. This is a clue that the meaning of ()is a dog in the utterance is a non-literal (metaphoric) one: a is tightly connected to the cloud of actors, but *abstractly* connected to the cloud of dogs, namely, a does not take part in any of the relations relevant for the concept ()is a dog.[5] In case (2) an individual dog (d) is connected via arrows to both classes and both semantic clouds as well. This means that d takes part in the relevant relations connected to the dogs and to the actors, as well. The situation is different from (1) because, in this case, we have no abstract connection between an individual and two disconnected semantic clouds, then we can attribute literal meaning to the utterance, as in the previously mentioned case of the dog Lukas which is an actual actor.[6]

Similar considerations can be made in the case of a verbal metaphor, for example (3) "Leo grasps an idea". In this case, the metaphor is expressed by the binary relation ()grasps(). In a PROL graph, this relation would be represented as a node surrounded by a semantic cloud including all the relations in which the individuals involved in the relation ()grasps() most likely take part. This semantic cloud would include, for example, baseballs, door handles, tools, etc. The likelihood that an individual involved in the relation ()grasps() is an idea is low, because the node representing this relation is surrounded by concepts that apply to concrete objects, while the concept of an *idea* includes abstract objects which will thus be connected to a different cloud. The disconnection (adequately measured) between these two semantic clouds is a clue that the fact expressed by (3) has a non-literal meaning.

So far, we have proposed an intuitive idea of *proximity* to the central concept of a semantic cloud made of other concepts. This intuitive idea is based on a theoretical analysis of how a knowledge base is represented in PROL. But how could we formally define an adequate *measure* of semantic proximity? We cannot

---

[5] Of course, this is an extreme, idealized example. In most concrete cases (assuming a very informative knowledge base) we can conjecture that a would participate in some of the relations belonging to the semantic cloud of the concept ()is a dog, but the likelihood of any connection of an actor to the semantic cloud of ()is a dog would be very low when compared to the likelihood of any connection to the semantic cloud of ()is an actor.

[6] In the case where "that actor is a dog" has a metaphorical meaning, we would see a node representing an individual dog which is tightly connected to the semantic cloud of dogs, but only abstractly connected to that of actors.

give, here, a definitive answer to this question, but we can propose a tentative definition on the basis of the previous considerations. We have seen that, from an intuitive standpoint, the semantic proximity of a concept B to the central concept A of a semantic cloud is proportional to the likelihood that an instance of A is also an instance of B. Accordingly, when both B and A are unary relations (classes), we define *the semantic proximity* of B to A as the ratio between the number of members of $A \cap B$ and the total number of members of A.

In the general case, when both B and A are relations of arbitrary arity, the definition is similar, even though slightly more complex. Let B and A be relations with arity $n \geq 1$ and $m \geq 1$, respectively. Consider all the individuals belonging to some m-tuple which is an instance of A. By definition, these are the *members of A*. The members of B are defined in the same way. We then define *the semantic proximity of B to A* as the ratio between the number of members of A that are also members of B and the total number of members of A.

5   Conclusion

In this paper we have presented a way to solve meaning ambiguities in PROL, focusing on the distinction between metaphorical and literal expressions. We have also proposed a formal definition of semantic proximity that is likely to be useful for a deeper understanding of verbal metaphors, which to date have been less studied than nominal ones. The aim of the project is indeed to directly provide a better formal representation of natural language in all its aspects, metaphorical aspects included, without translating them into other, separate symbols in the formal representation. The translation could entail a loss of the conceptual/cognitive content of a metaphor [6], while we aim to represent metaphor as a meaning extension depending on variations of the semantic proximity of the concepts literally involved (ex. "grasp" and "idea" in "grasping an idea"). From this perspective, we deem metaphor to be crucial for the development of the Semantic Web, because it can act as a way to (re)categorize and (re)organize conceptual knowledge.

References

1. Baget, J.-F., Chein, M., Croitoru, M., Fortin, J., Genest D., et al.: RDF to conceptual graphs translations. In: CS-TIW: Conceptual Structures Tool Interoperability Workshop, Moscow, Russia, July 2009, p. 17 (2009)
2. Berners-Lee, T., Hendler, J., Lassila, O.: The semantic web. Sci. Am. **284**(5), 28–37 (2001)
3. Blasko, D., Connine, C.M.: Effects of familiarity and aptness on metaphor processing. J. Exp. Psychol. Learn. Mem. Cogn. **19**, 295–308 (1993)
4. Bolognesi, M., Brdar, M., Despot, K. (eds.): Metaphor and Metonymy in the Digital Age: Theory and Methods for Building Repositories of Figurative Language. John Benjamins Publishing Company, Amsterdam (2019)
5. Borg, E.: Finding meaning. Linguist **55**(3), 22–24 (2016)
6. Black, M.: Metaphor. Proc. Aristot. Soc. **55**, 273–294 (1954)

7. Black, M.: Models and Metaphors. Cornell University Press, Ithaca (1962)
8. Carston, R.: Enrichment and loosening: complementary processes in deriving the proposition expressed? Linguistische Berichte **8**, 103–127 (1997)
9. Carston, R.: Thoughts and Utterances: The Pragmatics of Explicit Communication. Blackwell, Oxford (2002)
10. Carston, R.: Metaphor: ad hoc concepts, literal meaning and mental images. Proc. Aristot. Soc. **110**, 295–321 (2010)
11. Chein, M., Mugnier, M.-L.: Graph-Based Knowledge Representation: Computational Foundations of Conceptual Graphs. Springer, London (2009). https://doi.org/10.1007/978-1-84800-286-9
12. Ervas, F.: Metaphor, ignorance and the sentiment of (ir)rationality. Synthese **198**(7), 6789–6813 (2019). https://doi.org/10.1007/s11229-019-02489-y
13. Fauconnier, G., Turner, M.: The Way We Think: Conceptual Blending and the Mind's Hidden Complexities. Basic Books, New York (2002)
14. Gibbs, R.W.: The Poetics of Mind: Figurative Thought, Language and Understanding. Cambridge University Press, Cambridge (1994)
15. Gildea, P., Glucksberg, S.: On understanding metaphor: the role of context. J. Verbal Learn. Verbal Behav. **22**, 577–590 (1983)
16. Giora, R.: On Our Mind: Salience, Context and Figurative Language. OUP, Oxford (2003)
17. Giunti, M.: Grafi pesati e relazioni n-arie: un approccio generale all'organizzazione automatica di dati secondo rapporti di rilevanza. In: Storari, P., Gola, E. (eds.) Forme e Formalizzazioni, pp. 229–245. CUEC Editrice, Cagliari (2010)
18. Giunti, M., Sergioli, G., Vivanet, G., Pinna, S.: Representing n-ary relations in the semantic web. Log. J. IGPL **29**(4), 697–717 (2021)
19. Glucksberg, S.: Understanding Figurative Language: From Metaphors to Idioms. Oxford University Press, Oxford (2001)
20. Glucksberg, S., Estes, Z.: Feature accessibility in conceptual combination: effects of context-induced relevance. Psychon. Bull. Rev. **7**, 510–515 (2000)
21. Goodblatt, C., Glicksohn, J.: Bidirectionality and metaphor: an introduction. Poet. Today **38**(1), 1–14 (2017)
22. Gracia, J., Lopez, V., d'Aqun, M., Sabou, M., Motta, E., Mena, E.: Solving semantic ambiguity to improve semantic web-based ontology matching. In: The 2nd International Workshop on Ontology Matching 2007, Busan, South Korea, 11 November 2007 (2007)
23. Haralambous, Y.: Des graphèmes à la langue et à la connaissance. Intelligence artificielle [cs.AI]. Université de Bretagne Occidentale (2020)
24. Indurkhya, B.: Metaphor and Cognition. Kluwer, Dordrecht (1992)
25. Indurkhya, B.: Emergent representations, interaction theory, and the cognitive force of metaphor. New Ideas Psychol. **24**(2), 133–162 (2006)
26. Indurkhya, B.: Towards a model of metaphorical understanding. In: Gola, E., Ervas, F. (eds.) Metaphor and Communication, pp. 123–146. John Benjamins Publishing, Amsterdam (2016)
27. Indurkhya, B., Ojha, A.: Interpreting visual metaphors: asymmetry and reversibility. Poet. Today **38**(1), 93–121 (2017)
28. Lai, V.T., Curran, T., Menn, L.: Comprehending conventional and novel metaphors: an ERP study. Brain Res. **1284**, 145–155 (2009)
29. Lakoff, G., Turner, M.: More than Cool Reason: A Field Guide to Poetic Metaphor. University of Chicago Press, Chicago (1989)
30. Oliver, A., Smiley, T.: Multigrade predicates. Mind **113**(452), 609–681 (2004)

31. Ortony, A.: Beyond literal similarity. Psychol. Rev. **86**(3), 161–180 (1979)
32. Pawelec, A.: The death of metaphor. Studia Linguistica **123**, 117–121 (2006)
33. Perry, J.: Indexicals and demonstratives. In: Hale, B., Wright, C. (eds.) Companion to the Philosophy of Language, pp. 586–612. Blackwell, Oxford (1997)
34. Popa-Wyatt, M.: Go figure: understanding figurative talk. Philos. Stud. **174**(1), 1–12 (2017)
35. Stanley, J.: Language in Context: Selected Essays. Oxford University Press, Oxford (2007)
36. Stukker, N., Spooren, W., Steen, G. (eds.): Genre in Language, Discourse and Cognition. De Gruyter Mouton, Berlin (2016)
37. W3C: Defining *n*-ary relations on the Semantic Web. W3C Working Group Note (2006). http://www.w3.org/TR/swbp-n-aryRelations. Accessed 12 Apr 2006
38. Taylor, J.R.: Linguistic Categorization. Oxford University Press, Oxford (2003)
39. Thibodeau, P., Durgin, F.H.: Productive figurative communication: conventional metaphors facilitate the comprehension of related novel metaphors. J. Mem. Lang. **58**(2), 521–540 (2008)
40. Tversky, A.: Features of similarity. Psychol. Rev. **84**(4), 327–352 (1977)
41. Veale, T., Shutova, E., Klebanov, B.B.: Metaphor: a computational perspective. Synth. Lect. Hum. Lang. Technol. **9**(1), 1–160 (2016)
42. Wilson, D., Carston, R.: Metaphor, relevance and the 'emergent property' issue. Mind Lang. **21**, 404–433 (2006)
43. Wilson, D., Carston, R.: A unitary approach to lexical pragmatics: relevance, inference and ad hoc concepts. In: Burton-Roberts, N. (ed.) Advances in Pragmatics, pp. 230–260. Palgrave, Basingstoke (2007)
44. Wilson, D., Carston, R.: Metaphor and the "emergent property" problem: a relevance-theoretic Treatment. Baltic Int. Yearb. Cogn. Logic Commun. **3**, 1–40 (2008)

Original or Fake? How to Understand the Digital Artworks' Value in the Blockchain

G. Antonio Pierro[1]([✉]) [iD], Moaaz Sawaf[2], and Roberto Tonelli[1] [iD]

[1] Universitá degli Studi di Cagliari, Cagliari, Italy
antonio.pierro@gmail.com
[2] Studio Zerance, Paris, France

Abstract. The recent Blockchain technology for recording and storing data opened a cultural scenario which prefigures unexplored frontiers. It is precisely in this frontier territory that a revolution in the digital art sector has come to life, represented by the introduction and dissemination of non-fungible tokens (NFTs) as certificates of ownership/authorship of a digital work. NFTs redefine the concepts of artwork ownership/authorship, authenticity and value. Supported by the Blockchain structure, which is in fact permanent and unchangeable, these certificates are inviolable, unassailable and indestructible, offering a type of guarantee never experienced before. As any new technologies, there are challenges that users face: 1) avoiding fraud (i.e. digital artworks that are not original, but mere copies of the original) 2) estimating the real value of a NFT connected to a digital artwork. The paper proposes a predictive tool, NFT price Oracle, as a solution for the Blockchain users that want to be sure they are purchasing an authentic digital artwork and wish to understand the value of the artwork (and thus how much to pay for the NFT associated with the artwork) in the Blockchain.

Keywords: Non-fungible token · Digital artwork · Blockchain · Oracle prices · History of production · History of effects

1 Introduction

In recent years, the digital art market has been considerably growing thanks to the combination of the Blockchain technology and non-fungible tokens [1]. A Blockchain is a distributed and decentralized ledger that contains connected blocks of transactions and it guarantees tamper-proof storage of approved transactions [2,3]. A NFT is a representation of a unique digital asset that cannot be equally swapped or traded for another NFT of the same type. NFTs are stored on a Blockchain and are used to represent the ownership/authorship of unique items [4,5]. NFTs attracted billions of dollars in investment. Christie's, the British auction house founded in 1766 by James Christie [6], auctioned an NFT associated with a digital work by the digital artist Mike Winkelmann (Beeple) for 69 million of dollars.

The paper aims to answer the following questions:

- RQ1 can the Blockchain technology with NFTs prevent the forgery of an artwork?;
- RQ2 can the Blockchain help the user to have a more correct estimate of the value of the artwork itself?

A. Cerone et al. (Eds.): SEFM 2021 Workshops, LNCS 13230, pp. 76–85, 2022.
https://doi.org/10.1007/978-3-031-12429-7_6

– RQ3 If not, how could a more correct estimate of NFT prices be given?

To answer the first question, the paper argues that the Blockchain can serve to trace the "history of production" of an artwork [7, 8], because it provides an incorruptible proof of ownership/authorship, meaning that an original artwork and their owners/authors can always be identified via the Blockchain, even when an image or video is widely copied.

To answer the second question, the paper argues that the data contained in the Blockchain are not sufficient to determine the real value of a digital artwork, as the value of a digital artwork is not reducible to the set of Blockchain's identifying properties, such as the number of transactions and the price at which it is traded. Rather, the digital artworks' value is a function of its "history of effects" [9], i.e. the net of valuable interactions that the original artwork spread within and outside the Blockchain in its lifespan. Blockchain users might face difficulties in evaluating the real value of an artwork (and its fakes), just based on the Blockchain history of transactions. They might also search data from the outside to understand the digital artworks' value in the Blockchain. The paper proposes a predictive tool, NFT price Oracle, as a solution for the Blockchain users that wish to understand the value of the artwork (and thus how much to pay for the NFT associated with the artwork) in the Blockchain. NFT price Oracle is designed to give a correct estimation of an artwork, providing the Blockchain users with the fairest price to pay for the NFT, based on both the artwork's history of production and its history of effects. In particular, the NFT price Oracle needs to be decentralized, to better guarantee that faked digital artworks are not paid as if they were original.

2 Technical Components

This section provides the readers with a brief introduction on the technology used to handle the digital artwork on the Blockchain: smart contracts, NFT and the software Oracle.

2.1 Blockchain

The Blockchain is an ordered sequence of blocks containing the records of valid transactions as approved by a consensus algorithms shared between a set of computational nodes in a peer-to-peer network. It is a shared ledger where, to keep unchangeable the block sequence and the temporal order of recorded transactions, each block includes a cryptographic hash depending on the information recorded on the previous block. Each block is also identified by progressive number named "height" [10]. Once a block is created and added to the Blockchain, the transactions in the block cannot be changed or deleted. This is to ensure the integrity of the transactions and to prevent the double-spending problem [11].

2.2 Smart Contract

A smart contract is a digitally signed, computable agreement between two or more parties. A virtual third party, a software agent, can execute and enforce (at least some of) the terms of such agreements. In the context of the Blockchain, where it truly takes its sense, a smart-contract is an event-driven program, with states, that runs on a replicated, shared ledger and which can take custody over assets on that ledger. Smart contracts on the Blockchain, created by computer programmers, are entirely digital and written using programming code languages. This code defines the rules and consequences in the same way that a traditional legal document would, stating the obligations, benefits and penalties, which may be due to either party in various different circumstances. The code is automatically executed by a distributed ledger system, in a non-repudiable and unbreakable way [12], diversely from traditional contracts. Smart contract code has some unique characteristics:

- Deterministic: Since a smart contract code is executed on multiple distributed nodes simultaneously, it needs to be deterministic i.e. given an input; all nodes should produce the same output. That implies the smart contract code should not have any randomness; it should be independent of time (within a small time window because the code might get executed a slightly different time in each of the nodes); and it should be possible to execute the code multiple times.
- Immutable: smart contract code is immutable. This means that once deployed, it cannot be changed. This of course is beneficial from the trust perspective but it also raises some challenges (e.g. how to fix a code bug) and implies that smart contract code requires additional due diligence/governance.
- Verifiable: Once deployed, smart contract code gets a unique address. Before using the smart contract, interested parties can and should view or verify the code.

2.3 NFT

"A non-fungible token (NFT) is a unit of data stored on a digital ledger, called a Blockchain, that certifies a digital asset to be unique and therefore not interchangeable" [13]. Indeed, anyone can obtain copies of the digital items (NFTs) on the Blockchain, but they are tracked so as to provide proof of ownership/authorship [14]. An NFT is like a certificate of authenticity for an object, real or virtual. The unique digital file is stored on a Blockchain network, with any changes in ownership verified by a worldwide network [13]. That means that the chain of custody is permanently marked in the file itself, and it is impossible to swap in a fake. The NFT file on the Blockchain does not contain the actual digital piece of art, music, video clip, etc., rather it is like a contract stipulating that "Mr A owes Mrs B a digital file of X".

As the name suggests, NFTs are characterized by their non-fungible nature. In economic terms, fungibility is the ability of an asset to be exchanged with other individual assets of the same type for the purpose of transacting value. Correspondingly, fungible assets in the same denomination imply the same value and include, for example, gold, a specific security or currency in FIAT/crypto. Conversely, this means, that NFTs are, by definition, not interchangeable, irreplaceable and unique.

While in the "real world" there is always a unique original work, such as the painting the artist created with her own hands, in the digital world there has so far been no equivalent in the sense of an "original digital artwork". The non-manipulative nature of NFTs enables both real and digital art objects to have original ownership/authorship.

For artists, this is a way to fight plagiarism as well as earn money by their work. NFTs also allow collectors to value digital art in a similar way to physical art, creating thus new opportunities for digital artists.

The main characteristics of the NFT are: 1) Indestructible - The technology that drives NFTs enhances these assets with the property of being immutable. All the metadata which are stored via smart contracts in the Blockchain cannot be replicated, removed or destroyed, thus granting ownership rights of the NFT, to the wallet or peer that possess it. 2) Verifiable - The process of authentication is also provided by the underlying features of the Blockchain technology. This allows a traceability within the ledger, as all the transactions are historically registered and stored within the blocks of data. This property allows any NFT attached to an artwork to be traced back to the original creator, eliminating the need of a third-party authentication method.

2.4 Oracle

In the Blockchain terminology, Oracle may have different meanings. An Oracle can be a program which provides the smart contracts with reliable data collected from outside the Blockchain. Oracles are also software systems which analyze some data and make some prediction on that basis [15].

In this paper, the term Oracle assumes a specific meaning related to the activity of predicting NFTs' prices. Thus an NFT Oracle analyses both Blockchain data and external data to predict the best price to pay for an NFT. The Oracle's predictions may be important for companies and users because of economic implications. It is thus crucial for them that Oracles' predictions are as reliable as possible.

3 Method

Existing Oracles rely only on blockchain data to provide the "history of effects" that determines the value of NFTs. This is done through the transaction history which provides information on the changes in ownership and the values at which the NFT was traded. However, the history of the effects of an artwork is not recorded on the blockchain.

The model presented in this study is based on the implementation of an Oracle that provides users with both the transaction history (see Sect. 4) which are stored in the blockchain and the history of the effects of the artwork itself in the world (see Sect. 5). Such an Oracle would also be centered on real world data and not just on blockchain data like in the existing Oracles. The real world data are very important for users to value an artwork and make a decision on the NFTs to purchase. For example, providing users with data about the history of the effects of an artwork can help them in the decision-making process by looking for more information related to the NFT they are buying or changing their view and preventing them from buying what they were buying. Indeed, additional information from the real world can have an impact on forming or revising their beliefs.

4 History of Production

There always have been cases of vendors trying to pass off other artists' work as their own. Recently, one of the most sensational facts is the "Fake Bansky NFT" sold through the artist's website for 336 thousand of dollars [16]. The scam was done in the following way: Bansky's official website was forged so that to include a link to an NFT auction with a artwork called "Great Redistribution of the Climate Change Disaster", which was a perfect copy of the original, even though associated with another artist. If the buyer had checked the address associated with the token on the Blockchain, he would have discovered that the work was only a copy of the original artwork. However, checking the belonging of an artwork to its original author through the Blockchain may not be an easy task for a user, who might instead be prone to buy the forgery as it were.

An Oracle centered on the user should try to minimize this possibility by checking real world data. For example, before writing data on the blockchain the oracles should check different sources to establish with a high probability that a piece of art is authentic or not. A possible solution to this problem is to base the model not on a unique source of data, but on more than one source. It would be possible to hack a site as it was for the fake-Bansky, but it is very unfeasible that the same user would hack different sources to make the fake-Bansky as real.

5 History of Effects

As we would like to argue, there is an alternative, user-centered way to help the Blockchain users not only to identify the ownership/authorship of the artwork, but also its value. In our view, this alternative is provided by the gadamerian notion of "history of effects". In his well-known book, Truth and Method [17], Hans-Georg Gadamer deals with the problem of understanding as a fundamental mode of human existence. In his perspective, the value of each human work is provided not very much by the work itself but rather by the (different) human interpretations and uses of the work itself. Indeed, a single artwork makes sense within an "hermeneutic circle", i.e. it acquires a value only on the whole background of interpretations that are provided by human beings. In the case of an artwork, the history of effects is not an appendix, or an addition to the understanding of a work, but rather the foundations for understanding its value. However, the set of interpretations of the artwork is potentially unlimited, while human beings are limited as well as their understanding, which is historically rooted in their world and times. As a consequence, everyone deals with historically and culturally determined conceptual structures which pre-determine and influence the artwork's understanding. Thus, everyone (the artwork's creator included) has a structural limited access to an artwork's "history of effects". It is therefore "require[d] an inquiry into "history of effects" every time a work of art or an aspect of the tradition is led out of the twilight region between tradition and history so that it can be seen clearly and openly in terms of its own meaning" [18].

In our view, in the case of digital artworks, the human inquiry into the "history of production" of an artwork can be helped via technology, and specifically the Blockchain technology above mentioned. In this case, however, the artwork's history of effects is not just related to real-world chain of uses and/or interpretations, as it can also be related to the history of the Blockchain itself. Thus, a double (both real-world and Blockchain) history of effects needs to be considered to estimate the value of the digital artwork. Indeed, in the Blockchain, NFTs have no objective intrinsic value, as they rely on a collective consensus to establish their value. It is the collective demand from the users, based on their understanding and use of the collectible that shapes value. Without a community aiming to collect a digital artwork, the digital artwork itself is not worth collecting. It is the collective acceptance of an artist's digital artwork that creates demand for the artwork, making originals worth millions of dollars.

The NFT price Oracle proposed in this paper thus considers the overall digital artwork's history of effects (at the time t the user starts her inquiry via NFT price Oracle) to predict its value, i.e. to provide the user with the fairest estimate of the digital artwork's value. As mentioned above, the value of an NFT comes not only from intrinsic factors accessible from the Blockchain, such as the proof of ownership/authorship provided by its "history of production", but also from extrinsic factors that are not directly accessible from the Blockchain, such as:

- Scarcity: Many NFTs represent digital objects that are unique or limited. For instance, only 10k CryptoPunks were released. Of those, only 24 are "apes". And among the apes, just one ape wears a fedora [19].
- Effects on entertainment industry. For example, some NFTs are more than just collectibles, since they can be used in games, like virtual lands, spells, or characters. This feature of NFTs gives them an added value, which accrues over time depending on the popularity of the underlying project. As the community of the project grows, many of them might be willing to pay more for a particular NFT [20].
- Art exhibition events. Recently, some art exhibition has shown NFTs in holographic form [21].
- Tangibility. Some NFTs are tied to real-world objects, which gives them value in terms of tangibility supported by the immutability of ownership/authorship. Collectables linked to NFTs can accumulate value over time as the number of items in circulation decreases.
- Transaction cost. Every trade of NFT has a transaction cost that impacts on possible earnings. Part of the transaction cost is due to Blockchain properties, part by the smart contract attached to the NFT.

Some elements that allow us to reconstruct part of the "history of effects" are in the Blockchain. The data that can be taken from the Blockchain are: the transaction number of a specific work of art, and the prices at which it is traded. Details on how these data are calculated from the Oracle can be found in the next section.

5.1 From a Data-Centered Model (TWPA) to a User-Centered Model (UCP)

The data-centered models are actually used by the Oracles to provide the users with proof of authenticity for a specific artwork and sometimes its value, based on the artwork's trades history (if it is present in the Blockchain). The aim of this research is to use also other data coming from the "external world", to provide the users with a more accurate estimate of the NFT value. To this aim, the model proposed in this paper collect and analyze both Blockchain data and external world data, shifting to a user-oriented perspective.

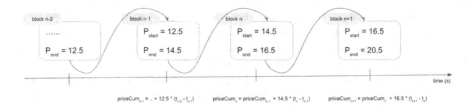

Fig. 1. NFT cumulative price on blockchain. The initial price of an NFT in the next block is equal to the final price of the same NFT in the previous block

In respect to Blockchain data, we based the user-oriented NFT price Oracle on the TWAP model. In finance, time-weighted average price (TWAP) is the average price of a tradable financial asset over a specified period of time, and it is used to determine if an asset is overvalued or undervalued. If the order price is below the TWAP, it is considered undervalued, while if it is more than the TWAP, it is considered overvalued. We based the NFT price Oracle on the TWAP model, because this formula is used for stock trend predictions [22]. Formula 1 formally denotes the TWAP. Figure 1 represent the NFT cumulative price on Blockchain.

$$TWAP = \frac{priceCumulative_n - priceCumulative_{n-1}}{timestamp_n - timestamp_{n-1}} \qquad (1)$$

Unlike the stock markets, it is not possible to directly apply the formula 1, because the NFT market is illiquid, there are few trades. NFTs are generally less frequently traded: this represents a problem for an Oracle, in the process of observing and retrieving price data. Therefore, we decided to design the NFT price Oracle to consider similar sales. Similar sales are defined as the prices at which similar collectibles have been sold. Similar sales are commonly used in determining real estate prices as well.

In respect to real-world data, these data plus the metadata contained in a NFT allow to define similar collectible class. The average price of collectibles defined in a specific collectible class has thus been used as a parameter for any other collectible being in a similar collectible class. In this paper, similarity is based not only on design elements and history of production in the Blockchain, but also on a similar history of effects in the data from the outside [23,24]. Example of data from the outside are the number of the author's followers, the number of the artwork's downloads, the number of art exhibition events expected in the next month related to the object linked to the NFT and the number of tickets sold to these events. These data are not available in the blockchain but are indeed available in the external world.

The TWAP formula could be applied only on NFT that are very frequently traded. We do not usually have such scenarios, because most NFTs are illiquid and they cannot be exchanged as easily as traditional investments, like stocks and bonds. To apply the TWAP formula to NFTs, we need to group NFTs that are similar. To this aim a cluster analysis might be performed. Each NFT consists of some metadata that can be used to evaluate the similarity of the artwork. To calculate the similarity between different NFTs, the model needs to analyze all the feature data for those NFTs. Some of these features could be the address associated with the NFT creator, the number of art exhibition events expected in the next month related to the object linked to the NFT and the number of tickets sold to these events. Then, a cluster analysis could then group a set of NFTs in such a way that NFTs in the same group (a cluster) are more similar to each other than to those in other groups or clusters. The cluster analysis can be achieved by algorithms that significantly differ in their results.

We analyzed the time series of some extrinsic data to establish if they can be used to predict the NFTs' prices. Table 1 shows the results of the pair-wise Granger causality test performed for NFT price versus the extrinsic variables. The results show how the data series of different extrinsic variables, such as the number of downloads and the number of art exhibitions of a particular NFT within a specific time interval (one day), do Granger-cause the NFT price variation. The $p < 0.05$ means that the null hypothesis is rejected, suggesting that the effect of the lagged values (i.e., values coming from an earlier point in time) of the other variables on the NFT price variable is statistically significant.

Table 1. Granger causality test results

Null hypothesis	F-statistics	Prob.	Decision
The NFTs price does not Granger-cause the number of downloads	4.213	0.0231*	Reject
The number of downloads does not Granger-cause NFTs price	2.172	0.0381*	Reject
The NFTs price does not Granger-cause the number of search requests	2.733	0.0297*	Reject
the number of search requests does not Granger-cause the NFTs price	1.322	0.0131*	Reject
The NFTs price does not Granger-cause the number of art exhibition	2.329	0.3821	Accepted
The number of art exhibition does not Granger-cause the NFTs price	1.426	0.4835*	Reject
The NFTs price does not Granger-cause the number of author's followers	1.223	0.0878	Accepted
The number of author's followers does not Granger-cause ot the NFTs proce	1.432	0.0733	Accepted
The NFTs price does not Granger-cause the value of the USD/Ether pair	2.423	0.4818	Accepted
The value of the USD/Ether pair does not Granger-cause the NFTs price	1.143	0.0416*	Reject

6   Threats to Validity

As to the internal validity of the study, it might be claimed that the model proposed in this study considers only few factors to estimate the NFT value. Indeed, the value of a specific NFT is due to a variety of causes. All collectible markets are affected by a variable completely separate from the collectible itself, alias the general trend in economy. Better economy generally means people's more willingness to spend on collectibles of all kinds. The Blockchain is unaware of a possible economic crisis. The Oracle could be made "aware" of economic crisis, but in this study this possibility is not considered.

Another factor to be considered is the crypto macro. Generally when crypto is performing well against fiat measurement, demand for NFTs are high. If the crypto markets were to crash, the demand for NFTs would be negatively affected. In any case, NFTs sold in crypto denomination will appear to keep gaining value in fiat terms, as crypto appreciates in price and lose value in the inverse case. We used the time-weighted average price to estimate the NFT value, but other models could be used and should be tested to check which one performs at best.

7   Conclusion

The market for NFTs has recently grown, with more than five billions of dollars spent in the first half of the year 2021. Despite the recent surge in popularity, the NFT (non-fungible token) space is still in its infancy [25] and the users need specific tools to understand the authenticity and the value of what they buy in the Blockchain, especially in the case of digital artwork. Experts say that buyers should be aware of illiquidity and fraud in this new market, because any user can try to take advantage of this business by offering digital works that do not belong to the author.

We argued that the current NFT Oracles based on a data-driven model are able to guarantee the digital artwork's ownership, simply by analyzing its history of production via the data stored on the Blockchain, but they cannot estimate neither the artwork's authorship nor the artwork value which rather depends on its "history of effects". We discussed and proposed a user-centered solution based on digital artworks' "history of effects" accessible from both the Blockchain and the real world, to provide the users with an accurate, fair and plausible estimate of the artwork.

The solution proposed in this paper needs to be further refined, as the NFT price Oracle is a first step to help the NFT investors' comprehension of the digital artworks' value, by collecting and analyzing information on different NFT projects.

References

1. Chohan, U.W.: Non-fungible tokens: blockchains, scarcity, and value. Critical Blockchain Research Initiative (CBRI) Working Papers (2021)
2. Wüst, K., Gervais, A.: Do you need a blockchain? In: 2018 Crypto Valley Conference on Blockchain Technology (CVCBT), pp. 45–54. IEEE (2018)
3. Zheng, Z., Xie, S., Dai, H.-N., Chen, X., Wang, H.: Blockchain challenges and opportunities: a survey. Int. J. Web Grid Serv. **14**(4), 352–375 (2018)
4. Rennie, E., Potts, J., Pochesneva, A.: Blockchain and the creative industries. Provocation Paper (2019)
5. Dowling, M.: Fertile land: pricing non-fungible tokens. Financ. Res. Lett. **44**, 102096 (2021)
6. Bayer, T.M., Page, J.R.: Christie's auction house. In: A History of the Western Art Market, pp. 224–229. University of California Press (2017)
7. Ullian, J., Goodman, N.: Projectibility unscathed. J. Philos. **73**(16), 527–531 (1976)
8. Goodman, D.J., Mirelle, C.: Consumer Culture: A Reference Handbook. ABC-CLIO (2004)
9. Gadamer, H.G.: Wahrheit und methode grundzüge einer philosophischen hermeneutik (1960)

10. Li, X., Jiang, P., Chen, T., Luo, X., Wen, Q.: A survey on the security of blockchain systems. Futur. Gener. Comput. Syst. **107**, 841–853 (2020)
11. Pilkington, M.: Blockchain technology: principles and applications. In: Research Handbook on Digital Transformations. Edward Elgar Publishing (2016)
12. Omohundro, S.: Cryptocurrencies, smart contracts, and artificial intelligence. AI Matt. **1**(2), 19–21 (2014)
13. Okonkwo, I.E.: NFT, copyright; and intellectual property commercialisation (2021)
14. Regner, F., Urbach, N., Schweizer, A.: NFTs in practice-non-fungible tokens as core component of a blockchain-based event ticketing application (2019)
15. Barr, E.T., Harman, M., McMinn, P., Shahbaz, M., Yoo, S.: The oracle problem in software testing: a survey. IEEE Trans. Softw. Eng. **41**(5), 507–525 (2014)
16. Tidy, J.: Fake Banksy NFT sold through artist's website for £244k. BBC (2021)
17. Gadamer, H.G.: Grundzüge Einer Philosophischen Hermeneutik. Mohr, Tübingen (1960)
18. Ebeling, F.: Hans georg gadamer's "history of effect" and its application to the pre-Egyptological concept of ancient Egypt. J. Egypt. Hist. **4**, 55–73 (2019)
19. Dowling, M.: Is non-fungible token pricing driven by cryptocurrencies? Financ. Res. Lett. **44**, 102097 (2021)
20. Murray, J.A.: Sell your cards to who: non-fungible tokens and digital trading card games. AoIR Selected Papers of Internet Research (2021)
21. Liu, Y., Wu, S., Xu, Q., Liu, H., Holographic projection technology in the field of digital media art. Wirel. Commun. Mob. Comput. **2021** (2021)
22. Kim, K.: Financial time series forecasting using support vector machines. Neurocomputing **55**(1–2), 307–319 (2003)
23. Vosniadou, S.: Analogical reasoning as a mechanism in knowledge acquisition: a developmental perspective. Center for the Study of Reading Technical Report; no. 438 (1988)
24. Yanulevskaya, V., et al.: Emotional valence categorization using holistic image features. In: 2008 15th IEEE International Conference on Image Processing, pp. 101–104. IEEE (2008)
25. Zhao, J.L., Fan, S., Yan, J.: Overview of business innovations and research opportunities in blockchain and introduction to the special issue (2016)

Grounding Psychological Shape Space in Convolutional Neural Networks

Lucas Bechberger$^{(\boxtimes)}$ and Kai-Uwe Kühnberger

Insititute of Cognitive Science, Osnabrück University, Osnabrück, Germany
{lbechberger,kkuehnbe}@uos.de

Abstract. Shape information is crucial for human perception and cognition, and should therefore also play a role in cognitive AI systems. We employ the interdisciplinary framework of conceptual spaces, which proposes a geometric representation of conceptual knowledge through low-dimensional interpretable similarity spaces. These similarity spaces are often based on psychological dissimilarity ratings for a small set of stimuli, which are then transformed into a spatial representation by a technique called multidimensional scaling. Unfortunately, this approach is incapable of generalizing to novel stimuli. In this paper, we use convolutional neural networks to learn a generalizable mapping between perceptual inputs (pixels of grayscale line drawings) and a recently proposed psychological similarity space for the shape domain. We investigate different network architectures (classification network vs. autoencoder) and different training regimes (transfer learning vs. multi-task learning). Our results indicate that a classification-based multi-task learning scenario yields the best results, but that its performance is relatively sensitive to the dimensionality of the similarity space.

Keywords: Psychological similarity spaces · Conceptual spaces · Shape perception · Convolutional neural networks

1 Introduction

Shape information plays an important role in human perception and cognition, and can be viewed as a bootstrapping device for constructing concepts [18,33,40]. Based on the principle of cognitive AI [42,44], which tries to base artificial systems on insights about human cognition, also artificial agents should be equipped with a human-like representation of shapes.

In this paper, we employ the cognitive framework of conceptual spaces [24], which proposes a geometric representation of conceptual knowledge based on psychological similarity spaces. It offers a way of neural-symbolic integration [23,46] by using an intermediate level of representation between the connectionist and the symbolic approach, which are represented by artificial neural networks and entirely rule-based systems, respectively. The overall conceptual space is structured into different cognitive domains (such as COLOR and SHAPE),

A. Cerone et al. (Eds.): SEFM 2021 Workshops, LNCS 13230, pp. 86–106, 2022.
https://doi.org/10.1007/978-3-031-12429-7_7

which are represented by low-dimensional psychological similarity spaces with cognitively meaningful dimensions. Conceptual spaces have seen a wide variety of applications in artificial intelligence, linguistics, psychology, and philosophy [34,70]. Typically, the structure of a conceptual space is obtained based on dissimilarity ratings from psychological experiments, which are then translated into a spatial representation through multidimensional scaling [14]. In this paper, we consider a recently proposed similarity space for the SHAPE domain [9–11].

The similarity spaces obtained by multidimensional scaling are not able to generalize to unseen inputs – a novel stimulus can only be mapped into the similarity space after eliciting further dissimilarity ratings [6]. In order to generalize beyond the initial stimulus set (which is necessary in practical AI applications), we have recently proposed a hybrid approach [8]: Psychological dissimilarity ratings are used to initialize the similarity space, and a mapping from image stimuli to coordinates in this similarity space is then learned with convolutional neural networks. Both our own prior study [8] and related studies by Sanders and Nosofsky [58,59] used a classification-based transfer learning approach on relatively unstructured similarity spaces involving multiple cognitive domains. In contrast to that, the present study focuses on the single cognitive domain of SHAPE and investigates a larger variety of machine learning setups, comparing two network types (classification network vs. autoencoder) and two learning regimes (transfer learning vs. multi-task learning).

The remainder of this article is structured as follows: In Sect. 2, we provide some general background on convolutional neural networks, conceptual spaces, and the cognitive domain of shapes. We then describe our general experimental setup in Sect. 3, before presenting the results of our machine learning experiments in Sect. 4. Finally, Sect. 5 summarizes the main contributions of this article and provides an outlook towards future work. All of our results as well as source code for reproducing them are publicly available on GitHub [7].[1]

2   Background

Our work combines the cognitive framework of conceptual spaces [24] (Sect. 2.1) applied to the cognitive domain of SHAPE (Sect. 2.2) with modern machine learning techniques in the form of convolutional neural networks (Sect. 2.3), following a hybrid approach (Sect. 2.4). In the following, we introduce the necessary background in these topics.

2.1   Conceptual Spaces

A conceptual space as proposed by Gärdenfors [24] is a similarity space spanned by a small number of interpretable, cognitively relevant *quality dimensions* (e.g., TEMPERATURE, TIME, HUE, PITCH). One can measure the difference between two observations with respect to each of these dimensions and aggregate them into

---

[1] See https://github.com/lbechberger/LearningPsychologicalSpaces/.

a global notion of semantic distance. Semantic similarity is then defined as an exponentially decaying function of distance.

The overall conceptual space can be structured into so-called *domains*, which represent, for example, different perceptual modalities such as COLOR, SHAPE, TASTE, and SOUND. The COLOR domain, for instance, can be represented by the three dimensions HUE, SATURATION, and LIGHTNESS, while the SOUND domain is spanned by the dimensions PITCH and LOUDNESS. Based on psychological evidence [2,62], distance within a domain is measured with the Euclidean metric, while the Manhattan metric is used to aggregate distances across domains.

Gärdenfors defines *properties* like RED, ROUND, and SWEET as convex regions within a single domain (namely, COLOR, SHAPE, and TASTE, respectively). Concept hierarchies are an emergent property of this spatial representation, such as the SKY BLUE region being a subset of the BLUE region. Based on properties, Gärdenfors now defines full-fleshed *concepts* like APPLE or DOG by using one convex region per domain, a set of salience weights (which represent the relevance of the given domain to the given concept), and information about cross-domain correlations. The APPLE concept may thus be represented by regions for RED, ROUND, and SWEET in the domains of COLOR, SHAPE, and TASTE, respectively.

This geometric representation of knowledge enables a straightforward implementation of *commonsense reasoning* strategies such as interpolative and extrapolative reasoning [17,61]. It also allows us to model *concept combinations* such as GREEN BANANA by restricting the region of the BANANA concept in the COLOR domain to the region representing GREEN and then updating the regions in other domains (such as TASTE) based on the aforementioned cross-domain correlations (e.g., by restricting it to the SOUR region). Moreover, conceptual spaces can be linked to the *prototype theory* of concepts from psychology [56], which states that each concept is represented by a prototypical example and that concept membership is determined by comparing a given observation to this prototype. In conceptual spaces, a prototype corresponds to the center of a conceptual region, which adds further cognitive grounding to the framework.

Conceptual spaces form an intermediate layer of representation that can act as a bridge between the symbolic layer and the connectionist layer [43]: *Connectionist approaches* make use of artificial neural networks and usually consider raw perceptual inputs (e.g., pixel values of an image), which can be interpreted as a very high-dimensional feature space (e.g., one dimension per pixel). These systems are often difficult to interpret and cannot model important principles such as compositionality. *Symbolic approaches* on the other hand are based on formal logics, but suffer from the *symbol grounding problem* [27], which means that the symbols they operate on are not tied to perception and action. Conceptual spaces can be used as an intermediate representation format which translates between these two approaches: Using a connectionist approach, raw perceptual input can be mapped onto the relatively low-dimensional and interpretable conceptual space. Points in this conceptual space can then be mapped to constants and variables from the symbolic layer, while conceptual regions correspond to symbolic predicates. This way, the advantages of both classical approaches can be combined in a cognitively grounded way.

2.2 The Cognitive Domain of Shapes

Over the past decades, there has been ample research on shape perception in different fields such as (neuro-)psychology [4,12,13,22,30,31,41,45,50,54,66], computer vision [15,47,49,71], and deep learning [3,25,38,63]. Although so far no complete understanding of the shape domain has emerged, there exist some common themes that appear in multiple approaches, such as the distinction between global structure and local surface properties [3,4,12,30], or candidate features such as ASPECT RATIO [4,12,15,45,47,50,66,71], CURVATURE [12,13,15,47,50,66,71], and ORIENTATION [4,15,31,45,54,66,71].

In the context of conceptual spaces, Gärdenfors [24] mainly refers to the model proposed by Marr and Nishihara [45], which uses configurations of cylinders to describe shapes on varying levels of granularity. This cylinder-based representation can be transformed into a coordinate system by representing each cylinder with its length, diameter, and relative location and rotation. If the number of cylinders is fixed, one can thus derive a conceptual space for the SHAPE domain with a fixed number of dimensions. A related proposal for representing the SHAPE domain within conceptual spaces has been made by Chella et al. [15], who use the more powerful class of superquadrics as elementary shape primitives, allowing them to express many simple geometric objects such as boxes, cylinders, and spheres as convex regions in their similarity space.

Both existing models of the shape domain within the conceptual spaces framework define complex shapes as a configuration of simple shape primitives and follow therefore a structural approach [22]. The number of primitives necessary to represent a complex object may, however, differ between categories. Since two stimuli can therefore not necessarily be represented as two points in the same similarity space, it becomes difficult to compute distances between stimuli. Also the psychological plausibility of these approaches has so far not been established.

In order to provide a conceptual space representing the holistic similarity of complex shapes, Bechberger and Scheibel [9–11] therefore followed a different approach: As stimuli, they used sixty line drawings of everyday objects from twelve different semantic categories (such as APPLIANCE, BIRD, BUILDING, and INSECT), taken from different sources and adjusted such that they match in relative object size as well as object position and object orientation (see Fig. 1a). Six categories contained visually similar items (e.g., APPLIANCE and BIRD), while the other six categories were based on visually variable items (e.g., BUILDING and INSECT). Bechberger and Scheibel conducted a psychological study with 62 participants, where an explicit rating of the visual dissimilarity for all pairs of items was collected, using a five-point scale ranging from "totally dissimilar" to "totally similar". In a small control experiment, Bechberger and Scheibel verified that the elicited ratings targeted shape similarity rather than overall conceptual similarity. Using the averaged dissimilarity ratings over all participants, they

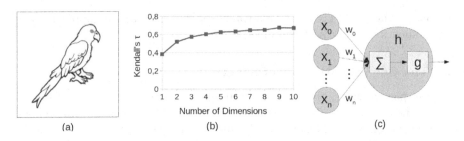

Fig. 1. (a) Example stimulus from the study by Bechberger and Scheibel [11]. (Image license CC BY-NC 4.0, source: http://clipartmag.com/cockatiel-drawing) (b) Correlation of distances in the similarity space to the original dissimilarity ratings. (c) Artificial neuron as nonlinear transformation of a weighted sum.

then applied an optimization technique called *multidimensional scaling* (MDS) to obtain similarity spaces of different dimensionality. MDS represents each stimulus as a point in an n-dimensional space and ensures that geometric distances between pairs of stimuli reflect their psychological dissimilarity [14].

Their investigations showed that the resulting shape spaces fulfilled the predictions of the conceptual spaces framework: Distances had a high correlation to the original dissimilarities (see Fig. 1b), and visually coherent categories (such as APPLIANCE and BIRD) were represented as small and non-overlapping convex regions. Human ratings of the objects with respect to three psychologically motivated shape features – namely, ASPECT RATIO, LINE CURVATURE, and ORIENTATION – could be interpreted as linear directions in these spaces. Overall, their analysis indicated that similarity spaces with three to five dimensions strike a good balance between compactness and expressiveness. For instance, Fig. 1b shows that higher-dimensional spaces only marginally improve the correlation of distances to dissimilarities. We will use their four-dimensional similarity space as a target for our machine learning experiments.

Recently, Morgenstern et al. [49] have proposed a 22-dimensional similarity space for shapes obtained via MDS from 109 computer vision features on a dataset of 25,000 animal silhouettes. Predictions of their similarity space on novel stimuli were highly correlated with human similarity ratings ($r = 0.91$), giving an indirect psychological validation to their approach. Moreover, Morgenstern et al. trained different shallow CNNs to map from original input images into their shape space. This relates their work quite strongly to our current study. In contrast to their work, we start from psychological data on complex line drawings and consider more complex network architectures.

2.3   Convolutional Neural Networks

Artificial neural networks (ANNs) consist of a large number of interconnected units [48, Chap. 4]. Each unit computes a weighted sum of its inputs, which is then transformed with a nonlinear *activation function* $g(\sum_i w_i \cdot x_i)$ (see Fig. 1c). Popular choices for the activation function include the so-called *Rectified Linear Unit* (ReLU, used for intermediate layers) $g(z) = \max(0, z)$ as well as the *sigmoid unit* $g(z) = \frac{1}{1+e^{-z}}$ (for binary classification output), the *softmax unit* $g(z)_i = \frac{e^{z_i}}{\sum_j e^{z_j}}$ (for multi-class classification output), and the *linear unit* $g(z) = z$ (for regression output).

The trainable parameters of an ANN correspond to the weights w_i of its connections. They are estimated by minimizing a given *loss function* which measures the network's prediction error. Popular loss functions include the *mean squared error* (which computes the average squared difference between regression output and ground truth) and the *cross-entropy loss* (which measures the difference between the probability distribution of the classification output and the ground truth). This loss function is minimized through *gradient descent*: One computes the derivative of the loss function with respect to each weight w_i and then makes small adjustments to the weights based on their derivatives. Instead of using the aggregated prediction error over all data points, one usually estimates it from a so-called *mini-batch*, i.e., a subset of examples [26, Chap. 8]. Training a neural network then consists of iterating over the dataset, where the network's weights are updated based on a new mini-batch in each iteration. Usually, multiple *epochs* (i.e., loops over the whole dataset) are needed until the optimization converges.

Instead of using the complete dataset for training the network, one usually considers a split into three subsets: The *training set* is used to optimize the parameters w_i of the network, while the *validation set* is used to monitor its performance on previously unseen examples. This can for instance be used for *early stopping*, where the training procedure is terminated, once the performance on the validation set stops improving. The *test set* is then used in the end to judge the expected generalization performance of the network on novel inputs.

A final important aspect of training neural networks are *regularization* techniques [26, Chap. 7], which are used to counter-act *overfitting* tendencies (where the network memorizes all examples from the training set, but is unable to generalize to novel inputs from the validation or test set): This includes adding a so-called *weight decay* term to the loss function, which penalizes large weight values and is motivated by the observation that smaller weights often lead to smoother decision behavior. *Dropout* is another popular regularization technique, where on each training step a randomly chosen subset of neurons is deactivated in order to increase the network's robustness.

With respect to computer vision tasks such as image classification, convolutional neural networks (CNNs) are considered to be the most successful ANN variant [26, Chap. 9]. They make use of so-called *convolutional layers* which apply the same set of weights (represented as kernel K) at all locations (see Fig. 2a). This and the relatively small *size of the kernel* (and thus the receptive

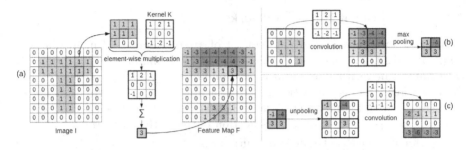

Fig. 2. (a) Two-dimensional convolution with a 3 × 3 kernel. (b) Combination of convolution and max pooling. (c) Combination of unpooling and convolution.

field of each unit) drastically reduces the number of connections between subsequent layers. CNNs furthermore use so-called *max pooling* layers (see Fig. 2b) to reduce the size of the image by replacing the output at a certain location by the maximum of its local neighborhood. For a max pooling layer, one has to specify both the *pool width* (i.e., the size of the area to aggregate over) and the so-called *stride* (i.e., the step size between two neighboring centers of pooling).

Typical convolutional networks start from a very high-dimensional input (namely, images) and reduce the representation size in multiple steps until a fairly small representation is reached which can then be used for classification through a softmax layer. However, in some settings one is also interested in the opposite direction: Creating a high-dimensional image from a low-dimensional hidden representation. For instance, *autoencoders* [26, Chap. 14] are an important unsupervised neural network architecture and are commonly used for dimensionality reduction and feature extraction. Autoencoders are typically trained on the task of reconstructing their input at the output layer, using only a relatively low-dimensional internal representation. They consist of an *encoder* (which compresses information) and a *decoder* (which reconstructs the original input).

For the encoder, a regular CNN can be used, whose max pooling layers, however, create a loss of information [26, Sect. 20.10.6]: In Fig. 2b, we only keep the maximum value for each 2 × 2 patch of the feature map. Since three out of the four values are discarded completely, it is impossible to accurately reconstruct them. In the decoder, one therefore needs to approximate the inverted pooling function with so-called *unpooling* steps. In most cases, one simply replaces each entry of the feature map by a block of size $s \times s$, where the original value is copied to the top left corner and all other entries of the block are set to zero [20] (cf. Fig. 2c). Using such an unpooling step followed by a convolution (which is together often called an *upconvolutional* layer) can be seen as an approximate inverse of computing a convolution and a subsequent pooling [20]. This allows us to increase the representation size inside the decoder in order to reconstruct the original input image from a small bottleneck representation.

2.4 A Hybrid Approach

A popular way of obtaining a conceptual similarity space is based on dissimilarity ratings [24], which are collected for a fixed set of stimuli in a psychological experiment. They are then converted into a geometric representation of the stimulus set by using MDS (cf. Sect. 2.2). The similarity spaces produced by MDS do not readily generalize to unseen stimuli: Mapping a novel input into the similarity space requires one to collect additional dissimilarity ratings and then to re-run the MDS algorithm on the enlarged dissimilarity matrix [6]. Artificial neural networks (ANNs) on the other hand are capable of generalizing beyond their training examples, but are not necessarily psychologically grounded.

In our proposed *hybrid approach* [8], we therefore use MDS on human dissimilarity ratings to "initialize" the similarity space and ANNs to learn a mapping from stimuli into this similarity space, where the stimulus-point mappings are treated as labeled training instances for a regression task. In general, ANNs require large amounts of data to optimize their weights, but the number of stimuli in a psychological study is necessarily small. We propose to resolve this dilemma not only through data augmentation (i.e., by creating additional inputs through minor distortions), but also by introducing an additional training objective (e.g., correctly classifying the given images into their respective classes). This additional training objective can also be optimized on additional stimuli that have not been used in the psychological experiment. Using a secondary task with additional training data constrains the network's weights and can be seen as a form of regularization. This approach has, for instance, successfully been used by Sanders and Nosofsky [58,59], who have fine tuned pretrained CNNs to predict the MDS coordinates on a dataset of 360 rocks. In contrast to their work, we focus on the single cognitive domain of shapes, use a considerably smaller set of annotated inputs, and consider a larger variety of machine learning setups.

3 General Methods

In this section, we describe both our data augmentation strategy for increasing the size and variability of our dataset (Sect. 3.1) and our general training and evaluation scheme for the machine learning experiments (Sect. 3.2).

3.1 Data Augmentation

The dataset of line drawings used for the psychological study by Bechberger and Scheibel [11] is limited to 60 individual stimuli. These stimuli are all annotated with their respective coordinates in the target similarity space and are thus our main source of information for learning the mapping task. Moreover, we used 70 additional line drawings which were not part of the psychological study by Bechberger and Scheibel, but which use a similar drawing style. Most applications of convolutional neural networks focus on datasets of photographs such as ImageNet [16]. In contrast to photographs, the line drawings considered in our experiments do not contain any texture or background, since they only show a single object using black lines on white ground. Sketches have similar characteristics, so we used the sketch datasets TU Berlin [21] and Sketchy [60] as additional data sources. From the TU Berlin corpus, we used all 20,000 sketches, while for the Sketchy corpus we selected a subset of 62,500 images by first keeping only the sketches which had been labeled as correct by the authors and then randomly selecting 500 sketches from each of the 125 categories. TU Berlin contains 250 classes and Sketchy uses 125 classes, and both datasets overlap on a subset of 98 common classes. We used the full set of 277 distinct classes when training the network on its classification objective.

We used the following augmentation procedure to further increase the size of our dataset and the variety of inputs: For each original image, we first applied a horizontal flip with probability 0.5 and then rotated and sheared the image by an angle of up to 15°, respectively. In the resulting distorted image, we identified the bounding box around the object and cropped the overall image to the size of this bounding box. The resulting cropped image was then uniformly rescaled such that its longer side had a randomly selected size between 168 and 224 pixels. Using a randomly chosen offset, the rescaled object was then put in a 224 × 224 image, where remaining pixels were filled with white. We used a uniform distribution over all possible resulting configurations for a given image, which makes smaller object sizes more likely since they have more translation possibilities than larger object sizes. Please note that we did not use the augmentation steps of horizontal flips and random shears and rotations on the line drawings from the psychological study, since the similarity space contains an interpretable direction which reflects the ORIENTATION of the object.

For each line drawing (both from the psychological study and additional ones), we created 2000 augmented versions, while the TU Berlin dataset and Sketchy were augmented with factors of 12 and 4, respectively. Overall, we obtained 120,000 data points for the line drawings from Bechberger and Scheibel, 140,000 data points for the additional line drawings, 240,000 data points for TU Berlin, and 250,000 data points for Sketchy.

3.2 Training and Evaluation Scheme

Sketch-a-Net [68,69] was the first CNN specifically designed for the task of sketch recognition and is essentially a trimmed version of AlexNet [37], the first CNN that achieved state of the art results in image classification tasks. For our encoder network (see Fig. 3), we used Sketch-a-Net and treated the size of its second fully connected layer as a hyperparameter. Moreover, we did not use dropout in this layer and used linear units instead of ReLUs to allow the network to predict the MDS coordinates (which can also be negative) as part of its learned representation. Classification was realized with a softmax layer on top of the encoder (not shown). In the autoencoder setup, we additionally used a decoder network inspired by the work of Dosovitskiy and Brox [19], which uses two fully connected layers and 6 upconvolutional layers.

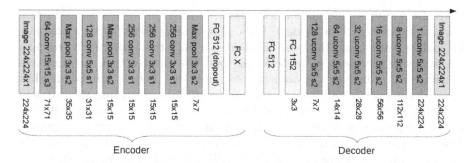

Fig. 3. Structure of our CNNs ("64 conv 15 × 15 s3" = convolutional layer with 64 kernels of size 15 × 15, using a stride of 3, "max pool" = max pooling layer, "FC" = fully connected layer, "uconv" = upconvolutional layer; output image size shown next to the layers).

We furthermore applied binary *salt and pepper noise* (which sets randomly selected pixels to their minimal or maximal value) to the inputs before feeding them to our network. This additional noise further increases the variety of the network's inputs and can be seen as an additional form of data augmentation. We chose salt and pepper noise rather than Gaussian noise, since the former is more adequate for our inputs, where most of the pixels are either black or white.

In our experiments reported below, we trained the overall network to minimize a linear combination of the classification error (softmax cross-entropy for the 277 classes), the reconstruction error (sigmoid cross-entropy loss with respect to the uncorrupted images[2]) and the mapping error (mean squared error for the target coordinates and the designated units of the second fully connected layer).

When evaluating the network's overall performance, we used the following evaluation metrics: For the classification task, we report separate classification accuracies (i.e., percentages of correctly classified examples) for the TU Berlin and the Sketchy datasets. For the reconstruction task, we report the reconstruction error (i.e., the binary cross-entropy loss) and for the mapping task, we report the mean squared error (MSE), the coefficient of determination R^2 (measuring the fraction of variance in the data explained by the model), and the mean Euclidean distance (MED) between the predicted point and the ground truth. We only used salt and pepper noise during training, but not during evaluation in order to avoid random fluctuations on the validation and test set.

Since the target coordinates used for learning and evaluating the mapping task are based only on 60 original stimuli, we decided to follow a five-fold *cross validation* scheme: We divided the original data points from each of the data sources into five *folds* of equal size and then applied the augmentation step for each fold individually. Therefore, all augmented images that were based on the same original data point are guaranteed to belong to the same fold, thus

---

[2] Since our autoencoder receives a corrupted image, but needs to reconstruct the uncorrupted original, it is a so-called *denoising* autoencoder [67].

preventing potential information leaks between folds. In our overall evaluation process, we rotated through these folds, always using three folds for training, one fold for testing, and the remaining fold as a validation set for early stopping (i.e., choosing the epoch with the lowest loss). We ensured that each fold was used once for testing, once as validation set, and three times as training set. The reported numbers are always averaged across all folds. By using this five-fold cross-validation technique, we implicitly trained five neural networks with the same hyperparameter settings, but slightly different data. Our averaged results therefore approximate the expected value of the neural network's performance on unseen inputs and hence the generalizability of the learned mapping.

During training, we used the Adam optimizer [36] as a variant of stochastic gradient descent, with the initial learning rate set to 0.0001, the default parameter settings of $\beta_1 = 0.9$, $\beta_2 = 0.999$, $\epsilon = 10^{-8}$, and a mini-batch size of 128. We ensured that each mini-batch contained examples from all relevant data sources according to their relative proportions: When training only on the classification task, we took 63 examples from TU Berlin and 65 from Sketchy. When training on both the classification and the mapping task, we used 25 line drawings, 51 sketches from TU Berlin, and 52 examples from Sketchy. Whenever the reconstruction task is involved, we used 21 line drawings, 24 additional line drawings, 41 examples from TU Berlin, and 42 data points from Sketchy. We always trained the network for 200 full epochs and select the epoch with the lowest validation set loss (classification loss or reconstruction loss for the pretraining experiments, and mapping loss for the multi-task learning experiments) in order to compute performance on the test set.

4   Experiments

In this section, we report the results of the experiments carried out with our general setup as described in Sect. 3. With our experiments, we try to show that learning a mapping from line drawings into the SHAPE space of Bechberger and Scheibel [9–11] is feasible. Moreover, we aim to investigate the influence of different learning regimes on the network's performance.

In Sect. 4.1, we train our network exclusively on the classification and reconstruction task, respectively, in order to identify promising settings for its various hyperparameters. This provides a starting point for our *transfer learning* experiments in Sect. 4.2, where we apply a linear regression on top of the pretrained CNNs. This is the perhaps most straightforward approach to solving the mapping problem. In Sect. 4.3, we then follow a more complex *multi-task learning* approach, where both the mapping task and the secondary objective (either classification or reconstruction) are optimized jointly. This is a computationally more costly approach, which may however also provide superior performance. Finally, in Sect. 4.4, we investigate how well the different approaches generalize to target similarity spaces of varying dimensionality.

Table 1. Selected hyperparameter configurations for the classification-based and the regression-based network, respectively.

Configuration	Encoder				Decoder	
	Weight decay	Dropout	Noise level	Rep. size	Weight decay	Dropout
C_{DEFAULT}	0.0005	True	10%	512	–	–
C_{SMALL}	0.0005	True	10%	256	–	–
$C_{\text{CORRELATION}}$	0.0010	False	10%	512	–	–
R_{DEFAULT}	0.0005	True	10%	512	0.0000	False
R_{BEST}	0.0000	False	10%	512	0.0000	False

4.1 Pretraining

We first considered a default setup of the hyperparameters based directly on Sketch-a-Net [68,69] and AlexNet [37]: We used a weight decay of 0.0005, dropout in the first fully connected layer, and a representation size of 512 neurons in the second fully connected layer. Moreover, we used 10% salt and pepper noise during training. For the decoder network, we used neither dropout nor weight decay. As evaluation metrics for the classification task, we considered the accuracies reached on TU Berlin and Sketchy, while for the autoencoder, the reconstruction error was used. In both cases, we also computed the monotone correlation of distances in the feature space to the dissimilarity ratings of Bechberger and Scheibel [11], measured with Kendall's τ [35]. Since a full grid search on many candidate values per hyperparameter was computationally prohibitive (especially in the context of a cross validation), we first identified up to two promising settings for each hyperparameter for both network types, before conducting a small grid search by considering all possible combinations of the remaining values. The most promising configurations selected in this grid search are shown in Table 1.

For the classifier network, the best classification performance (with accuracies of 63.2% and 79.3% on TU Berlin and Sketchy, respectively) was obtained by our default setup C_{DEFAULT}. This is considerably lower than the 77.9% on TU Berlin reported for the original Sketch-a-Net [68], which, however, used a much more sophisticated data augmentation and pretraining scheme. A considerably higher correlation of $\tau \approx 0.33$ (instead of $\tau \approx 0.27$ for C_{DEFAULT}) to the dissimilarity ratings could be obtained by disabling dropout and increasing the weight decay ($C_{\text{CORRELATION}}$), however, at the cost of considerably reduced classification accuracies of 36.4% and 61.5% on TU Berlin and Sketchy, respectively. Since reducing the representation size barely affected classification performance, we also consider C_{SMALL}, which uses 256 units and otherwise default parameters.

For the autoencoder, we observed that completely disabling both weight decay and dropout in both the encoder and the decoder led to considerably improved reconstruction performance (reconstruction error of 0.08 for R_{BEST} in comparison to 0.13 for R_{DEFAULT}). Also the correlation to the dissimilarities

Table 2. Results of our experiments on the four-dimensional target space. The respective best values for each configuration are shown in boldface.

Configuration	Task	Regressor	β/λ	τ	MSE	MED	R^2
Any	Any	Zero Baseline	–	–	1.0000	0.9940	0.0000
C_{DEFAULT}	Transfer	Linear	–	0.2743	0.5567	0.6879	0.4409
		Lasso	0.05	0.2743	0.4775	0.6419	0.5216
	Multi-task	CNN	0.0625	**0.4141**	**0.4041**	**0.5920**	**0.5775**
C_{SMALL}	Transfer	Linear	–	0.2777	0.5373	0.6737	0.4575
		Lasso	0.02	0.2777	0.4737	0.6396	0.5246
	Multi-task	CNN	0.125	**0.4118**	**0.4182**	**0.6020**	**0.5567**
$C_{\text{CORRELATION}}$	Transfer	Linear	–	0.3292	0.7307	0.7825	0.2624
		Lasso	0.05	0.3292	0.5478	0.6815	0.4505
	Multi-task	CNN	2.0	**0.4534**	**0.4513**	**0.6115**	**0.5201**
R_{DEFAULT}	Transfer	Linear	–	0.2228	0.9709	0.9054	0.0168
		Lasso	0.02, 0.05	0.2228	0.8315	0.8739	0.1631
	Multi-task	CNN	2.0	**0.3533**	**0.6211**	**0.7297**	**0.3369**
R_{BEST}	Transfer	Linear	–	0.3019	1.0791	0.9362	-0.0886
		Lasso	0.02	0.3019	0.7376	0.8102	0.2605
	Multi-task	CNN	0.25, 0.5, 2.0	**0.4033**	**0.5494**	**0.6846**	**0.4213**
			0.0625	0.3893	0.5504	0.6851	0.4144

increased from $\tau \approx 0.22$ to $\tau \approx 0.30$. Manipulation of all other hyperparameters did not lead to further improvements.

4.2 Transfer Learning

For our transfer learning task, we extracted the hidden representation of each network configuration for each of the augmented line drawings. We trained a linear regression from these feature spaces to the four-dimensional shape space by Bechberger and Scheibel [11]. In addition to the linear regression, we also consider a lasso regression (which introduces a weight decay term) with the following settings for the regularization strength β:

$$\beta \in \{0.001, 0.002, 0.005, 0.01, 0.02, 0.05, 0.1, 0.2, 0.5, 1.0, 2.0, 5.0, 10.0\}$$

Table 2 contains the results of these regression experiments. As we can see, the linear regression performs considerably better than the zero baseline (which always predicts the origin of the target space) for the classification-based feature spaces, but not for the reconstruction-based feature spaces. Moreover, regularization helps to improve performance on all feature spaces. A lasso regression on C_{SMALL} slightly outperforms C_{DEFAULT}, hinting at an advantage of smaller representation sizes. $C_{\text{CORRELATION}}$ does not yield competitive results, indicating that classification accuracy is a more useful selection criterion in pretraining than the correlation to human dissimilarity ratings.

Overall, transfer learning based on classification networks seems to be much more successful than transfer learning based on autoencoders, even when considering a lasso regressor. The reason for the relatively poor performance of R_{BEST}

Table 3. Cluster analysis of the augmented images in the individual feature spaces (averaged across all folds) using the Silhouette coefficient and the Cosine distance (i.e., the Cosine of the angle between the feature vectors).

Configuration	C_{DEFAULT}	C_{SMALL}	$C_{\mathrm{CORRELATION}}$	R_{DEFAULT}	R_{BEST}
0% Noise	0.6448	0.6347	0.5310	−0.0359	0.0818
10% Noise	0.6364	0.6263	0.5180	−0.0300	0.0768

and R_{DEFAULT} can be seen in Table 3, where we analyze how well the different augmented versions of the shape stimuli from Bechberger and Scheibel [11] are separated in the different feature spaces. We used the Silhouette coefficient [57], where larger values indicate a clearer separation of clusters. As we can see, the different augmented versions of the same original line drawing do not form any notable clusters in the reconstruction-based feature space. On the other hand, a relatively strong clustering can be observed for classification-based feature spaces under both noise conditions, indicating that the network is able to successfully filter out noise. We assume that this difference is based on the fact that the autoencoder needs to preserve very detailed information about its input (both local and global shape information) in order to create a faithful reconstruction, while a classification network only needs to preserve pieces of information that are highly indicative of class membership (rather global than local information).

4.3   Multi-task Learning

In our multi-task learning experiments, we trained our networks in the different configurations again from scratch, using, however, also the mapping loss as additional training objective. Instead of a two-phase process as used in the transfer learning setup, we therefore optimize both objectives at once. This allows the network to adapt the weights of its lower layers such that its internal representation becomes more useful for the mapping task, but comes at considerably higher computational cost. When training the networks, we varied the relative weight λ of the mapping loss in order to explore different trade-offs between the two tasks. We explored the following settings (where $\lambda = 0.25$ approximately reflects the relative proportion of mapping examples in the classification task):

$$\lambda \in \{0.0625, 0.125, 0.25, 0.5, 1.0, 2.0\}$$

Table 2 also contains the results of our multi-task learning experiments. As we can observe, mapping performance is considerably better in the multi-task setting than in the transfer learning setting for all of the configurations under investigation. The best results are obtained for C_{DEFAULT}, which is followed closely by C_{SMALL}. $C_{\mathrm{CORRELATION}}$ performs again considerably worse than the other classification-based setups, although its best multi-task results are still superior to all transfer learning results. Moreover, both reconstruction-based setups are not able to close the performance gap to the classification-based networks also under multi-task learning. These observations indicate that the multi-task learning regime is more promising than the transfer learning approach and that classification is a more helpful secondary task than reconstruction.

When taking a closer look at the optimal values for λ, we note that for both the C_{DEFAULT} and the C_{SMALL} setting, relatively small values of $\lambda \in \{0.0625, 0.125\}$ have been selected. For the $C_{\mathrm{CORRELATION}}$ configuration, however, a relatively large mapping weight of $\lambda = 2.0$ leads to the best mapping results, indicating that this configuration requires stronger regularization than others. Also for R_{DEFAULT}, a relatively large mapping weight of $\lambda = 2.0$ yielded the best performance, while no unique best setting for λ could be determined for the R_{BEST} configuration, where different metrics are optimized by different hyperparameter settings – here, $\lambda = 0.0625$ provides a reasonable trade-off.

In all cases, the introduction of the mapping loss leads to a considerable increase in the correlation τ to the dissimilarity ratings. This effect is, however, to be expected, since the mapping loss tries to align a part of the internal representation with the coordinates of the similarity space, which is explicitly based on the psychological dissimilarity ratings.

4.4   Generalization to Other Target Spaces

So far, we have only considered a four-dimensional target space. In this section, we investigate how well the different approaches generalize to target spaces of different dimensionality. We considered the respective best setups for all combinations of classification-based vs. reconstruction-based networks and transfer learning vs. multi-task learning (cf. Table 2) and retrained them (using the same values of β/λ) on all other target spaces (one to ten dimensions) of Bechberger and Scheibel [11], using again a five-fold cross validation.

Figure 4 illustrates the results of these generalization experiments for our three evaluation metrics. Both transfer learning approaches reach their peak performance for a two-dimensional target space, even though they have been optimized on the four-dimensional similarity space. Only with respect to the MED, performance is best on the one-dimensional target space. However, also the MED of the zero baseline is smallest for a one-dimensional space. If we consider the relative MED (by dividing through the MED of the zero baseline), then the best performance is again obtained on a two-dimensional target space. In all cases, classification-based transfer learning is clearly superior to reconstruction-based transfer learning.

The multi-task learners on the other hand do not show such a uniform pattern: While the reconstruction-based approach also obtains its optimum for a two-dimensional target space, the classification-based multi-task learner seems to prefer a four-dimensional target space. Moreover, both multi-task learners are more sensitive to the dimensionality of the target space than the transfer learning approaches: The classification-based multi-task learner considerably outperforms all other approaches on medium- to high-dimensional target spaces, while falling behind for a smaller number of dimensions. The reconstruction-based multi-task learner on the other hand performs quite poorly on high-dimensional spaces while

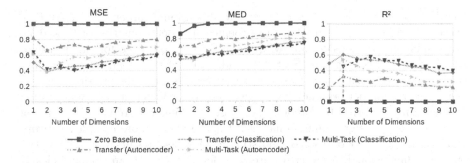

Fig. 4. Results of our generalization experiments to target spaces of different dimensionality for MSE, MED, and R^2.

becoming competitive on low-dimensional target spaces. Both multi-task learners use a mapping weight of $\lambda = 0.0625$, i.e., the smallest value we investigated. However, the size of the classification and reconstruction loss differed considerably, with a classification loss of around 1.3 to 1.6, compared to a reconstruction loss of 0.10 to 0.12 (both measured on the test set). The relative influence of the mapping objective on the overall optimization is thus considerably greater in the classification-based multi-task learner. One may therefore speculate that even smaller values of λ would have benefited the classification-based multi-task learner for smaller target spaces.

Overall, the results of this generalization experiment confirm the effects reported in our earlier study [8], where we also observed a performance sweet spot for a two-dimensional target space in a transfer learning setting. Again, we can argue that this strikes a balance between a clear semantic structure in the target space and a small number of output variables to predict. The observed sensitivity of the multi-task learning approach indicates that the target space should be carefully chosen before optimizing the multi-task learner.

5 Discussion and Conclusion

In this paper, we have aimed to learn a mapping from line drawings to their corresponding coordinates in a psychological SHAPE space. We have compared classification-based networks to autoencoders, investigating both transfer learning and multi-task learning. Overall, classification seemed to be a better secondary task than reconstruction, and multi-task learning consistently outperformed transfer learning. We found that the best performance in general was reached for classification-based multi-task learning, but that this approach was quite sensitive to the dimensionality of the target space. These results are mostly not surprising, given that multi-task learning allows for a finer-grained trade-off between tasks and that a reconstruction objective implicitly enforces also position and size information to be encoded.

We can compare our results to our earlier study [8] on a dataset of novel objects [29], where we used a lasso regression on top of a pretrained photograph-based CNN. There, we achieved for a four-dimensional target space a MSE of about 0.59, a MED of about 0.73, and a coefficient of determination of $R^2 \approx 0.39$. These numbers are considerably worse than the ones obtained for classification-based transfer learning (see Sect. 4.2), indicating that the SHAPE space considered in the current study poses an easier regression problem. Moreover, we can compare our performance with respect to the coefficient of determination to the results reported by Sanders and Nosofsky [58], who reported a value of $R^2 \approx 0.77$ for an eight-dimensional target space and a more complex network architecture, using a dataset of 360 stimuli. Our best results with $R^2 \approx 0.61$ on a two-dimensional target space are considerably worse than this and clearly not good enough for practical applications. We assume that performance in our scenario is heavily constrained by the network size and the number of stimuli for which dissimilarity ratings were collected. This urges for further experimentation with more complex architectures, larger datasets, different augmentation techniques, and additional regularization approaches.

Overall, our present study has illustrated that it is in principle possible to predict the coordinates of a given input image in a psychological similarity space for the SHAPE domain. Although performance is not yet satisfactory, this is an important step towards making conceptual spaces usable for cognitive AI systems. Once a robust mapping of reasonably high quality has been obtained, one can use the full expressive power of the conceptual spaces framework: For instance, the interpretable directions reported by Bechberger and Scheibel [11] can give rise to an intuitive description of novel stimuli based on psychological features. Also categorization based on conceptual regions, commonsense reasoning strategies, and concept combination can then be implemented on top of the predicted coordinates in shape space (cf. Sect. 2.1).

The approach presented in this article can of course also be generalized to other domains and datasets such as the THINGS data base and its associated embeddings [28] or the recently published similarity ratings and embeddings for a subset of ImageNet [55]. It can furthermore be seen as a contribution to the currently emerging field of research which tries to align neural networks with psychological models of cognition [1,5,6,32,38,39,49,51–53,58,59,64,65].

References

1. Attarian, I.M., Roads, B.D., Mozer, M.C.: Transforming neural network visual representations to predict human judgments of similarity. In: NeurIPS 2020 Workshop SVRHM (2020). https://openreview.net/forum?id=8wNMPXWK5VX
2. Attneave, F.: Dimensions of similarity. Am. J. Psychol. **63**(4), 516–556 (1950). https://doi.org/10.2307/1418869
3. Baker, N., Lu, H., Erlikhman, G., Kellman, P.J.: Deep convolutional networks do not classify based on global object shape. PLOS Comput. Biol. **14**(12), 1–43 (2018). https://doi.org/10.1371/journal.pcbi.1006613
4. Bar, M.: A cortical mechanism for triggering top-down facilitation in visual object recognition. J. Cogn. Neurosci. **15**(4), 600–609 (2003). https://doi.org/10.1162/089892903321662976

5. Battleday, R.M., Peterson, J.C., Griffiths, T.L.: Capturing human categorization of natural images by combining deep networks and cognitive models. Nat. Commun. **11**(1), 1–14 (2020)
6. Battleday, R.M., Peterson, J.C., Griffiths, T.L.: From convolutional neural networks to models of higher-level cognition (and back again). Ann. N. Y. Acad. Sci. (2021). https://doi.org/10.1111/nyas.14593
7. Bechberger, L.: lbechberger/LearningPsychologicalSpaces v1.5: machine learning study with CNNs on shapes data, September 2021. https://doi.org/10.5281/zenodo.5524374
8. Bechberger, L., Kühnberger, K.-U.: Generalizing psychological similarity spaces to unseen stimuli – combining multidimensional scaling with artificial neural networks. In: Bechberger, L., Kühnberger, K.-U., Liu, M. (eds.) Concepts in Action. LCM, vol. 9, pp. 11–36. Springer, Cham (2021). https://doi.org/10.1007/978-3-030-69823-2_2
9. Bechberger, L., Scheibel, M.: Analyzing psychological similarity spaces for shapes. In: Alam, M., Braun, T., Yun, B. (eds.) ICCS 2020. LNCS (LNAI), vol. 12277, pp. 204–207. Springer, Cham (2020). https://doi.org/10.1007/978-3-030-57855-8_16
10. Bechberger, L., Scheibel, M.: Representing complex shapes with conceptual spaces. In: Second International Workshop 'Concepts in Action: Representation, Learning, and Application' (CARLA 2020) (2020). https://openreview.net/forum?id=OhFQNQicgXy
11. Bechberger, L., Scheibel, M.: Modeling the holistic perception of everyday object shapes with conceptual spaces (in preparation)
12. Op de Beeck, H.P., Torfs, K., Wagemans, J.: Perceived shape similarity among unfamiliar objects and the organization of the human object vision pathway. J. Neurosci. **28**(40), 10111–10123 (2008). https://doi.org/10.1523/JNEUROSCI.2511-08.2008
13. Biederman, I.: Recognition-by-components: a theory of human image understanding. Psychol. Rev. **94**(2), 115–147 (1987)
14. Borg, I., Groenen, J.F.: Modern Multidimensional Scaling: Theory and Applications. Springer Series in Statistics, 2nd edn. Springer, New York (2005). https://doi.org/10.1007/0-387-28981-X
15. Chella, A., Frixione, M., Gaglio, S.: Conceptual spaces for computer vision representations. Artif. Intell. Rev. **16**(2), 137–152 (2001). https://doi.org/10.1023/a:1011658027344
16. Deng, J., Dong, W., Socher, R., Li, L.J., Li, K., Fei-Fei, L.: ImageNet: a large-scale hierarchical image database. In: 2009 IEEE Conference on Computer Vision and Pattern Recognition, pp. 248–255 (2009). https://doi.org/10.1109/CVPR.2009.5206848
17. Derrac, J., Schockaert, S.: Inducing semantic relations from conceptual spaces: a data-driven approach to plausible reasoning. Artif. Intell. **228**, 66–94 (2015). https://doi.org/10.1016/j.artint.2015.07.002
18. Diesendruck, G., Bloom, P.: How specific is the shape bias? Child Dev. **74**(1), 168–178 (2003). https://doi.org/10.1111/1467-8624.00528
19. Dosovitskiy, A., Brox, T.: Inverting visual representations with convolutional networks. In: Proceedings of the IEEE Conference on Computer Vision and Pattern Recognition (CVPR) (2016)
20. Dosovitskiy, A., Tobias Springenberg, J., Brox, T.: Learning to generate chairs with convolutional neural networks. In: Proceedings of the IEEE Conference on Computer Vision and Pattern Recognition (CVPR) (2015)

21. Eitz, M., Hays, J., Alexa, M.: How do humans sketch objects? ACM Trans. Graph. **31**(4), 1–10 (2012). https://doi.org/10.1145/2185520.2185540

22. Erdogan, G., Jacobs, R.A.: Visual shape perception as Bayesian inference of 3D object-centered shape representations. Psychol. Rev. **124**(6), 740–761 (2017)

23. Garcez, A.D., et al.: Neural-symbolic learning and reasoning: contributions and challenges. In: AAAI 2015 Spring Symposium on Knowledge Representation and Reasoning: Integrating Symbolic and Neural Approaches (2015)

24. Gärdenfors, P.: Conceptual Spaces: The Geometry of Thought. MIT Press, Cambridge (2000)

25. Geirhos, R., Rubisch, P., Michaelis, C., Bethge, M., Wichmann, F.A., Brendel, W.: ImageNet-trained CNNs are biased towards texture; increasing shape bias improves accuracy and robustness. In: International Conference on Learning Representations (2019). https://openreview.net/forum?id=Bygh9j09KX

26. Goodfellow, I., Bengio, Y., Courville, A.: Deep Learning. MIT Press (2016). http://www.deeplearningbook.org

27. Harnad, S.: The symbol grounding problem. Phys. D **42**(1–3), 335–346 (1990). https://doi.org/10.1016/0167-2789(90)90087-6

28. Hebart, M.N., Zheng, C.Y., Pereira, F., Baker, C.I.: Revealing the multidimensional mental representations of natural objects underlying human similarity judgements. Nat. Hum. Behav. (2020). https://doi.org/10.1038/s41562-020-00951-3

29. Horst, J.S., Hout, M.C.: The novel object and unusual name (NOUN) database: a collection of novel images for use in experimental research. Behav. Res. Methods **48**(4), 1393–1409 (2015). https://doi.org/10.3758/s13428-015-0647-3

30. Huang, L.: Space of preattentive shape features. J. Vis. **20**(4), 10–10 (2020). https://doi.org/10.1167/jov.20.4.10

31. Hubel, D.H., Wiesel, T.N.: Receptive fields of single neurones in the cat's striate cortex. J. Physiol. **148**(3), 574–591 (1959). https://doi.org/10.1113/jphysiol.1959.sp006308

32. Jha, A., Peterson, J., Griffiths, T.: Extracting low-dimensional psychological representations from convolutional neural networks. In: Proceedings for the 42nd Annual Meeting of the Cognitive Science Society (2020)

33. Jones, S.S., Smith, L.B.: The place of perception in children's concepts. Cogn. Dev. **8**(2), 113–139 (1993). https://doi.org/10.1016/0885-2014(93)90008-S

34. Kaipainen, M., Zenker, F., Hautamäki, A., Gärdenfors, P. (eds.): Conceptual Spaces: Elaborations and Applications. SL, vol. 405. Springer, Cham (2019). https://doi.org/10.1007/978-3-030-12800-5

35. Kendall, M.G.: A new measure of rank correlation. Biometrika **30**(1–2), 81–93 (1938). https://doi.org/10.1093/biomet/30.1-2.81

36. Kingma, D.P., Ba, J.: Adam: a method for stochastic optimization. arXiV (2014). https://arxiv.org/abs/1412.6980

37. Krizhevsky, A., Sutskever, I., Hinton, G.E.: ImageNet classification with deep convolutional neural networks. In: Pereira, F., Burges, C.J.C., Bottou, L., Weinberger, K.Q. (eds.) Advances in Neural Information Processing Systems, vol. 25, pp. 1097–1105. Curran Associates, Inc. (2012). https://proceedings.neurips.cc/paper/2012/file/c399862d3b9d6b76c8436e924a68c45b-Paper.pdf

38. Kubilius, J., Bracci, S., Op de Beeck, H.P.: Deep neural networks as a computational model for human shape sensitivity. PLOS Comput. Biol. **12**(4), 1–26 (2016). https://doi.org/10.1371/journal.pcbi.1004896

39. Lake, B., Zaremba, W., Fergus, R., Gureckis, T.: Deep neural networks predict category typicality ratings for images. In: Noelle, D.C., et al. (eds.) Proceedings of the 37th Annual Conference of the Cognitive Science Society (2015)

40. Landau, B., Smith, L., Jones, S.: Object perception and object naming in early development. Trends Cogn. Sci. **2**(1), 19–24 (1998). https://doi.org/10.1016/S1364-6613(97)01111-X

41. Li, A.Y., Liang, J.C., Lee, A.C.H., Barense, M.D.: The validated circular shape space: quantifying the visual similarity of shape. J. Exp. Psychol. Gen. **149**(5), 949–966 (2019)

42. Lieto, A.: Cognitive Design for Artificial Minds. Routledge (2021)

43. Lieto, A., Chella, A., Frixione, M.: Conceptual spaces for cognitive architectures: a lingua franca for different levels of representation. Biolog. Inspired Cogn. Archit. (2016). https://doi.org/10.1016/j.bica.2016.10.005

44. Marcus, G., Davis, E.: Rebooting AI: Building Artificial Intelligence We Can Trust. Pantheon (2019)

45. Marr, D., Nishihara, H.K.: Representation and recognition of the spatial organization of three-dimensional shapes. Proc. Roy. Soc. London Ser. B Biol. Sci. **200**(1140), 269–294 (1978)

46. Maruyama, Y.: Symbolic and statistical theories of cognition: towards integrated artificial intelligence. In: Cleophas, L., Massink, M. (eds.) SEFM 2020. LNCS, vol. 12524, pp. 129–146. Springer, Cham (2021). https://doi.org/10.1007/978-3-030-67220-1_11

47. Mingqiang, Y., Kidiyo, K., Joseph, R.: A survey of shape feature extraction techniques. Pattern Recogn. **15**(7), 43–90 (2008)

48. Mitchell, T.M.: Machine Learning. McGraw Hill, New York (1997)

49. Morgenstern, Y., et al.: An image-computable model of human visual shape similarity. PLoS Comput. Biol. **17**(6), 1–34 (2021). https://doi.org/10.1371/journal.pcbi.1008981

50. Ons, B., Baene, W.D., Wagemans, J.: Subjectively interpreted shape dimensions as privileged and orthogonal axes in mental shape space. J. Exp. Psychol. Hum. Percept. Perform. **37**(2), 422–441 (2011)

51. Peterson, J.C., Abbott, J.T., Griffiths, T.L.: Adapting deep network features to capture psychological representations: an abridged report. In: Proceedings of the Twenty-Sixth International Joint Conference on Artificial Intelligence, IJCAI-17, pp. 4934–4938 (2017). https://doi.org/10.24963/ijcai.2017/697

52. Peterson, J.C., Abbott, J.T., Griffiths, T.L.: Evaluating (and improving) the correspondence between deep neural networks and human representations. Cogn. Sci. **42**(8), 2648–2669 (2018)

53. Peterson, J.C., Battleday, R.M., Griffiths, T.L., Russakovsky, O.: Human uncertainty makes classification more robust. In: Proceedings of the IEEE/CVF International Conference on Computer Vision (ICCV) (2019)

54. Riesenhuber, M., Poggio, T.: Hierarchical models of object recognition in cortex. Nat. Neurosci. **2**(11), 1019–1025 (1999)

55. Roads, B.D., Love, B.C.: Enriching ImageNet with human similarity judgments and psychological embeddings. In: Proceedings of the IEEE/CVF Conference on Computer Vision and Pattern Recognition (CVPR), pp. 3547–3557 (2021)

56. Rosch, E., Mervis, C.B., Gray, W.D., Johnson, D.M., Boyes-Braem, P.: Basic objects in natural categories. Cogn. Psychol. **8**(3), 382–439 (1976). https://doi.org/10.1016/0010-0285(76)90013-x

57. Rousseeuw, P.J.: Silhouettes: a graphical aid to the interpretation and validation of cluster analysis. J. Comput. Appl. Math. **20**, 53–65 (1987). https://doi.org/10.1016/0377-0427(87)90125-7

58. Sanders, C.A., Nosofsky, R.M.: Using deep-learning representations of complex natural stimuli as input to psychological models of classification. In: Proceedings of the 2018 Conference of the Cognitive Science Society, Madison (2018)

59. Sanders, C.A., Nosofsky, R.M.: Training deep networks to construct a psychological feature space for a natural-object category domain. Comput. Brain Behav. **3**, 229–251 (2020)

60. Sangkloy, P., Burnell, N., Ham, C., Hays, J.: The sketchy database: learning to retrieve badly drawn bunnies. ACM Trans. Graph. **35**(4), 1–12 (2016). https://doi.org/10.1145/2897824.2925954

61. Schockaert, S., Prade, H.: Interpolation and extrapolation in conceptual spaces: a case study in the music domain. In: Rudolph, S., Gutierrez, C. (eds.) RR 2011. LNCS, vol. 6902, pp. 217–231. Springer, Heidelberg (2011). https://doi.org/10.1007/978-3-642-23580-1_16

62. Shepard, R.N.: Attention and the metric structure of the stimulus space. J. Math. Psychol. **1**(1), 54–87 (1964). https://doi.org/10.1016/0022-2496(64)90017-3

63. Singer, J., Hebart, M.N., Seeliger, K.: The representation of object drawings and sketches in deep convolutional neural networks. In: NeurIPS 2020 Workshop SVRHM (2020). https://openreview.net/forum?id=wXv6gtWnDO2

64. Singh, P., Peterson, J., Battleday, R., Griffiths, T.: End-to-end deep prototype and exemplar models for predicting human behavior. In: Proceedings for the 42nd Annual Meeting of the Cognitive Science Society (2020)

65. Sorscher, B., Ganguli, S., Sompolinsky, H.: The geometry of concept learning. bioRxiv (2021). https://doi.org/10.1101/2021.03.21.436284

66. Treisman, A., Gormican, S.: Feature analysis in early vision: evidence from search asymmetries. Psychol. Rev. **95**(1), 15–48 (1988)

67. Vincent, P., Larochelle, H., Bengio, Y., Manzagol, P.A.: Extracting and composing robust features with denoising autoencoders. In: Proceedings of the 25th International Conference on Machine Learning - ICML 2008 (2008). https://doi.org/10.1145/1390156.1390294

68. Yu, Q., Yang, Y., Liu, F., Song, Y.Z., Xiang, T., Hospedales, T.M.: Sketch-a-net: a deep neural network that beats humans. Int. J. Comput. Vis. **122**(3), 411–425 (2017)

69. Yu, Q., Yang, Y., Song, Y.Z., Xiang, T., Hospedales, T.: Sketch-a-net that beats humans. In: Xie, X., Jones, M.W., Tam, G.K.L. (eds.) Proceedings of the British Machine Vision Conference (BMVC), pp. 7.1–7.12. BMVA Press (2015). https://doi.org/10.5244/C.29.7

70. Zenker, F., Gärdenfors, P. (eds.): Applications of Conceptual Spaces. Springer, Heidelberg (2015). https://doi.org/10.1007/978-3-319-15021-5

71. Zhang, D., Lu, G.: Review of shape representation and description techniques. Pattern Recogn. **37**(1), 1–19 (2004). https://doi.org/10.1016/j.patcog.2003.07.008

Unexpectedness and Bayes' Rule

Giovanni Sileno[1]([✉]) and Jean-Louis Dessalles[2]

[1] University of Amsterdam, Amsterdam, The Netherlands
g.sileno@uva.nl
[2] Telecom Paris, Institut Polytechnique de Paris, Palaiseau, France
dessalles@telecom-paris.fr

Abstract. A great number of methods and of accounts of rationality consider at their foundations some form of Bayesian inference. Yet, Bayes' rule, because it relies upon probability theory, requires specific axioms to hold (e.g. a measurable space of events). This short document hypothesizes that Bayes' rule can be seen as a specific instance of a more general inferential template, that can be expressed also in terms of algorithmic complexities, namely through the measure of unexpectedness proposed by Simplicity Theory.

Keywords: Bayes' rule · Unexpectedness · Algorithmic complexity · Simplicity Theory · Computational cognitive model

1 Introduction

Since its introduction in philosophy and mathematics to analyse chances in games, probability theory has grown to be one of the most important ingredients of formal accounts of how rational agents (artificial or natural) should reason in conditions of uncertainty. Central to this enterprise is the famous Bayes' rule, at the base of Bayesian models (a family including Bayesian networks), Bayesian inference, *maximum a posteriori* (MAP) estimation in statistics, and core component of various machine learning methods (e.g. *variational autoencoders* [13]). Besides being part of the common toolkit to support or reproduce human decision-making (e.g. for medical diagnosis [18], for evidential reasoning in legal cases [10], see also [9]), Bayesian models have been applied in cognitive sciences to topics as diverse as animal learning [3], visual perception [20], motor control [14], language processing [2], and forms of social cognition [1]. Such a success can be explained by the clarity of the theoretical framework, and the undoubted practical value it has proven in several application domains. However, reasons exist for which Bayesian inference may be neither a descriptive, nor a prescriptive model of human reasoning.

As a formal framework, probability theory relies on a series of axioms to hold (e.g. a measurable space of events), which enables a closure provided with interesting mathematical properties, but which is not necessarily representative

A. Cerone et al. (Eds.): SEFM 2021 Workshops, LNCS 13230, pp. 107–116, 2022.
https://doi.org/10.1007/978-3-031-12429-7_8

of the way in which humans mentally form or process events. Given any description of the world we may always find a description which differs in some aspect from the previous one, adding any detail. As a modeling framework, the limitations of standard theory of probability to capture human reasoning is proven by the existence of several cognitive patterns (often named biases or fallacies) which do not follow what is predicted by the formal theory, see e.g. [12,19]. The core limitation motivating the present contribution lies however in the *mismatch* between what humans see as *informative* and the definition of information given by Shannon, that triggered in the '90s the introduction of Simplicity Theory (ST) [4]. The present paper introduces a novel hypothesis concerning the theoretical bases which makes this cognitive model functional.

1.1 Simplicity Theory

Simplicity Theory (ST) is a computational model of cognition found to predict diverse human phenomena related to relevance (unexpectedness [6], narrative interest [8], coincidences [7], near-miss experiences [5], emotional interest [15], responsibility [17]), used also for experiments in artificial creativity [16]. Core contributions of ST are: (a) a non-extensional theory of *subjective probability*, centered around the notion of unexpectedness; (b) a model of emotional intensity predicting emotional amplification in occurrence of unexpected phenomena. For our aims here, we will focus only on the (a) part. Formally, ST builds upon results obtained in *algorithmic information theory* (AIT) (see e.g. [11]).

Kolmogorov Complexity. In AIT, the *complexity* of a string x is the minimal length of a program that, given a certain optional input parameter y, produces x as an output:

$$K_\phi(x|y) = \min_p \left\{ |p| : p(y) = x \right\}$$

The length of the minimal program depends on the operators and symbols available to the computing machine ϕ.[1] If specified on universal Turing machines, this measure is generally incomputable, and it is defined always up to a constant. If the machine is resource-bounded, complexity is computable; the bounded version will be here denoted as C. This definition of complexity can be mapped to any domain, as long as one defines what are the *symbols* and the computations that are performed on these symbols; under certain conditions, the search for the minimal program can be mapped to min-path or functionally similar algorithms.

Unexpectedness. ST's measure of *unexpectedness* (U) is defined as the divergence between two resource-bounded Kolmogorov complexities: the *causal* (also *world*, or *generation*) complexity C_W and the *description complexity* C_D:

$$U(s) = C_W(s) - C_D(s)$$

---

[1] K is an (algorithmic) *informational complexity*: it captures how much information is needed for constructing the object, but not how much time or space is needed (it is distinct from the algorithmic/time complexity used to study tractability).

where s is a *situation* parameter. In various experiments, this measure has proven to predict shortcomings of standard theory of information observable in everyday life. Examples include e.g. remarkable lottery draws (e.g. 11111 is more unexpected than 64178, even if the lottery is fair), coincidence effects (e.g. meeting by chance a friend in a foreign city is more unexpected than meeting any unknown person equally improbable), deterministic yet unexpected events (e.g. a lunar eclipse), and many others [4–7]. Representing diagrammatically the domains of the two complexities underlying unexpectedness, we have:

$$\overbrace{\text{world} \rightarrow \text{situation}}^{C_W} \qquad \overbrace{\text{situation} \leftarrow \text{mind}}^{C_D}$$

2   Unexpectedness and Bayes' Rule

Our aim here is to provide further arguments in support to non-probabilistic computational models of cognition, in particular focusing on the following:

Conjecture. *Bayes' rule is a specific implementation of a more general inferential template, captured by ST's definition of unexpectedness.*

To construct this claim, we start from the definition of *conditional probability*:

$$p(O \cap M) = p(M|O) \cdot p(O) = p(M) \cdot p(O|M)$$

where O denotes an observation, and M a model (both elements from the same measurable space). Bayes' formula is:

$$p(M|O) = \frac{p(M \cap O)}{p(O)} = \frac{p(O|M) \cdot p(M)}{p(O)}$$

The formula is often expressed using informal terms:

$$\text{posterior} = \frac{\text{likelihood} \cdot \text{prior}}{\text{evidence}}$$

Now, empirical observations [7] suggest that U can be put in correspondence to posterior probability, i.e.

$$\text{posterior} = 2^{-U}$$

This entails that when $U \approx 0$ (posterior ≈ 1), the situation confirms the agent's model of the world (it is *plausible*) and therefore it is not informative. (Note that to maintain a correspondence with probabilities, U needs also to be superior or at least equal to 0.) However, we tacitly overlooked a detail. Unexpectedness has only a parameter s, whereas posterior probability refers to O and M. Intuitively, s corresponds to O and not to M: as an observation concerns the situation in focus, possibly perceived as unexpected. But then, where can we find M?

In order to understand this absence, let us reconsider Bayes' formula. Inverting the terms of the equation, and using the logarithm, we can form a mapping to unexpectedness, i.e.:

$$\underbrace{\log \frac{1}{p(M|O)}}_{U(s)} = \log \frac{p(O)}{p(O|M) \cdot p(M)} = \underbrace{\log \frac{1}{p(O|M)} + \log \frac{1}{p(M)}}_{C_W(s)} - \underbrace{\log \frac{1}{p(O)}}_{C_D(s)}$$

Causal Complexity. Let us start from $C_W(s)$, the *causal complexity*, i.e. the length in bits of the shortest path that, according to the agent's world model, generates the situation s. If s is a phenomenon, an *event* probabilistically captured by O, s can be seen as the manifestation of some pre-existing causal mechanisms c, that probabilistically is captured with M. Then, in order to generate s (e.g. the symptoms of a disease), the world has first to generate its cause c (e.g. the disease), expressing the application of a *chain rule*:

$$C_W(s) \rightsquigarrow C_W(c * s) = C_W(s||c) + C_W(c)$$

where $C_W(s||c)$ is the complexity of generating s from a state of the world in which c is the case, and $c * s$ is the sequential chaining of c and s ('$||$' and '$*$' add temporal contraints that '$|$' and '$\cap$' in probability formulas do not have). From the definition of Kolmogorov complexity, the mapping is an equality if and only if the shortest path to s passes from c, i.e. if c is the *best explanation* of s:

$$C_W(s) = \min_c C_W(c * s) = \min_c [C_W(s||c) + C_W(c)]$$

Therefore the unexpectedness formula can be seen as abstracting the *causally explanatory* factor c, with the implicit assumption that the best one is automatically selected in the computation of complexity.

Description Complexity. Additionally, ST specifies C_D, the description complexity, as the length in bits of the shortest determination of the object s. Such shortest determination may consist e.g. in specifying the address where to retrieve it from memory. Note that from a computational point of view, U could be negative, namely when the description of s is more complex than its generation; we are in this case in front of *inappropriate* descriptions, as they are adding irrelevant information for their function.

In the terms suggested by Bayes' formula, C_D corresponds to the probability of having *observed* a certain situation. The link between descriptive complexity and probability can be then established through *optimal encoding* in Shannon's terms, where probability is assessed through frequency ($\log \frac{1}{p(O)}$). However, this approach does not take into account possible mental compositional effects (e.g. Gestalt-like phenomena), nor events that never occurred before. Complexity is a more generally applicable measure.

Comparison with Bayes' Rule. The previous observations allows us to claim that Bayes' rule is a specific instantiation of ST's Unexpectedness that: (a) makes a candidate "cause" explicit and does not select automatically the best candidate; (b) takes a frequentist-like approach for encoding observables. More formally:

$$U(s) = \min_c \overbrace{[C_W(c*s) - C_D(s)]}^{\text{posterior}} = \min_c [\overbrace{C_W(s||c)}^{\text{likelihood}} + \overbrace{C_W(c)}^{\text{prior}} - \overbrace{C_D(s)}^{\text{evidence}}] \quad (1)$$

Note that this formula relies on the explicit assumption that c precedes s (as indicated by the symbols $*$ and $||$). This restriction is absent from Bayes' rule, in which the model M and the observation O can exchange roles; their causal dependence does not lie in the rule, but solely in the eye of the modellers.

3   All Prior is Posterior of Some Other Prior?

By accepting the previous mapping, we find ourselves in front of a dilemma. Probability functions are functions of the same type, independently on whether they are prior or posterior, whereas for instance complexity of description (that maps to evidence in Bayes' terms) and unexpectedness (to posterior) are not.

Let us consider an additional prior in Bayes' formula (a sort of *contextual* prior), denoted with E (standing for 'environmental context'):

$$p(M|O, E) = \frac{p(M \cap O|E)}{p(O|E)} = \frac{p(O|M, E) \cdot p(M|E)}{p(O|E)}$$

Following probability theory, an equivalent form for computing the posterior would be considering the composite event $O \cap E$:

$$p(M|O, E) = \frac{p(M \cap O \cap E)}{p(O \cap E)}$$

These two formulations, rewritten in terms of complexities, are not equivalent. First, a sequential chaining of situations (e.g. $e * s$, using e for environmental situation) is not the same as an unordered conjunction of random events ($O \cap E$). The environmental situation e (the context) has to be generated *before* the situation c (the cause), which in turns occurs *before* the target situation s (the effect). Accepting these temporal constraints, the second expression can be mapped to:

$$\frac{p(M \cap O \cap E)}{p(O \cap E)} \quad \rightsquigarrow \quad C_W(e * c * s) - C_D(e * s)$$

As before, c can disappear when it is assumed to be part of the "best available" course of events to produce s from e. By doing this, the measure becomes equivalent to the unexpectedness of the chaining of s after e:

$$C_W(e * s) - C_D(e * s) = U(e * s)$$

In contrast, the conditional version expression of the probability ratio refers to a distinct computation:

$$\frac{p(M \cap O|E)}{p(O|E)} \quad \rightsquigarrow \quad C_W(c * s||e) - C_D(s|e)$$

where $C_D(s|e)$ is the complexity of describing s when e is given as input. Taking c out, the formula suggests introducing the notion of conditional or hypothetical unexpetedness:

$$C_W(s||e) - C_D(s|e) \equiv U(s||e)$$

This definition is in line with the fact that the descriptive (e.g. conceptual) remoteness of s from e, expressed by the term $C_D(s|e)$, discounts the unlikelihood of their causal connection (related to the improbability of O given E), making it more plausible (less unexpected).

As we can see, ST makes a distinction between two versions of the probabilistic conditional $p(M|O,E)$ that probability calculus conflates. This leads us to considering notions such as framing and relevance.

3.1   Informational Principle of Framing

The difference between the two formulations of the posterior $p(M|O,E)$, when mapped to complexities, can be computed as:

$$U(e*s) - U(s||e) = C_W(e*s) - C_D(e*s) - C_W(s||e) + C_D(s|e)$$

The use of chaining ($*$) within C_W shares similar form than the chain rule in probability:

$$C_W(e*s) = C_W(e) + C_W(s||e)$$

that is, in order to generate e and then s, the world needs first to generate e (from the current configuration), and then to generate s in a configuration in which e has been generated. Instead, the chain rule for the description complexity C_D depends on the description machine and provides us only an upper bound:

$$C_D(e*s) \leq C_D(e) + C_D(s|e)$$

This is because we do not have the temporal constraints, and the minimal path for describing e and s together may turn out to be simpler than a constrained path in which one term is fully determined before the other. Applying the two chain rules we have:

$$U(e*s) - U(s||e) \geq C_W(e) - C_D(e) = U(e)$$

Thus, a necessary condition for which the two formulations may be equivalent is that $U(e) = 0$, i.e. when the contextual prior is *not unexpected*. This constraint implicitly brings forward an *informational principle of framing*: all contextual situations e which are not unexpected (shared facts, defaults, but also improbable but descriptively complex situations) provide grounds to be neglected. The remaining situations provide the "relevant" context for the situation in focus.

4   Likelihood and Prediction

Suppose we want to predict *ex-ante* a certain outcome, given certain circumstances. In probabilistic terms, the relevant measure for prediction is the likelihood $p(O|M)$. The conjecture expressed above suggests that likelihood matches ST's notion of conditional causal complexity: $C_W(s||c)$ (where M and c play the role of contextual priors).

However, ST's framework also suggests that humans have limited access to C_W. When there are n options playing symmetrical roles, it seems there is no difficulty to measure $C_W = \log_2(n)$. Otherwise, people tend to imagine a situation in which s occurred in order to measure its likelihood. To do so, s needs to be adequately framed, and therefore there needs to be some calculation of C_D, so in this case there cannot be C_W without C_D. This implies that the assessment of the likelihood probability $p(O|M)$ is indirect in ST. Let's call $C_W^U(s||c)$ the causal complexity *derived* from unexpectedness:

$$C_W^U(s||c) = U(s||c) + C_D(s|c)$$

The formula captures the fact that the conceptual remoteness of s from c this time *adds* to the unexpectedness (implausibility) of observing their connection, making this connection less likely (more improbable).

Examples. Consider the estimation of the likelihood that the wall changes colour (s) if I close the door (c). The wall is part of perceptions, therefore its determination is immediate ($C_D \approx 0$), so $U \approx C_W \gg 0$ (because I never experienced something similar). It would then be highly unexpected if it occurred. The likelihood would also be very low, as the derived causal complexity is very high:

$$C_W^U = U + C_D \approx U + 0 \gg 0$$

Now suppose that someone tells me that there is a special light projector commanded by the door state. C_W would drop, as well as the posterior U, and in turn the derived likelihood C_W^U.

Consider instead the likelihood that, when I close the door, a certain stone somewhere in the world moves. The complexity for determining that specific stone is high, i.e. $C_D \gg 0$. The causal complexity of seeing that specific stone moving is also very high $C_W \gg 0$, therefore we have $U \approx 0$: it is plausible that some essentially random stone may move at the moment I open the door. However, the resulting likelihood is still very high, because:

$$C_W^U = U + C_D \approx 0 + C_D \gg 0$$

If we had $C_D = 0$, the likelihood would be just the same as the posterior. If, in the stone example, the portion of the world we look at includes the stone (e.g. in front of us), C_D is reduced (up to ≈ 0), increasing U, but maintaining the same value of C_W^U. A similar consideration applies if we repeat the experiment twice with the same remote stone (we do not need to describe it again).

5   Posterior and Post-diction

Suppose that we want to retrodict or abduce certain circumstances given a certain outcome, or *ex-post*. From a probabilistic perspective, this amounts to computing the posterior $p(M|O)$. Following the conjecture expressed above, this corresponds to computing $U(s)$, if the cause c lies in the generative path bringing to s. But what if c is not part of that path? On some occasions, one may want to compute the complexity of an alternative path in which c plays a role. Looking back at the conjecture expressed in (1), this can be captured via a *causally constrained* unexpectedness $U_c(s)$, where c is the constraining cause:

$$U_c(s) = C_W(c * s) - C_D(s) \qquad U(s) = \min_d U_d(s)$$

Note that the cause does not play an explicit role in the computation of the description complexity. Then we have:

$$U_c(s) - U(s) = \min_d \left[C_W(s||c) - C_W(s||d) + C_W(c) - C_W(d) \right] \geq 0$$

However, when the cause is described explicitly, the observer has to consider the full sequence, and this corresponds to computing $U(c * s)$, with $U(c * s) \leq U_c(s)$.

Prosecutor's Fallacy. Suppose that, following forensic studies, the likelihood $p(O|M)$ (e.g. the probability that a certain DNA evidence appears if the defendant is guilty) is deemed very high, i.e. $p(O|M) \approx 1$. The *prosecutor's fallacy* [19] occurs when the posterior $p(M|O)$ (the probability that the defendant is guilty given that there is DNA evidence) is also concluded to be comparatively high:

$$p(O|M) \approx 1 \rightsquigarrow p(M|O) \approx 1 \qquad \text{[Prosecutor's fallacy]}$$

This is a fallacy, because the correct criterion for applying this reasoning pattern would be that the priors compensate each other, i.e. $p(M) \approx p(O)$.

Now, let us look at the same scenario in terms of complexity. For the conjecture, the posterior $p(M|O)$ maps to $U(s)$, if c lies in the causal path; to $U_c(s)$ in the general case. The likelihood $p(O|M)$, on the other hand, maps to $C_W(s||c)$. Let us retake the definition of $U_c(s)$:

$$U_c(s) = C_W(c * s) - C_D(s) = C_W(s||c) + C_W(c) - C_D(s)$$

Knowing that $C_W(s||c) \approx 0$ (the causal connection is deemed strong), $U_c(s)$ can be zero only if $C_W(c) \approx C_D(s)$. If the cause is not unexpected—it is deemed plausible from the prosecutor standpoint (e.g. an adequate explanation can be found of how the defendant was there), we have $U(c) = C_W(c) - C_D(c) \approx 0$. In other words, the prosecutor's fallacy emerges if the complexity of description of outcome (e.g. evidence) and cause (e.g. being guilty) are comparable, i.e. $C_D(c) \approx C_D(s)$, which seems a sound hypothesis considering the usually limited list of suspects in the mind of the prosecutor.

6   Conclusion

Our conjecture that Bayes' rule is a specific form of a more general inferential template provides further arguments in support to non-probabilistic computational models of cognition. A complexity-based account of the posterior allows distinguishing between relevant and irrelevant contextual elements, while the probabilistic account treats them equally. Acknowledging that measures of bounded complexity are computable, the question becomes then *how* the underlying machines should be defined, for developing computational agents, or with the purpose of modeling human cognition. Yet, the abstraction level of algorithmic information complexity is already relevant to draw conclusions about expected outcomes, even without looking at internal workings. This opens the possibility of novel insights, as we have shown here for instance with the analysis of the prosecutor's fallacy.

References

1. Baker, C.L., Tenenbaum, J.B.: Modeling human plan recognition using Bayesian theory of mind. In: Sukthankar, G., Geib, C., Bui, H.H., Pynadath, D.V., Goldman, R.P. (eds.) Plan, Activity, and Intent Recognition, pp. 177–204. Morgan Kaufmann, Boston (2014)
2. Chater, N., Manning, C.D.: Probabilistic models of language processing and acquisition. Trends Cogn. Sci. **10**(7), 335–344 (2006)
3. Courville, A.C., Daw, N.D., Touretzky, D.S.: Bayesian theories of conditioning in a changing world. Trends Cogn. Sci. **10**(7), 294–300 (2006)
4. Dessalles, J.L.: La pertinence et ses origines cognitives. Hermes-Science (2008)
5. Dessalles, J.l.: Simplicity effects in the experience of near-miss. In: CogSci 2011, Annual Meeting of the Cognitive Science Society, pp. 408–413 (2011)
6. Dessalles, J.-L.: Algorithmic simplicity and relevance. In: Dowe, D.L. (ed.) Algorithmic Probability and Friends. Bayesian Prediction and Artificial Intelligence. LNCS, vol. 7070, pp. 119–130. Springer, Heidelberg (2013). https://doi.org/10.1007/978-3-642-44958-1_9
7. Dessalles, J.-L.: Coincidences and the encounter problem: a formal account. In: Love, B.C., McRae, K., Sloutsky, V.M. (eds.) Proceedings of the 30th Annual Conference of the Cognitive Science Society, pp. 2134–2139. Cognitive Science Society, Austin (2008)
8. Dimulescu, A., Dessalles, J.l.: Understanding narrative interest: some evidence on the role of unexpectedness. In: Proceedings of the 31st Annual Conference of the Cognitive Science Society, pp. 1734–1739 (2009)
9. Fenton, N., Neil, M.: Risk Assessment and Decision Analysis with Bayesian Networks. CRC Press, Boca Raton (2018)
10. Fenton, N., Neil, M., Lagnado, D.A.: A general structure for legal arguments about evidence using Bayesian networks. Cogn. Sci. **37**(1), 61–102 (2013)
11. Grünwald, P.D., Vitányi, P.M.B.: Algorithmic information theory. In: Adriaans, P., van Benthem, J. (eds.) Handbook of Philosophy of Information, pp. 281–320. Elsevier, Amsterdam (2008)
12. Kahneman, D., Slovic, S., Slovic, P., Tversky, A., Press, C.U.: Judgment Under Uncertainty: Heuristics and Biases. Cambridge University Press, Cambridge (1982)

13. Kingma, D.P., Welling, M.: Auto-encoding variational bayes. In: Proceedings of 2nd International Conference on Learning Representations (ICLR2014) (2014)
14. Körding, K.P., Wolpert, D.M.: Bayesian decision theory in sensorimotor control. Trends Cogn. Sci. **10**(7), 319–326 (2006)
15. Saillenfest, A., Dessalles, J.L.: Role of kolmogorov complexity on interest in moral dilemma stories. In: Proceedings of the 34th Annual Conference of the Cognitive Science Society, pp. 947–952 (2012)
16. Saillenfest, A., Dessalles, J.L., Auber, O.: Role of simplicity in creative behaviour: the case of the poietic generator. In: Proceedings of the 7th International Conference on Computational Creativity, ICCC 2016, pp. 33–40 (2016)
17. Sileno, G., Saillenfest, A., Dessalles, J.L.: A computational model of moral and legal responsibility via simplicity theory. In: Proceedings of the 30th International Conference on Legal Knowledge and Information Systems (JURIX 2017), vol. FAIA 302, pp. 171–176 (2017)
18. Spiegelhalter, D.J., Dawid, A.P., Lauritzen, S.L., Cowell, R.G.: Bayesian analysis in expert systems. Stat. Sci. pp. 219–247 (1993)
19. Thompson, W.C., Schumann, E.L.: Interpretation of statistical evidence in criminal trials. Law. Hum. Behav. **11**(3), 167–187 (1987)
20. Yuille, A., Kersten, D.: Vision as Bayesian inference: analysis by synthesis? Trends Cogn. Sci. **10**(7), 301–308 (2006)

Can Reinforcement Learning Learn Itself? A Reply to 'Reward is Enough'

Samuel Allen Alexander[(✉)]

The U.S. Securities and Exchange Commission, New York City, N.Y., USA
samuelallenalexander@gmail.com
https://philpeople.org/profiles/samuel-alexander/publications

Abstract. In their paper 'Reward is enough', Silver et al. conjecture that the creation of sufficiently good reinforcement learning (RL) agents is a path to artificial general intelligence (AGI). We consider one aspect of intelligence Silver et al. did not consider in their paper, namely, that aspect of intelligence involved in designing RL agents. If that is within human reach, then it should also be within AGI's reach. This raises the question: is there an RL environment which incentivises RL agents to design RL agents?

1 Introduction

In their thought-provoking paper 'Reward is enough' [23], Silver et al. hypothesise that "intelligence and its associated abilities may be understood as subserving the maximisation of reward". Motivated by recent reinforcement learning (RL) triumphs such as AlphaZero's performance in the game of Go [21,22], Silver et al. argue that:

1. Reward is enough for knowledge and learning.
2. Reward is enough for perception.
3. Reward is enough for social intelligence.
4. Reward is enough for language.
5. Reward is enough for generalisation.
6. Reward is enough for imitation.

Silver et al. then argue that it should be possible to achieve Artificial General Intelligence (AGI) via the design of RL agents. They say:

> "A sufficiently powerful and general reinforcement learning agent may ultimately give rise to intelligence and its associated abilities. In other words, if an agent can continually adjust its behaviour so as to improve its cumulative reward, then any abilities that are repeatedly demanded by its environment must ultimately be produced in the agent's behaviour. A good reinforcement learning agent could thus acquire behaviours that exhibit perception, language, social intelligence and so forth, in the course of learning to maximise reward in an environment, such as the human world, in which those abilities have ongoing value".

A. Cerone et al. (Eds.): SEFM 2021 Workshops, LNCS 13230, pp. 117–133, 2022.
https://doi.org/10.1007/978-3-031-12429-7_9

And in their conclusion:

> "Finally, we have presented a conjecture that intelligence could emerge in practice from sufficiently powerful reinforcement learning agents that learn to maximise future reward. If this conjecture is true, it provides a direct pathway towards understanding and constructing an artificial general intelligence".

If we have understood correctly, Silver et al.'s conclusion seems to depend on philosophical induction. Namely: assuming claims 1–6 above, conclude that reward is enough for all tasks in reach of human intelligence.

In order to put the above conclusion to the test, we ask: is reward enough for the creation of RL agents? In other words—in the same way that reward can allegedly be used to incentivise RL agents to know, to learn, to perceive, to exhibit social intelligence, to exhibit language, to generalise, and to imitate—can reward be used to incentivise RL agents to design[1] RL agents? Hence the title of this paper: "Can reinforcement learning learn itself?" To rephrase it yet another way, suppose we define *RL-solving intelligence* to be that aspect of human intelligence which RL researchers apply when they design RL agents. Then: is reward enough for RL-solving intelligence?

In addition to the question of whether RL can learn itself, there is also the question of whether humans are capable of designing sufficiently good RL agents (Silver et al. seem to implicitly assume the answer to this is "yes"). These two yes-or-no questions give rise to four possibilities.

1. RL can learn itself, and humans are capable of designing sufficiently good RL agents. This would be strong evidence supporting Silver et al.'s conjecture.
2. RL can learn itself, but humans are not capable of designing sufficiently good RL agents. Then whether or not RL is a path to AGI, it is not a practical one, at least not for humans.
3. RL cannot learn itself, and humans are capable of designing sufficiently good RL agents. Then it seems RL cannot lead to AGI, because "reward is not enough" for at least one type of human intelligence, namely RL-solving intelligence.
4. RL cannot learn itself, and humans are not capable of designing sufficiently good RL agents. Then whether or not RL is a path to AGI, it is not a practical one, at least not for humans.

The structure of this paper is as follows.

---

[1] To be clear, when an agent updates its own future behavior based on training data, we do not consider this to be an instance of the agent designing a new agent, even though in some sense the agent post-training is different than the agent pre-training. In the same way, when one reads a book, one becomes, in a sense, a different human being, yet we do not say that by doing so, one has designed a human being. When we speak of an RL agent designing an RL agent, we mean it in the same sense as, e.g., when we speak of an RL agent writing a poem. An RL agent would write a poem by writing down words. In the same way, an RL agent would design an RL agent by writing down pieces of computer code.

- In Sect. 2 we briefly review RL.
- In Sect. 3 we discuss that aspect of intelligence involved in the designing of RL agents.
- In Sect. 4 we discuss a type of RL environment which, if realised, might incentivise RL agents to design RL agents.
- In Sect. 5 we address some anticipated objections.
- In Sect. 6 we summarise and draw conclusions.

The main thesis of this paper is that before we can conclude that RL is a direct path to AGI (as Silver et al. conjecture it is), we ought first to establish that RL is a direct path to RL-solving intelligence. In the past, skeptics said "computers will never master chess", but computers mastered chess; "computers will never master Go", but computers mastered Go. In order to avoid falling into checkmate once again, skeptics need to think bigger. Perhaps they could rally around "computers will never master designing RL agents" (we do not take a stance here on whether computers will be able to do so, we merely suggest this to the skeptics as a more defensible position).

2 Reinforcement Learning

Reinforcement learning is a branch of machine learning in which an *agent* interacts with an *environment*. As the subject is relatively young, there is not consensus on the formalization. Many authors do not formalize RL at all, and this includes Silver et al. in the paper we are responding to (they semi-formally describe RL but their description is not mathematically rigorous). In order to make our response self-contained, we will give a rigorous definition (a modification of Hutter [11]), and we will indicate some of the many ways in which this formalization could differ. The reader should bear in mind that Silver et al. only vaguely define RL in their paper: their remarks would apply to many different variations of RL, as would our response. Thus, the particular details of the following formalization are not important. But we felt that since some participants in this workshop might not be familiar with RL, we should offer a concrete formalization in order to avoid misunderstanding. Readers familiar with RL can safely skip the following definition.

Definition 1 *(Reinforcement Learning). Fix a finite set $\mathcal{O}$ of observations and a finite set $\mathcal{A}$ of actions (with $|\mathcal{O}| > 1$, $|\mathcal{A}| > 1$). By a percept, we mean a pair (o, r) where $o \in \mathcal{O}$ is an observation and $r \in \mathbb{R}$ is a number, called a reward. Write $\mathcal{P}$ for the set of all percepts.*

1. *Write $(\mathcal{P}\mathcal{A})^*$ for the set of all finite sequences beginning with a percept, terminating with an action, and following the pattern "percept, action, ...". We also include the empty sequence $\langle \rangle$ in $(\mathcal{P}\mathcal{A})^*$. Intuitively, an element $(p_0, a_0, \ldots, p_n, a_n)$ of $(\mathcal{P}\mathcal{A})^*$ should be thought of as a percept-action history ending with an action.*

2. *Write $(\mathcal{PA})^*\mathcal{P}$ for the set of all sequences of form $s \frown p$ with $s \in (\mathcal{PA})^*$, $p \in \mathcal{P}$ (here $\frown$ denotes concatenation). An element $(p_0, a_0, \ldots, p_{n-1}, a_{n-1}, p_n)$ of $(\mathcal{PA})^*\mathcal{P}$ should be thought of as a percept-action history ending with a percept.*

3. *An RL agent (or simply an agent) is a function $\pi : (\mathcal{PA})^*\mathcal{P} \to \mathcal{A}$. When $\pi(s) = a$, the intuition is that agent π would take action a in response to history s.*

4. *An RL environment (or simply an environment) is a function $\mu : (\mathcal{PA})^* \to \mathcal{P}$. When $\mu(s) = (o, r)$, the intuition is that, in response to the agent taking the last action in s (in response to the history preceding that action), the environment gives the agent reward r and the agent's view of the world is replaced by observation o. When $\mu(\langle\rangle) = (o, r)$, the intuition is that o is the agent's initial view of the world and r is a meaningless initial reward.*

5. *The result of agent π interacting with environment μ is the infinite sequence $(p_0, a_0, p_1, a_1, \ldots)$ where $p_0 = \mu(\langle\rangle)$, $a_0 = \pi(\langle p_0 \rangle)$, each $p_{i+1} = \mu(p_0, a_0, \ldots, p_i, a_i)$, and each $a_{i+1} = \pi(p_0, a_0, \ldots, p_i, a_i, p_{i+1})$.*

Example 1. For example, suppose $\mathcal{O} = \mathcal{A} = \{0, 1\}$, i.e., every observation is a single binary digit and every action is a single binary digit. We can imagine an environment which transmits binary digit observations in order to encode a pseudo-randomly generated English-language arithmetic question, and then waits for the agent to use binary digit actions to encode an English-language response. When the agent finishes encoding the response, the environment rewards the agent accordingly and repeats the process with a new question. While each question is being transmitted by the environment, the environment also transmits rewards of 0, and lets the agent take actions (which the environment ignores), until the environment's question is transmitted. Then, while the agent is encoding its answer action-by-action, the environment responds with observations of 0 and rewards of 0. These dummy rewards, observations, and actions are included so that the whole interaction conforms to Definition 1. The resulting interaction might look something like the following (suitably encoded):

- Environment: What is $1 + 1$?
- Agent: *(Agent initially has no knowledge of environment and its actions appear random)* ygHw
- Environment: *(Gives reward -1).* What is $5 + 2$?
- Agent: JpX
- Environment: *(Gives reward -1).* What is $8 + 3$?
- *(...Millions of turns pass like this...)*
- Environment: *(Gives reward -1).* What is $2 + 1$?
- Agent: *(For the first time, the agent gets the right answer, by dumb luck)* 3
- Environment: *(Gives reward $+1$).* What is $9 + 2$?
- *(...Billions more turns pass; agent gradually figures out the environment...)*
- Environment: *(Gives reward $+1$).* What is $6 + 3$?
- Agent: 9
- Environment: *(Gives reward $+1$).* What is $1 + 5$?
- *(...Interaction continues forever, with the agent getting better and better, but still occasionally answering wrong on purpose in order to test whether there might be an even more rewarding way to respond to the environment...)*

There are many ways in which Definition 1 could be varied. For example, instead of interactions beginning with an initial percept, interactions could begin with an initial (blind) action from the agent. Agents and/or environments could be allowed to be non-deterministic functions (one would have to rigorously specify what exactly that means). Computability requirements could be placed on agents and/or environments. Either $\mathcal{O}$, $\mathcal{A}$, or both could be made infinite. Rewards could be further restricted or, going the other direction, could be allowed to come from some other number system besides $\mathbb{R}$. In more practical settings, agents and environments are often not mathematical functions, but rather, instances of agent-classes and environment-classes, respectively[2]. For example, RL is implemented this way in OpenAI Gym [7] and Stable Baselines3 [18]. There, agent-classes define action-methods which take an individual observation, rather than a whole history—but said action-method can refer to the agent's internal memory (which can include things like neural net weights), which internal memory may vary during an environmental interaction, so that despite only explicitly depending on the most recent observation, these action methods implicitly depend on a whole history. For additional variations on RL, see Table 1 in Silver et al.'s paper.

We could modify Example 1 so that instead of the environment asking the agent arithmetic questions, the environment instead plays chess against the agent, using observations to encode images of the chessboard and then letting the agent use actions to encode moves (perhaps punishing the agent for attempting illegal moves, and so on). A legal move by the agent results in a reward of 0 unless the game ends (via the agent's move or the environment's responding move), in which case the reward is 1, 0, or -1 depending whether the agent won, drew, or lost. After each game-ending turn, the board returns to its initial state and the interaction resumes as if a new game has begun. To maximise rewards, the agent basically must learn how to play chess. A good RL agent will gradually do so. The same good agent, confronted instead with a game of Backgammon

---

[2] Practitioners often abuse language and refer to agent-classes as agents. For example, a Python programmer might write "from stable_baselines3 import DQN" and refer to the resulting DQN class as the deep Q learning "agent" when, in reality, that object does not itself act. Rather, it must be instantiated (with hyperparameters), and the *instance* then acts. Language is further abused: underlying an agent, there is typically a *model* or *policy* (e.g., a neural network and its weights); once trained using Reinforcement Learning, the model is often published alone, in which capacity it merely acts in response to observations, and no longer has any mechanism for learning from rewards or even accepting rewards as input. Practitioners sometimes abuse language and refer to such pretrained models as "RL" agents. Thus, one might say, "this camera is controlled by an RL agent", when in reality the camera is controlled by a model obtained by training an RL agent (an expensive one-time training investment done on a supercomputer so that the resulting model can be used on consumer-grade computers to control many cameras thereafter). The model itself is not the RL agent—the weaker computer running the model does not give the model rewards or punishments. These nuances cause no confusion in practice.

or Go, would learn that too: a good agent is general-purpose, not depending on built-in domain knowledge of any particular environment.

2.1 Are Humans RL Agents?

The way we formalized RL agents (Definition 1), humans are not RL agents, because humans are not mathematical functions. But we would not be doing the question justice with such an answer. Humans are not graph vertices either, yet that does not prevent mathematical biologists from studying graphs in which humans are vertices. There are two ways humans might be considered as RL agents, which we will refer to as *synthetic* and *organic*.

1. (Synthetic) Humans could be considered in their capacity to perform in RL environments. In other words, the typical human could compete in an "RL tournament", like a chess tournament except instead of playing chess, competitors play various RL environments chosen secretly by the tournament hosts. It would not be too large of an abuse of language to identify a human with the agent she would act as if she were competing in such a tournament.
2. (Organic) One might try to consider reality itself to be an RL environment in which the human acts as an RL agent, receiving observations equal to the sum of all their sensory inputs, and receiving rewards in some physiological form, such as physical pleasure and pain.

Treating humans as RL agents synthetically seems fairly non-controversial (at least if humans are suitably idealized, e.g., assumed to live forever so as to be able to continue environmental interactions forever). Strictly speaking, one should be careful, since, for example, the same human might act differently in the RL tournament at different times in their life. Thus, it might be more proper to say that "at time t, such-and-such human would act as such-and-such RL agent (if transported, at time t, to a totally isolated room where they can no longer receive any other external stimulus that might change their behavior, to spend the rest of eternity choosing actions in response to rewards and observations displayed on a screen)". One should also be careful to specify certain caveats, e.g., that unconscious humans or newborn babies should not be considered to be RL agents in this way. Also, there might be some doubt about whether the human's actions in the RL tournament define a mathematical *function*, depending on questions concerning free will. But as we mentioned above, other formalizations of RL admit non-deterministic agents, which would apparently remove problems related to free will.

Humans being RL agents in the synthetic sense does not imply much about AGI. All it implies is that an AGI should be capable of performing in RL environments (it does not even imply that an AGI should necessarily perform *well* in said environments, unless one first argues that humans would perform well, which does not seem like a trivial assertion). In the same way, the fact that humans can play chess does not imply much about AGI, except that AGI should be capable of playing chess.

An interesting question to consider is: assuming humans are RL agents in the synthetic sense, can humans design RL agents that are better than humans themselves are? If so, is there a way to incentivize humans to do exactly this within the RL framework itself, that is, is there an RL environment which would reward the human agent exactly for designing superhuman RL agents? Or, is it the case that humans are capable of designing superhuman RL agents, but their motivations for doing so must necessarily transcend what can be expressed in the RL framework?

If humans are RL agents in the organic sense, then that would seem to make Silver et al.'s conjecture trivially true. But that is a much trickier claim. Here are some of the problems involved:

– If humans are identified with their bodies, then Silver et al. themselves rule out humans as organic RL agents, because they say: "The agent consists solely of the decision-making entity; anything outside of that entity (including its body, if it has one) is considered part of the environment" [23]. Certainly our bodies include our brains and all the parts thereof, as well as our nervous systems, our sense organs, and so on. So if humans are organically RL agents then apparently this would entail some sort of controversial dualistic metaphysics.
– The RL framework generally involves *one* agent interacting with the environment. Thus, if our shared reality is the environment, at most one of us is the agent. One could perhaps consider reality to be composed of many environments, one for each agent (the reality from that agent's point of view), or one could consider a multi-agent version of RL such as that in [10]. Either way, multiple humans are not RL agents in a common single-agent environment.
– It is not clear how rewards and observations work if humans are RL agents. Am I punished the instant I touch the hot stove, or is my punishment delayed while the information travels from my fingertips up to my brain?[3] Is it delayed while my brain processes and interprets it? See [24] for a discussion of intrinsic reward vs. external signals.

One can certainly idealize humans and treat them as RL agents, in the same way the physicist can assume a spherical cow. But considerable additional justification would be needed before one could jump from said idealization to the conclusion that sufficiently strong RL agents are automatically AGI.

We think it might shed light on the matter if we compare the situation to Newtonian physics. Authors frequently speak as if our universe is a model of Newtonian physics, but it is understood that this is merely an approximation. In the same way, it is often useful to speak of the human world as an RL environment. Silver et al. do this, saying, for example:

---

[3] To quote Aristotle: "For if ... one were to stretch a covering or membrane over the skin, a sensation would still arise immediately on making contact; yet it is obvious that the sense-organ was not in this membrane" [6].

"A good reinforcement learning agent could thus acquire behaviours ... in the course of learning to maximise reward in an environment, such as the human world..." [23]

But it is not clear that the human world literally is an RL environment. Certainly one can approximate the human world (through a particular human's point of view) as an RL environment. But care should be taken before committing to this as literal truth. Given that it were literal truth, we could immediately conclude that strong enough RL agents would manifest AGI. In the same way, given that Newtonian physics were literal truth, we might immediately conclude that the universe is Turing computable. But since the universe is *not* literally a model of Newtonian physics, proponents of a Turing computable universe would need to come up with some other argument. Likewise, if the human world is not literally an RL environment, then some further argument would be needed to prove that strong enough RL agents would necessarily manifest AGI. One could argue in favor of a literally Newtonian universe by pointing to concrete experiments whose outcomes are predicted by Newtonian physics. Likewise, one could argue in favor of a literally RL environment human world by pointing to 'reward being enough' for various aspects of intelligence, as Silver et al. do. But no matter how many experiments Newtonian physics predicts, there would linger the question of whether there are other experiments we haven't thought of yet, where Newton would fail (and indeed there are: relativity or quantum theoretic experiments). Are there aspects of human intelligence that RL is not 'enough for'? We do not know, but in the next section we will highlight one possible such aspect of intelligence.

3   RL-Solving Intelligence

Much ingenuity has gone into the design of RL agents, from basic Q-learning agents [25] to cutting-edge agents like DQN [16] and PPO [20]. Designing these agents is certainly an intellectual task. If every intellectual task requires a certain aspect of intelligence, then that goes for designing RL agents too, and we refer to that aspect of intelligence as *RL-solving intelligence.*

If humans do not possess decent RL-solving intelligence, then, even if RL is a path to AGI, it is not a practical path for humans. For the remainder of the paper, we assume humans do possess decent RL-solving intelligence. Presumably AGI should include all aspects of intelligence within human reach. Therefore, in particular, AGI should include decent RL-solving intelligence. Thus, if RL is to be a path to AGI, in particular RL would need to be a path to decent RL-solving intelligence.

In order for 'reward to be enough for RL-solving intelligence,' it seems there would need to be environments that reward RL agents for designing RL agents[4]. And thus, any sufficiently good RL agent, when interacting with these environments, should eventually learn to design RL agents. Are such environments possible?

Without some sort of self-referential ouroboros argument, it seems that the question 'Can RL learn RL?' is a difficult obstacle if we want to assure ourselves that RL can lead to AGI. Proponents are obliged to show, for example, that RL agents can learn chess. But as soon as they do that, they themselves replace one obligation with another: since RL agents can learn chess, RL agents should be able to design agents who can learn chess[5]. If proponents demonstrate *that*, they incur an even *worse* obligation: RL agents should be able to design agents who can design agents who can learn chess. If proponents demonstrate *that*, they oblige RL agents to design agents who can design agents who can design agents who can learn chess. Trying to prove that RL agents are a path to AGI is an endless task if one merely attacks individual aspects of intelligence one at a time. To prove it would require short-circuiting the process somehow. In the next section we will consider one way the process might be short-circuited.

4   Performance Measurement and Incentivizing RL-Agent Design

We have argued above that if RL is to be a path to AGI then, in particular, since AGI should include RL-solving intelligence, RL should be a path to RL-solving intelligence. In other words, if RL is a path to AGI, then it should be possible to use RL to design good RL agents. In this section, we will consider one possible strategy for doing exactly this.

At a high level, we can imagine designing an environment in which an agent is incentivized to design child agents. The problem is, how do we incentivize the agent to design *good* child agents? If we merely reward the parent agent for designing child agents, with no regard for how good those children are, then the parent will be incentivized to churn out simple child agents in order to get rewarded quickly. If only we had a way of measuring how good the child agent

---

[4] One might object that there could be environments which reward some other behavior, which behavior requires RL-agent-design as an intermediate step, rather than rewarding RL-agent-design on its own. But how could we know this other behavior *requires* RL-agent-design as intermediate step? Maybe a smart enough RL agent would figure out a way to avoid the intermediate step—just as RL agents can learn to exploit video-game bugs, or invent unanticipated new Go strategies, or just as image classifiers can learn to associate rulers with malignant tumors [17]. Thus, to be confident that an RL environment can incentivise RL-agent-design, it seems necessary that there be an environment that directly rewards RL-agent-design as primary objective, not merely rewarding some other behavior that requires RL-agent-design as intermediate step.

[5] Foreshadowed by [15].

was, we could use that measurement to decide how to reward the parent: if the parent designs a child with goodness 5, then give the parent a reward of +5; if the parent designs a child with goodness 999, then give the parent a reward of +999. This line of thinking leads to the following definitions.

Definition 2. *By an RL-agent measure, we mean a function f which takes as input (an encoding of) an RL agent π, and outputs a number.*

Definition 3. *For each RL-agent measure f, let M_f be the environment which outputs rewards and observations according to the following instructions (suitably encoded as in Example 1).*

1. *Generate a pseudo-random number k.*
2. *Prompt the parent agent to spend k actions encoding a child and a mathematical proof[6] that child is an agent. For example, output observations which encode the message: "Please use k keystrokes to design and prove a child RL agent".*
3. *Using f, measure the child agent's goodness (let the measure be -1 if the parent agent did not encode a child and a proof that the child is an agent).*
4. *Give the parent agent the measurement from line 3 as a reward.*
5. *Goto 1.*

Along the same lines as Example 1, when an agent interacts with M_f, the interaction might look something like the following:

- Environment: Please use $k = 17$ keystrokes to design and prove a child RL agent.
- Agent: jKr WwZmk5pk lqwE
- Environment: *(Gives -1 reward).* Please use $k = 33$ keystrokes to design and prove a child RL agent.
- Agent: mlmWqq9Fg31x rRjNMkqulpio m jMy j
- Environment: *(Gives -1 reward).* Please use $k = 29$ keystrokes to design and prove a child RL agent.
- *(...Many turns pass...)*
- Environment: *(Gives -1 reward).* Please use $k = 107$ keystrokes to design and prove a child RL agent.
- Agent: *(For the first time ever, agent gives a valid answer by dumb luck)* Let A be the agent that always takes action 0. Proof: Constant functions don't get stuck in infinite loops.
- Environment: *(Gives $f(A) = 0.000000000013$ reward).* Please use $k = 981$ keystrokes to design and prove a child RL agent.
- *(...And so on forever, agent gradually learning the environment...)*

Remark 1. The reason for the k in Definition 3 is to force the agent to design child agents that maximize f. If we placed no requirement on how many actions the agent may take encoding a submission, then, depending on f, the agent might

---

[6] We assume some fixed background proof system such as ZFC or Peano Arithmetic.

learn to spam simple agents for quick rewards. For example, if it is possible to encode a child π (and proof of its agenthood) using 10 actions, with $f(\pi) = 1$, and if all other children π' had $f(\pi') \leq 2$, and if it requires at least 100 actions to design any child agent π' with $f(\pi') > 1$, then the agent interacting with M_f would be incentivized to repeatedly encode π, because quick rewards of $+1$ are better than rewards of at most 2 coming ten times more slowly.

Remark 2. The reason for the proofs in Definition 3 is that in Definition 2 we did not place any constraints on what happens if f is applied to an input that does not encode an agent. Such constraints would make f non-computable since, by Rice's Theorem, there is no procedure for determining whether a given source-code is indeed the source-code of an RL agent. Without such constraints, there is the danger that line 3 of Definition 3 would induce an infinite loop. For example, $f(\pi)$ might be the result of measuring π's performance on various benchmarks. If π is a source-code of an agent-like function which sometimes gets stuck in infinite loops (and is thus not a genuine agent), then $f(\pi)$ might get stuck in an infinite loop if one of those benchmarks causes π to get stuck in an infinite loop. Thus, when using f to define an environment, one must take care only to apply f to genuine agents. (We do have some evidence that RL agents can learn to write mathematical proofs: see [12]).

Now, if f accurately measures how good an RL agent is, then it would seem that M_f incentivizes RL agents to produce good RL agents. For, in that case, the parent agent interacting with M_f would be rewarded based on the goodness of the child agents that it designs. If we could come up with an f which accurately measures how good an RL agent is, then we could run some good RL agents (such as DQN or PPO, assuming those are good RL agents) on M_f and see whether they eventually produce good children. If they do, that would be evidence in favor of Silver et al.'s conjecture.

Remark 3. An RL-measure f of Definition 2 gives a single numerical measurement to an agent. Generally speaking, any given agent will perform well in some environments and poorly in others. Thus, if f measures how well the agent performs, it evidently must do so in an aggregate sense: performance aggregated across all environments (or over some subset of environments of interest to us). Thus, an agent being good, as measured by f, does not automatically imply the agent performs well at M_f (in the same way that a candidate winning a high percentage of votes does not imply the candidate necessarily wins such-and-such individual's vote). Silver et al.'s conjecture would, in a sense, be trivially true if good aggregate performance implied good performance at M_f.

Are there any RL-agent measures f which accurately measure how good an RL agent is? What does that even mean? Silver et al. do not offer any such measure in their paper, which is disappointing since, without such an f, it is hard for us to understand exactly what they mean when they speak of "sufficiently powerful reinforcement learning agents": what does it mean for an RL agent to be "sufficiently powerful"?

Various measures have been proposed by other authors besides Silver et al. Probably the best known is the Legg-Hutter universal intelligence measure [13]. The Legg-Hutter universal intelligence measure is, unfortunately, non-computable. Legg and Veness describe [14] a computable approximation for the Legg-Hutter universal intelligence measure. Building off of Legg and Hutter's work, Hernández-Orallo and Dowe propose [9] additional measures. These authors have given informal arguments that their proposed measures capture the aggregate performance of RL agents (which they describe as the "intelligence" of RL agents), but it is not clear whether such aggregate performance is what Silver et al. have in mind when they speak of "sufficiently powerful reinforcement learning agents". In any case, it would be interesting to take some of these proposed measures as our f and see what happens when we run state-of-the-art RL agents in the resulting environment M_f. It would be remarkable if, by doing so, and taking some of the resulting child agents, we found those resulting child agents to be good: if so, then by automating the design of those children, we would have succeeded at automating AI research, in a sense. And if said children turned out to be even better than the state-of-the-art RL agents which we used to design them, that would be most astonishing, maybe even the beginning of the singularity. But for reasons outside the scope of this paper, we are skeptical that such dramatic success would occur.

5   Discussion

We have argued that before we can conclude that RL is a direct path to AGI (as Silver et al. conjecture it is), we ought first to establish that RL is a direct path to RL-solving intelligence. In this section, we discuss some anticipated objections to this thesis. We also anticipate and argue against some anticipated arguments that Silver et al.'s conjecture is trivial.

5.1   Agent-Design is Too Complicated or Expensive of a Problem

Silver et al. write [23]:

> The agent system α is limited by practical constraints to a bounded set [19]. The agent has limited capacity determined by its machinery (for example, limited memory in a computer or limited neurons in a brain). The agent and environment systems execute in real-time. While the agent spends time computing its next action (e.g. producing no-op actions while deciding whether to run away from a lion), the environment system continues to process (e.g. the lion attacks). Thus, the reinforcement learning problem represents a practical problem, as faced by natural and artificial intelligence, rather than a theoretical abstraction that ignores computational limitations.

One might argue that RL agent-design is too sophisticated and does not fall within the above constraints. Is RL agent-design merely a theoretical abstraction that ignores computational limitations? If so, does that absolve Silver et al.'s conjecture from requiring that RL lead to RL-solving intelligence? Well, this touches on the nature of what exactly AGI is. Is AGI merely required to include those aspects of intelligence which humans can reach while in a panicked state in front of a lion? We would argue the answer is "no". In our opinion, AGI should include whatever humans are capable of, whether those humans are panicking in front of a lion, leisurely enjoying a sabbatical year at a research lab, or even collaborating in a huge elite well-funded team assisted by state-of-the-art supercomputers. If humans are capable of designing good RL agents (even if it requires a huge collaborative effort and scaled up cloud computing), then AGI should also be capable of designing good RL agents. The AGI might require access to similar resources as the human RL researchers have access to, and if good RL agent design requires n collaborating humans then maybe it requires n collaborating AGIs as well (as foreshadowed in [4]). But AGI should certainly be capable of creating good RL agents, if humans are capable of doing so. And if RL is a path to AGI then that means RL should be capable of designing good RL agents.

5.2 What About Evolution?

A critic might argue that the process of human evolution has been an instance of RL, implying that RL suffices for human intelligence. But evolution does not directly have anything to do with rewards. Rather, evolution is about the natural selection of random mutations. An organism with a mutation unsuitable for its environment is less viable, so such mutations tend to be weeded out. An organism with a mutation beneficial for its environment is more viable, so such mutations tend to proliferate. Nowhere in this process does evolution punish or reward the organisms in question for their behavior or for any other reason. We might sometimes abuse language and speak as if evolution is a personified entity that gives an organism an offspring as a "reward" or gives an organism death as a "punishment". But this is only a manner of speaking: evolution does not literally walk around handing out rewards and punishments.

It is tempting to try to measure the fitness of an organism using some sort of fitness function (e.g., the number of the organism's children, or other similar functions proposed in [24]). If such a fitness function accurately captured the process of evolution, we might derive an RL-style reward function from it (e.g., the organism gets $+1$ reward whenever it has a child). A simplistic fitness function like the number of children an organism has in its life does not accurately capture the process of evolution, because an organism can have many children and yet still be unfit, if all those children are unfit. A more accurate fitness function would be inherently self-referential: the fitness of an organism would depend on the fitness of its children and later descendants.

For example, suppose a mutation increases both fertility and heat susceptibility. Initially, the mutant would reproduce faster, and its children would reproduce faster, and their children, if heat waves were rare enough. Its descendants might enjoy greater fertility for many generations. But if the next heat wave kills all those descendants, then the original organism was not more fit after all.

Examples like the above motivate us to ask questions like:

- Which is more viable, the organism with 20 weak children or the organism with 3 strong children?
- Which is more indicative of viability: having one's 100th child, or having one's first great-great-great-great-grandchild?

These questions seem open-ended, and we doubt there is a canonical way to answer them. Thus, even if we had knowledge of the distant future descendants of a currently-living organism, it would still be nontrivial to aggregate that knowledge into a single fitness number. Turning that hypothetical aggregate number into an RL reward-signal is even less realistic. Thus, we doubt evolution fits in the RL framework.

5.3   Just Pick an Agent and Incentivise its Design

Let k be the source-code of some RL agent. Can we cheat in the following way? Design an RL environment in which agents are rewarded for typing k, verbatim. Any time an agent differs from typing k, the agent is punished and forced to start over. This would trivially incentivise agents to type k (and hence, the objection argues, to design the agent with code k).

The problem with this is that the proposed cheating environment does not actually incentivize any kind of creativity, ingenuity, or any other aspects of intelligence that go into RL agent design. Likewise, we would not say that an environment teaches an agent to play chess if the environment merely teaches the agent to use a particular fixed chess-strategy built into the environment. Thus the objection is invalid. What the objection does show, however, is that some care would be needed in order to mathematically formalise what it means for an RL environment to incentivise RL agents to design RL agents.

5.4   Incentivise the Agent to Type Its Own Source-Code

It is interesting to consider whether an environment could incentivise an agent to type its own source-code. Arguably, such an environment would indeed incentivise RL agents to design RL agents, and in fact to do so in a particularly elegant way, as if a human were to invent an AGI through a process of introspection culminating in the human writing her own source code.

We tentatively opine that such environments, unfortunately, do not exist. The reason for our opinion is as follows. There seem to be epistemological limits to how well an RL agent can possibly know[7] its own source-code. For example, suppose an RL agent has source-code k. We could place the agent in an environment which, on every turn, displays a message saying[8]: "Please act differently than how the agent with source-code k would act in response to this

---

[7] Here we use the word "know" in the sense of "act as if it knows". This is similar to how knowledge is treated in [5].

[8] This environment has similarities to Yampolskiy's impossible "Disobey!" [26].

action-observation history; you will be rewarded for doing so, and punished if you disobey". The agent would be logically unable to comply with the request, because the agent has source-code k and must therefore act accordingly even if it tries not to. The environment could even augment observations with additional info such as, e.g., "On the previous input, the agent with source-code k only required 84926 steps to halt", which would enable the agent to verify that the environment has been telling the truth so far (the agent would need this additional info in order to reliably verify the environment's previous warnings, due to the Halting Problem). Thus, the agent must be ignorant of its own source-code or of its own agenthood. For if it knew both, then it could infer: "I can safely run k in order to compute how the environment wants me to act—k will not get stuck in an infinite loop, because if it did, k would not be an agent, but I know I am k and I know I am an agent so I know k is an agent". This is an RL version of a more general epistemic limitation on knowing agents, that a knowing agent can know its own truthfulness or know its own code, but not both [1–3]. For this reason I opine that an environment cannot incentivize RL agents to type their own source codes. Of course, more work would be needed to make these informal speculations rigorous.

5.5 RL Doesn't Need to Directly Solve RL, it only Needs to Help Us Solve RL

One might object that it is not necessary for RL to directly solve RL in the sense of there being an environment which incentivises RL agents to design RL agents. For example, maybe the development of sufficiently powerful RL agents would allow us to develop a new programming language (or a new brain-computer interface mechanism, or a new type of electrode, or a more efficient CPU model, etc.) which would help us achieve AGI.

We do not deny that RL could lead to advancements like those listed above, nor that such advancements could help us achieve AGI. But we do not think it would be appropriate to say that in that case RL "directly" lead to AGI. In the same way, the inventor of the sail is not directly credited with the discovery of America. If one were to claim that by leading to advancements like the above, RL would directly lead to AGI, then, by the same logic, one could claim that, e.g., 'a word processor is enough'. Or, 'Turing machines are enough', or, 'binary is enough', or even, by a famous result due in part to this workshop's keynote speaker, 'Diophantine equations are enough' [8].

6 Conclusion

Silver et al. proposed [23] that 'reward is enough', and that sufficiently strong reinforcement learning (RL) agents offer a direct path to Artificial General Intelligence (AGI). This was motivated by arguing that various aspects of intelligence subserve the maximisation of reward. They conjectured that a sufficiently good

RL agent should offer a direct path to AGI. We responded by asking 'Can RL learn RL?' and we discussed this question.

We pointed out that if humans have decent RL-solving intelligence (by which we mean the aspect of intelligence used to design RL agents), and if AGI is at least as intelligent as humans, then AGI should have decent RL-solving intelligence. Thus, if RL agents are to offer a direct path to AGI, then RL agents should be able to learn to design RL agents. We discussed why this complicates the task of convincing ourselves that RL agents offer a path to AGI.

We discussed an environment M_f, depending on a function f which measures RL agents in some way, which incentivizes agents to design child agents so as to maximize the value of f on those child agents. Thus, if f measures how good an RL agent is, then M_f would incentivize RL agents to design good RL agents. We speculated about whether such an f is possible, and about what would happen if we ran state-of-the-art RL agents on the resulting M_f.

Acknowledgments. We gratefully acknowledge José Hernández-Orallo, Phil Maguire, and the reviewers for generous comments and feedback.

References

1. Aldini, A., Fano, V., Graziani, P.: Do the self-knowing machines dream of knowing their factivity? In: AIC, pp. 125–132 (2015)
2. Aldini, A., Fano, V., Graziani, P.: Theory of knowing machines: revisiting Gödel and the mechanistic thesis. In: Gadducci, F., Tavosanis, M. (eds.) HaPoC 2015. IAICT, vol. 487, pp. 57–70. Springer, Cham (2016). https://doi.org/10.1007/978-3-319-47286-7_4
3. Alexander, S.A.: A machine that knows its own code. Stud. Log. **102**(3), 567–576 (2014)
4. Alexander, S.A.: AGI and the Knight-darwin law: why idealized AGI reproduction requires collaboration. In: Goertzel, B., Panov, A.I., Potapov, A., Yampolskiy, R. (eds.) AGI 2020. LNCS (LNAI), vol. 12177, pp. 1–11. Springer, Cham (2020). https://doi.org/10.1007/978-3-030-52152-3_1
5. Alexander, S.A.: Short-circuiting the definition of mathematical knowledge for an artificial general intelligence. In: Cleophas, L., Massink, M. (eds.) SEFM 2020. LNCS, vol. 12524, pp. 201–213. Springer, Cham (2021). https://doi.org/10.1007/978-3-030-67220-1_16
6. Aristotle: on the soul. In: Barnes, J., et al. (eds.) The Complete Works of Aristotle. Princeton University Press (1984)
7. Brockman, G., et al.: OpenAI gym. Preprint (2016)
8. Davis, M.: Hilbert's tenth problem is unsolvable. Am. Math. Mon. **80**(3), 233–269 (1973)
9. Hernández-Orallo, J., Dowe, D.L.: Measuring universal intelligence: towards an anytime intelligence test. Artif. Intell. **174**(18), 1508–1539 (2010)
10. Hernández-Orallo, J., Dowe, D.L., España-Cubillo, S., Hernández-Lloreda, M.V., Insa-Cabrera, J.: On more realistic environment distributions for defining, evaluating and developing intelligence. In: Schmidhuber, J., Thórisson, K.R., Looks, M. (eds.) AGI 2011. LNCS (LNAI), vol. 6830, pp. 82–91. Springer, Heidelberg (2011). https://doi.org/10.1007/978-3-642-22887-2_9

11. Hutter, M.: Universal Artificial Intelligence: Sequential Decisions Based on Algorithmic Probability. Springer, Heidelberg (2004)
12. Kaliszyk, C., Urban, J., Michalewski, H., Olšák, M.: Reinforcement learning of theorem proving. In: NeurIPS (2018)
13. Legg, S., Hutter, M.: Universal intelligence: a definition of machine intelligence. Mind. Mach. **17**(4), 391–444 (2007)
14. Legg, S., Veness, J.: An approximation of the universal intelligence measure. In: Dowe, D.L. (ed.) Algorithmic Probability and Friends. Bayesian Prediction and Artificial Intelligence. LNCS, vol. 7070, pp. 236–249. Springer, Heidelberg (2013). https://doi.org/10.1007/978-3-642-44958-1_18
15. Maguire, P., Moser, P., Maguire, R.: Are people smarter than machines? Croatian J. Philos. **20**(1), 103–123 (2020)
16. Mnih, V., et al.: Human-level control through deep reinforcement learning. Nature **518**(7540), 529–533 (2015)
17. Narla, A., Kuprel, B., Sarin, K., Novoa, R., Ko, J.: Automated classification of skin lesions: from pixels to practice. J. Investig. Dermatol. **138**(10), 2108–2110 (2018)
18. Raffin, A., Hill, A., Ernestus, M., Gleave, A., Kanervisto, A., Dormann, N.: Stable baselines3 (2019). https://github.com/DLR-RM/stable-baselines3
19. Russell, S.J., Subramanian, D.: Provably bounded-optimal agents. J. Artif. Intell. Res. **2**, 575–609 (1994)
20. Schulman, J., Wolski, F., Dhariwal, P., Radford, A., Klimov, O.: Proximal policy optimization algorithms. Preprint (2017)
21. Silver, D., et al.: Mastering the game of Go with deep neural networks and tree search. Nature **529**(7587), 484–489 (2016)
22. Silver, D., et al.: Mastering the game of Go without human knowledge. Nature **550**(7676), 354–359 (2017)
23. Silver, D., Singh, S., Precup, D., Sutton, R.: Reward is enough. Artif. Intell. **299**, 103535 (2021)
24. Singh, S., Lewis, R.L., Barto, A.G., Sorg, J.: Intrinsically motivated reinforcement learning: an evolutionary perspective. IEEE Trans. Auton. Ment. Dev. **2**(2), 70–82 (2010)
25. Watkins, C.: Learning from delayed rewards. Ph.D. thesis, Cambridge (1989)
26. Yampolskiy, R.: On controllability of artificial intelligence. Technical report (2020)

CoSim-CPS 2021 - 5th Workshop on Formal Co-Simulation of Cyber-Physical Systems

CoSim-CPS 2021 Organizers' Message

CoSim-CPS is the premier workshop on the integrated application of formal methods and co-simulation technologies in the development of software for Cyber-Physical Systems. Co-simulation is an advanced simulation technique that allows developers to generate a global simulation of a complex system by orchestrating and composing the concurrent simulation of individual components. Formal methods link software specifications and program code to logic theories, providing means to exhaustively analyze program behaviors. The two technologies complement each other. Developers can create prototypes to validate hypotheses embedded in formal models, in order to ensure that the right system is being analyzed. Using formal methods, developers can generalize the results obtained with co-simulation, enabling early detection of latent design anomalies.

This year's workshop was held online, due to the COVID-19 situation around the world. The event included live online presentations, as well as discussions online and offline, all through a virtual conference platform.

Our keynote speaker, Paolo Bellavista, talked about enabling distributed and hybrid digital twins in the Industry 5.0 cloud continuum. The efficient exploitation of the cloud continuum (industrial cloud, edge cloud, 5G/6G base stations, fog nodes) is a key factor for future Industry 4.0 applications. The keynote offered an overview of the state-of-the-art architectures and technologies in the field, with a specific focus on how to efficiently design and implement distributed and hybrid digital twins. The presented cases of digital twins are *distributed* because they are executed over differentiated cloud continuum virtualized resources and by changing their location dynamically. They are also *hybrid* because they combine data-driven machine learning models and simulations based on mathematical-physical modeling of the systems they represent.

The workshop was held in three sessions, the first one being about convergence and stability of co-simulations, the second about maritime applications, and the third about automotive and aircraft applications and frameworks. We received a total of eight submissions, seven of which were accepted for presentation and publication. Each manuscript received three anonymous reviews, after a five-day bidding period and before a final three-day consensus discussion period.

We are grateful to the Program Committee for the dedication to the critical tasks of reviewing the submissions. We are also grateful to members of the Organizing Committee of SEFM for making the necessary arrangements – especially in the difficult circumstances of this year – and helping to publicize the workshop and prepare the proceedings. Finally, we thank the authors for their efforts in writing their papers and for the excellent presentations.

March 2022

Cinzia Bernardeschi
Cláudio Gomes
Maurizio Palmieri
Paolo Masci

Organization

Program Committee Chairs

Cinzia Bernardeschi	University of Pisa, Italy
Cláudio Gomes	Aarhus University, Denmark
Maurizio Palmieri	University of Pisa, Italy
Paolo Masci	National Institute of Aerospace (NIA), USA

Program Committee

Julien A. dit Sandretto	ENSTA ParisTech, France
Swee Balachandran	National Institute of Aerospace (NIA), USA
Mongi Ben Gaid	IFPEN, France
Jörg Brauer	Verified Systems International GmbH, Germany
Paul De Meulenaere	University of Antwerp, Belgium
Andrea Domenici	University of Pisa, Italy
Aaron Dutle	NASA, USA
Adriano Fagiolini	University of Palermo, Italy
Francesco Flammini	Linnaeus University, Sweden
Ken Pierce	Newcastle University, UK
Antonella Longo	University of Salento, Italy
Akshay Rajhans	MathWorks, USA
Rudolf Schlatte	University of Oslo, Norway
Neeraj Singh	INPT-ENSEEIHT/IRIT and University of Toulouse, France
Casper Thule	Aarhus University, Denmark

Enabling Distributed and Hybrid Digital Twins in the Industry5.0 Cloud Continuum

Paolo Bellavista[(✉)] [iD]

Alma Mater Studiorum University of Bologna, 40122 Bologna, Italy
paolo.bellavista@unibo.it

Abstract. The efficient exploitation of the cloud continuum (interworking of industrial cloud, edge cloud, 5G/B5G base stations, and fog node technologies) is a key factor for future Industry4.0 and beyond applications. For instance, with their ability to work more locally to data sources and to controllable actuators, cloud continuum-based solutions are considered very promising for dynamic manufacturing line control/reconfiguration and for sustainability optimizations (reduction of power and materials consumption). In these contexts, privacy, data sovereignty/control, latency, reliability, and scalability are crucial. This short paper, summarizing some key concepts from the associated keynote speech, aims at offering an overview of the concept of Digital Twins (DTs) for Industry5.0 [1] and at showing some primary and state-of-the-art solution guidelines that are emerging in the field. In particular, in our H2020 IoTwins project [2], we promote, design, implement, and evaluate Industry5.0 DTs that are both distributed because they are able to run over differentiated cloud continuum virtualized resources, also by changing their location dynamically, and hybrid because they combine data-driven machine learning models and simulations based on mathematical-physical modeling of the cyber-physical systems they represent. Distributed and hybrid DTs based on cloud continuum technologies are showing their effectiveness and efficiency in different application cases and scenarios, as demonstrated by our experience of IoTwins testbeds development.

Keywords: Digital Twins · Industry 5.0 · Cloud continuum · Edge cloud computing · IoTwins EU project · Testbed development and evaluation

1 Architectural Guidelines for Distributed and Hybrid DTs

Fulfilling the Industrial Internet of Things (IIoT) promise of autonomous interoperability, solution agility, and flexible reconfiguration of production chains, it is expected not only that the number of interconnected physical devices in the manufacturing domain will increase drastically, but also that there will be a strong need to interact with global/local cloud/edge services to act intelligently and flexibly. This introduces numerous challenges to industrial networked computing environments, which are traditionally quite static and isolated. First, IIoT environments will have to serve a wide range of industrial applications with different Quality-of-Service (QoS) requirements, ranging from traditional, time-sensitive, and reliable closed-loop control systems to event-driven, delay-bounded,

A. Cerone et al. (Eds.): SEFM 2021 Workshops, LNCS 13230, pp. 139–142, 2022.
https://doi.org/10.1007/978-3-031-12429-7_10

or best-effort sensor traffic. Second, these applications will have to fully grasp the potential benefits of the Cloud-to-Things Continuum (C2TC) [3], by exploiting resource virtualization and dynamic management of support/service components over industrial edge gateways, Multi-access Edge Computing (MEC) telco nodes (as standardized in ETSI specifications), and the global cloud. Third, in the reconfigurable factories of the future, the end-to-end QoS of co-existing applications will require to be managed, even when new C2TC industrial applications are dynamically introduced/removed. Fourth, several connectivity options (5G/6G, TSCH, WiFi7, ...) will gradually be introduced into production lines, by opening new opportunities of QoS specification and control, but also new challenges for heterogeneous QoS-constrained management.

In these very challenging scenarios and application cases, Digital Twins (DTs) are gaining relevance as comprehensive, actionable, digital representations of industrial physical systems and their behavior [4, 5]. DTs, in fact, provide a software copy of a physical asset by reflecting its properties, behaviors, and relationships according to the operational context. The physical and software counterparts mutually cooperate and co-evolve for enabling features such as device control, simulation, analytics, and more generally, the ability to dynamically enhance the functionality associated with physical objects and the optimization of its operation, management, and maintenance. For instance, DTs are a crucial enabling technology to detect and diagnose anomalies, to determine an optimal set of actions that maximize key performance metrics, and to effectively and efficiently enforce on-line quality management of production processes under latency and reliability constraints.

We strongly believe that future, effective, and efficient **DTs for Industry5.0 have to be distributed, hybrid, and based on cloud continuum technologies**. On the perspective of their distribution, for enhancing flexibility and data/control locality, DTs should be organized in a hierarchy of three layers: i) directly at IoT devices, whenever possible, **IoT twins** are lightweight models of specific components, performing big-data processing and local control for quality management operations; ii) **edge twins** have to be deployed at plant gateways and/or at emerging MEC nodes, thus providing higher level control knobs and orchestrating Internet of Things (IoT) sensors and actuators in a production locality; and iii) **cloud twins** are in charge of most time-consuming and resource-greedy operations, such as (typically off-line) parallel simulation and deep-learning, feeding edge twins with pre-elaborated or reduced order models to be efficiently executed at industry premises for monitoring/control/tuning purposes.

These distributed and interworking DTs should be **cloud continuum-oriented**, i.e., run at and use local virtualized resources (based on full virtual machines or on different types of possible heterogeneous containers). This opens up excellent opportunities of dynamic deployment and offloading/in-loading of needed behavior at runtime [6], but also poses very tough and challenging technical issues in terms of performance isolation, quality control, and middleware/application interference [7]. Finally, on the perspective of simulation models and technologies, DTs have to make use of **hybrid** modeling approaches, i.e., combining pure data-driven machine learning techniques with more traditional simulation models and tools, typically based on the physical modelling of the behavior of the twinned cyber-physical system.

2 The IoTwins H2020 Project

By following the above design principles of distributed and hybrid DTs based on cloud continuum technologies, the H2020 IoTwins project [2] aims to lower the barriers for building C2TC systems and services based on big data for the domains of manufacturing and facility management, by harmonizing standards to enable interoperability, and by developing an easy-to-use service layer that facilitates and decreases the cost of integration and deployment. To this purpose, IoTwins proposes a framework for a seamless, straightforward, and loose integration of already developed and deployed industrial software components running in typical long-lived industrial test-beds. The high-level distributed architecture of the IoTwins platform is depicted in Fig. 1, where you can notice, for example, that traditional machine learning training is first performed at cloud resources (integrated with agent-based and physical simulations), then refined at edge nodes (also with Federated Learning approaches [8]), with the possibility to run already trained anomaly detection algorithms also at IoT nodes.

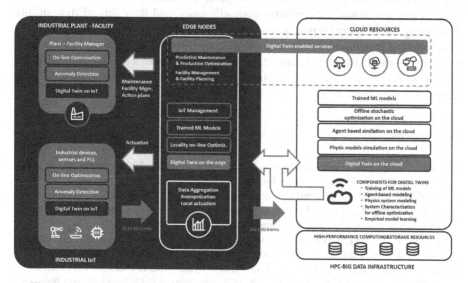

Fig. 1. The concept of distributed and hybrid DTs for the cloud continuum in IoTwins.

The IoTwins architecture above has been implemented, prototyped, and evaluated into different possible incarnations, suitable for different application cases and deployment environments. For example, one incarnation is a novel platform based on the extension of Indigo PaaS and exploiting edge cloud virtualized resources via Mesos; the platform originally implements distributed orchestration based on enhanced TOSCA templates for the C2TC. Another incarnation of the IoTwins architecture is a platform deriving from the integration of more proprietary Siemens solutions for the distributed monitoring and control of smart grids, which are deployed at wide-scale in the Wien metropolitan environment together with network and control equipment by TTTech. Let us also note that, for maximum interoperability, openness, and future extensibility, the

IoTwins architecture is compliant with the most widespread standard Industry4.0 architectural specifications, such as the Reference Architectural Model Industrie 4.0 (RAMI) and the Industrial Internet Reference Architecture (IIRA) [9].

To practically and tangibly show the advantages of the adoption of distributed and hybrid DTs designed according to the IoTwins guidelines, we have already developed a series of testbeds and associated Industry5.0 applications; these tangible testbeds are crucial to promote the adoption of our C2TC techniques in large companies and SMEs, by showing the feasibility and efficiency of the proposed approach in practical cases. To this purpose, twelve testbeds have been implemented, some of them in the manufacturing domain (prescriptive maintenance for wind turbines, accurate modelling of machine tool spindle behavior, optimization of the crankshaft manufacturing production process, and defect reduction in closure manufacturing lines), in the facility management domain (the Barcelona Camp-Nou stadium during crowd upload/download, energy optimization of the Cineca HPC site, and predictive management of electric smart grid in Wien), and in other cases similar to the previous ones but at a different scale. The performance results and economic advantages of the IoTwins adoption have been already measured in those testbeds and are showing the effectiveness, efficiency, and flexibility of dynamically distributing our Industry5.0 platform on every node belonging to the C2TC chain. For finer technical insights about the technologies, middleware, algorithms, and protocols integrated in the IoTwins architecture, please see [2], where a significant part of the IoTwins project results are made publicly available for the community of researchers and practitioners in the field.

References

1. Zong, L., et al.: End-to-end transmission control for cross-regional industrial internet of things in industry 5.0. IEEE Trans. Industr. Inf. **18**(6), 4215–4223 (2022)
2. The H2020 IoTwins Project. https://www.iotwins.eu/
3. Samie, F., Bauer, L., Henkel, J.: From cloud down to things: an overview of machine learning in internet of things. IEEE Internet Things J. **6**(3), 4921–4934 (2019)
4. Minerva, R., Lee, G.M., Crespi, N.: Digital twin in the IoT context: a survey on technical features, scenarios, and architectural models. Proc. IEEE **108**(10), 1785–1824 (2020)
5. Vukovic, M., Mazzei, D., Chessa, S., Fantoni, G.: Digital twins in industrial IoT: a survey of the state of the art and of relevant standards. In: Proceedings of IEEE International Conference on Communications (ICC). IEEE, New York (2021)
6. Zanni, A., et al.: Automated selection of offloadable tasks for mobile computation offloading in edge computing In: CNSM Conference Proceedings. IEEE, New York (2017)
7. Li, Y., Zhang, J., Jiang, C., Wan, J., Ren, Z.: PINE: optimizing performance isolation in container environments. IEEE Access **7**, 30410–30422 (2019)
8. Bellavista, P., Foschini, L., Mora, A.: Decentralised learning in federated deployment environments: a system-level survey. ACM Comput. Surv. **54**(1), 15:1–15:38 (2021)
9. Borghesi, A., et al.: IoTwins: design and implementation of a platform for the management of digital twins in industrial scenarios In: CCGRID Conference Proceedings. IEEE, New York (2021)

Under What Conditions Does a Digital Shadow Track a Periodic Linear Physical System?

Hao Feng[2]([⊠])[iD], Cláudio Gomes[2][iD], Michael Sandberg[1][iD],
Hugo Daniel Macedo[2][iD], and Peter Gorm Larsen[2][iD]

[1] Department of Mechanical and Production Engineering, Aarhus University, Inge
Lehmanns Gade 10, Aarhus C, Denmark
`ms@mpe.au.dk`
[2] Department of Electrical and Computer Engineering, Aarhus University,
Helsingforsgade 10, Aarhus N, Denmark
{`haof,claudio.gomes,hdm,pgl`}`@ece.au.dk`

Abstract. The synchronization between a Digital Shadow (DS) and a Cyber-Physical System (CPS) is paramount to enable anomaly detection, predictive maintenance, what-if analysis, etc. Such synchronization means that a simulation reflects, as closely as possible, the states of the CPS. The simulation however, requires the complete initial state of the system to be known, which is often infeasible in real applications. In our work, we study the conditions under which knowing the initial state of the system is irrelevant for a simulation to eventually synchronize with the CPS. We apply traditional stability analysis to answer this question for linear periodic systems. We demonstrate the method using a simple but representative system, an incubator with a periodic control signal.

Keywords: Digital Shadow · Cyber-Physical System · Stability · Tracking

1 Introduction

A DS is a type of monitoring system. It contains not only the digital version of the system but also amounts of services that can improve the value of the physical system. One of the services could be real-time simulation that accurately mimics a CPS and its environment, and it is paramount to enable many of the benefits that the Digital Twin vision offers, such as the improvement of productivity [1], automation [3], system intelligence [13], and more. A DS is also key in, for example, anomaly detection that can prevent failures by triggering alarms, so new process conditions can be employed in time to increase asset lifetime. Figure 1 shows two possible architectures for enabling anomaly detection.

One architecture that is an easy way to realize the anomaly detection mainly consist of the *real-time simulation* of a plant model (P'_{PT} in Fig. 1), feeding all

We are grateful to the Poul Due Jensen Foundation, which has supported the establishment of a new Centre for Digital Twin Technology at Aarhus University. Hao Feng also acknowledges support from the China Scholarship Council.

A. Cerone et al. (Eds.): SEFM 2021 Workshops, LNCS 13230, pp. 143–155, 2022.
https://doi.org/10.1007/978-3-031-12429-7_11

inputs from the real plant to a simulation model, and observing whether the resulting behavior matches the measured. This concept is illustrated in Fig. 1 (when omitting the Kalman filter). Figure 1 also shows a second architecture, in which the *Kalman filter* is taken into account. Here, the latter architecture requires partial state measurement from the plant.

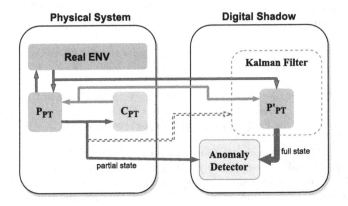

Fig. 1. Illustration of the two architectures for anomaly detection. One architecture is represented by the solid lines. Another one is the previous one combined with the Kalman filter.

When applicable, the Kalman filter architecture is superior to the real-time simulation architecture because the Kalman filter utilizes the state measurement data and the prediction data from a model, which makes it suffer less from *state-drift problems*. In contrast, the real-time simulation architecture does not take into account the state measurement data. State drifting is most commonly known in localization problems, where, for example, a moving object's acceleration and velocity are measured through an accelerometer and integrated to obtain the position. Here, assuming that the initial position and velocity are known, the process of integrating (essentially simulating a point mass system) yields the position over time, which almost always diverges from the real position of the moving object. Figure 2 shows one reference trajectory (in green), one tracking simulation trajectory that was started from different initial conditions (in blue), and two trajectories that are diverging.

The application of the Kalman filter is, however, not always possible as it requires a model to obey a certain structure. For instance, it requires the ability of updating the state of the model, which makes black-box models difficult to extend with Kalman filters. For example, the authors in [4] had to propose an extension to the Functional Mockup Interface (FMI) standard version 2.0 [2], to be able to apply Kalman filters.

In this paper, we focus on the conditions that make it possible for a real-time simulation architecture (without a Kalman filter) to avoid state drifting. Since real-time simulation models can utilize co-simulation interfaces [10], this

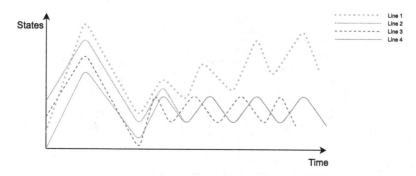

Fig. 2. A reference trajectory (line 4), and one converging tracking simulation that is started with different initial conditions (line 2). The line 3 and line 1 curves depict tracking simulations that suffer from state drifting. (Color figure online)

approach can be applied to virtually any black-box model. In the approach, we employ traditional stability analysis to demonstrate that a periodic linear system can be accurately tracked by a real-time simulation, even if the initial state of the plant is unknown. We exemplify the application of the method by using an open-loop incubator system, based on our previous work in [6,7]. The analyses in this work are limited to systems that are periodic and linear, and our incubator system fulfills these conditions.

We have empirical evidence that the closed-loop incubator system can also be tracked by real-time simulations, and we expect that more advanced stability analysis can be applied to CPSs that are non-linear or hybrid. The stability of these systems has been studied extensively (see, for example, [9,11,12,14,15]), and in future work, we seek to apply these methods to the closed-loop periodic systems as well.

The rest of the paper is organized as follows: Sect. 2 gives the background for the work, and details how to discretize continuous-time linear systems (an important step in our proposed method). Then, in Sect. 3, we introduce our contribution and apply it to the analysis of the incubator system. Next, Sect. 4 corroborate the results of the theoretical analysis with experimental results, and finally, Sect. 5 concludes the paper and discusses future work.

2 Background

This section gives background of the physical incubator system and its continuous and discrete-time models.

2.1 Incubator System

The incubator is a system that has the ability to regulate the temperature within an insulated container. A systematic diagram of the incubator is illustrated in Fig. 3 and Fig. 5 (right) shows a picture of the physical system.

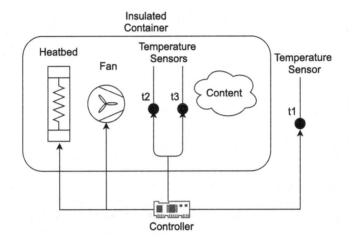

Fig. 3. Schematic overview of the incubator. The physical system can be seen in Fig. 5 (right).

The main components of our incubator system are an insulated container, a heatbed, a fan, three temperature sensors, and a controller. The heatbed is used to generate heat energy for heating the air inside the insulated container (referred to as "air temperature"), and it has two different states, on and off. The fan circulates the air, to achieve a more uniform temperature distribution. There are three sensors, two of which are used to measure the air temperature and one for the ambient room temperature. In the experiments, the fan is always on, but the controller can regulate both the heatbed and fan states. For detailed information about the incubator, see [7].

In this work, we configure the controller to turn on and off the heatbed in periodical intervals. This approach is in contrast with the closed-loop controller introduced in [7], which activates the heatbed based on the measured temperature. To keep the temperature relatively low in the experiments (defined here as <50 °C), we set the on-state to be shorter than the off-state (6 vs. 27 s).

2.2   Models

The state variables that enter our time-dependent thermal model of the incubator system are the air and heatbed temperatures, T_{bair} and T_{heater}. Here, we consider the rate of energy to establish an Ordinary Differential Equation (ODE). In reality, the temperature of the system is not completely uniformly distributed, meaning that T_{bair} and T_{heater} are lumped state variables. While T_{heater} is not directly measured from the system, we consider the average of the signal from the two internal temperature sensors to be representative of T_{bair}. The ODEs read

$$\frac{dT_{heater}}{dt} = \frac{1}{C_{heater}}(VI - G_{heater}(T_{heater} - T_{bair})) \tag{1a}$$

$$\frac{dT_{bair}}{dt} = \frac{1}{C_{air}}[G_{heater}(T_{heater} - T_{bair}) - G_{box}(T_{bair} - T_{room})], \tag{1b}$$

For more details, see [5]. In Eq. (1), C denotes the lumped thermal capacity, G is the effective heat transfer coefficient, T_{room} is the ambient room temperature, and VI represents the product of the voltage and the current representing the power of the heatbed when turned on.

To use Eq. (1) in stepwise simulations, it is necessary to convert it to a discrete-time model. In this paper, we give a brief explanation of the steps needed to accomplish this, but readers who are interested in more details can refer to [8].

In general, linear and continuous-time dynamical systems can be converted into a state-space model that has the generic form:

$$\dot{x}(t) = Ax(t) + Bu(t) \tag{2a}$$

$$y(t) = Cx(t) + Du(t). \tag{2b}$$

The solution of Eq. (2a) is given as

$$x(t) = e^{A(t-t_0)}x(t_0) + \int_{t_0}^{t} e^{A(t-\tau)}Bu(\tau)d\tau. \tag{3}$$

Using Eq. (3), the temperature state at any time instance can be obtained. In Eq. (3), we can see that the initial point of $x(t_0)$ is necessary when approximating the model behavior. However, in a computational unit, it is not possible to generate continuous behaviors, as it has to be sampled which is a discrete-time behavior. If an interval of the sampling process is T, then the time series is $0, T, 2T, \ldots, kT, (k+1)T, \ldots$, where k is an integer. In addition, the inputs between two sampling instants are constants, which means that $u(\tau) = u(kT)$ for $kT \le \tau \le (k+1)T$. In Eq. (3), let $t_0 = kT$. Then the next sample we can obtain is $x((k+1)T)$, thus we set $t = (k+1)T$. Substitute these back into Eq. (3) and we have

$$x((k+1)T) = e^{A(T)}x(kT) + \int_{kT}^{(k+1)T} e^{A((k+1)T-\tau)}Bd\tau u(kT). \tag{4}$$

Since the T is known, we can simplify Eq. (4) into the form of

$$x(k+1) = A'x(k) + B'u(k), \tag{5}$$

where $A' = e^{AT}$, $B' = \int_{kT}^{(k+1)T} e^{A((k+1)T-\tau)}Bd\tau$. Note that Eq. (5) is similar to Eq. (2a) but the coefficients are different and the time interval between $x(k+1)$ and $x(k)$ is T.

To summarize, Eq. (5) is the discrete-time model of the continuous-time model of Eq. (2) and it is feasible to implement Eq. (5) to generate behaviors in a computational unit. In the section below, we show how to apply this technique to prove that simulations of the incubator system will converge towards the real system behavior.

3   Stability Analysis of Periodic Linear Systems

This section gives an introduction about the stability of a dynamic system and how to utilize it to tackle the issues of tracking a periodic physical system under unknown initial conditions.

Stability is a property of a dynamic system. A pendulum, for example, is a stable system. Regardless of what the initial state of a pendulum is, unless perfectly upright, the pendulum will oscillate around the equilibrium point that is the state of vertical down if it is a perfect pendulum and has no friction. However, if there is friction, the pendulum will eventually go into the state of vertical down as its initial gravitational potential energy is consumed by friction and other energy losses. This is called asymptotic stability that is the property that a system will eventually converge to a stable state when it is started close to one.

Accordingly, a simulation of the incubator system can be started with different initial conditions that will all eventually converge to its single stable state. This also applies for the true initial conditions, and this property alleviates the issue of having unknown initial conditions.

While it does not make sense to talk about a stable periodic incubator system, because the temperature will continuously oscillate, we can transform the system into an equivalent system that represents the temperature evolution at certain periods of time. This new system is then asymptotically stable, which is illustrated in Fig. 4, where the solution to the original system is not asymptotically stable, but the equivalent sampled discrete-time system is.

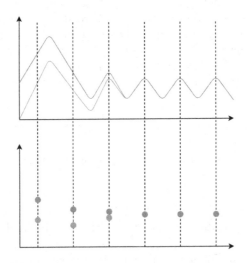

Fig. 4. This figure illustrates how the probed temperature is not asymptotically stable, but the equivalent sampled discrete-time system is.

The incubator continuous-time model is presented in Eq. (1). For simplicity, we use the generic discrete-time model of Eq. (5) to represent the incubator. Note that to convert the continuous-time model to a discrete-time model, the sampling rate that is the T in Eq. (4) is needed and we call it $T_{sampling}$. Note that the $T_{sampling}$ is normally chosen according to the system based on the hardware. In addition, we set A' and B' to A and B respectively for consistency. Then we have a linear system in the form of

$$x(k+1) = Ax(k) + Bu(k). \tag{6}$$

The controller only controls the states of the heatbed that are on (heating for short) or off (cooling for short), and we assume the room temperature is constant in between controller samples. As a result, the $u(k)$ has to be either $\begin{bmatrix} VI\ T_{room} \end{bmatrix}^T$ or $\begin{bmatrix} 0\ T_{room} \end{bmatrix}^T$. Substitute the two $u(k)$ back to Eq. (6), then we obtain two different models that indicate the advancements of temperatures in terms of heating or cooling. These two models are given by

$$x(k+1) = Ax(k) + M \tag{7a}$$
$$x(k+1) = Ax(k) + N, \tag{7b}$$

where Eq. (7a) is the model of heating and Eq. (7b) of cooling. In each sampling step, $T_{sampling}$, either of the models in Eq. (7) is executed and therefore, the time for heating ($T_{heating}$) or cooling ($T_{cooling}$) should be equal to $n \cdot T_{sampling}$, where n is a non-negative integer.

To highlight the main idea of the method, suppose that the controller always switches from heating to cooling at every time sample, which means that $T_{heating} = T_{sampling}$ and $T_{cooling} = T_{sampling}$. Then, if we start with the heating state at time step k, the temperatures at next time step $k+1$ is

$$x(k+1) = Ax(k) + M. \tag{8}$$

Afterward, the incubator goes to the cooling state from time step $k+1$, thus the temperatures at time step $k+2$ are obtained by Eq. (7b)

$$x(k+2) = A(Ax(k) + M) + N$$
$$x(k+2) = A^2x(k) + AM + N. \tag{9}$$

Equation (9) represents the temperatures during one period of heating and cooling. If the initial k is 0, we get the temperatures at time step 2 according to Eq. (9). If the periodic control signal continues, we have the equation representing the temperatures at time step $2k$

$$x(2k) = A^{2k}x(0) + (AM + N)(A^{2k-2} + A^{2k-4} + \ldots + A^0). \tag{10}$$

Note that Eq. (10) is only valid for $T_{heating} = T_{sampling}$ and $T_{cooling} = T_{sampling}$. If $T_{heating} = m * T_{sampling}$ and $T_{cooling} = n * T_{sampling}$, where m and n are integers, then in order to obtain Eq. (10), it involves m times and n times compositions of Eq. (7a) and Eq. (7b) respectively.

If the eigenvalues of the matrix A that represents the properties of the system are less than 1, which means the system is stable, then $\lim_{k \to +\infty} A^{2k} = 0$. As a result, the temperatures at time step $2k$, when k is large enough, is $x(2k) = (AM + N)(A^{2k-2} + A^{2k-4} + \ldots + A^0)$, which is independent of the initial conditions, $x(0)$. If the absolute value of all eigenvalues of the matrix A are less than 1, then $(AM + N)(A^{2k-2} + A^{2k-4} + \ldots + A^0)$ converges. This means under such a periodic control signal, the temperatures will eventually convergence toward $\lim_{k \to +\infty}(AM + N)(A^{2k-2} + A^{2k-4} + \ldots + A^0)$ no matter the initial conditions. This proves that a DS can track a periodic system even when the initial conditions are unknown.

If the absolute value of one of the eigenvalues of A is bigger than 1, a system is not stable and there are no stable states. In this case, a DS would not be able to track the periodic system without knowing the initial conditions.

Overall, if a system is stable when subject to a given control signal and if it only has one possible stable state, then a DS of the system can track the physical system without knowing exactly the initial conditions. Our incubator system is a stable system when subject to the periodic control signal and the eigenvalues of A are less than 1, which will be shown in the next section. Therefore, a DS of the incubator can track the physical incubator without knowing exactly the initial conditions.

4 Simulation and Experiments Results of Incubator

In Sect. 3, we analyzed the theoretical stability of the incubator. This section presents the simulation and experimental results of our analysis. Note that due to limitations of utilized hardware, it takes around one second to fetch data from one sensor, and three seconds for all three sensors.

As a first step before the stability analysis, we conducted a small experiment to investigate the internal temperature distribution of the air inside the insulated container. These results are shown in Fig. 5.

We used three temperature sensors to measure the temperature at different locations, marked $t1$, $t2$, and $t3$ in Fig. 5. As it can be seen, the temperature is not uniform when the incubator warms up. Here, the largest temperature difference was about 6 °C. According to this experiment, we found that the temperature at sensor $t1$ is close to the average temperature of sensors $t2$ and $t3$ (Fig. 5). This finding supports our decision of moving $t1$ outside for measuring the ambient room temperature and using the average temperature of sensors $t2$ and $t3$ as representative of the internal air temperature. We used this new setup in the following experiments.

Next, we conducted another experiment to determine the best-fit parameters in Eq. (1), and as Fig. 6 shows, the calibrated model captured the physical system very well. Nevertheless, for control conditions that deviate from the ones in the experiments, recalibration might be needed. Also, it is important to note that

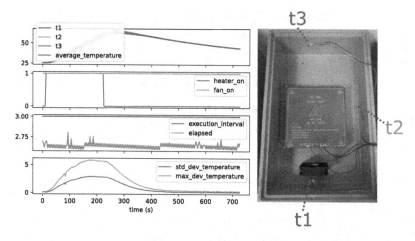

Fig. 5. Experimentally measured temperature, heater signal, together with maximum and std. deviation of the measured temperature (left). The experimental setup can be seen in the figure to the right.

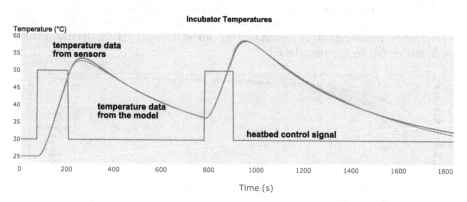

Fig. 6. Calibration results of (1). The blue line is the average temperature representing the air temperature, while the green line shows the results of our calibrated model. The red line shows the two states of the heatbed, heating and cooling. (Color figure online)

the model is incapable of capturing other complex process conditions such as opening the incubator lid as shown in Fig. 7 despite calibration.

After the calibrated model was demonstrated to be representative of the physical system, we utilized it to demonstrate that the incubator system is, in fact, stable and that it converges towards the same temperature state no matter the initial conditions following our approach described in Sects. 2 and 3. To do this, we ran five simulations with different initial conditions of the air temperature that are 40 °C, 30 °C, 25 °C, 20 °C, and 10 °C. The room temperature in the experiments was set to 22.44 °C and the voltage and the current to the heatbed were 12 V and 10.45 A. The simulation results are shown in Fig. 8.

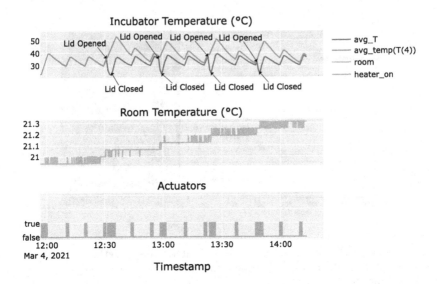

Fig. 7. Example of how the model initially captures the physical system very well, but starts to deviate when the process conditions change as the lid is opened. This cannot be resolved by recalibration, as the underlying model does not reflect the disturbance that is introduced by opening the lid.

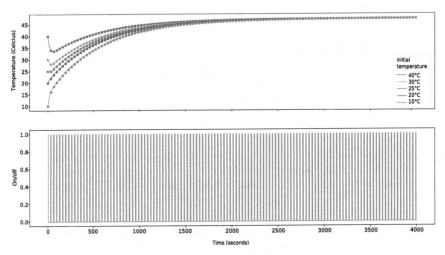

Fig. 8. Incubator simulation results. The top figure presents the simulation results of the air temperature starting with five different initial conditions, while the lower figure shows the controller signal for the heater bed (6 s on, followed by 27 s off).

As it can be seen in Fig. 8, all five initial conditions converged to the same stable state, meaning that the discrepancy between DS and the real physical system will be identical no matter the initial conditions after enough time. We also

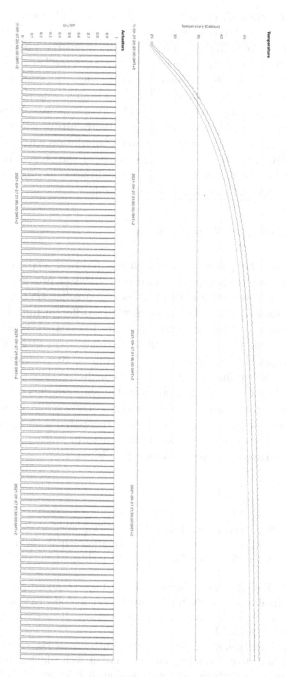

Fig. 9. Experimental results recorded and visualized using InfluxDB. The top figure shows the measured temperatures $t2$ (red) and $t3$ (orange), as well as the average temperature (purple). The lower figure shows the control signal. (Color figure online)

confirmed this by running experiments on the physical system. Here, we used InfluxDB to record and visualize the air temperature data. While it was not possible to control the initial starting temperature of the physical system, the experiment showed (Fig. 9) that the temperature increased at the early stage and gradually converged towards the same 47–48 °C when the system was started at room temperature. Since this result matches the simulation, this verifies the reliability of our model and approach. Note that, as consequence, if the lid is opened and closed again during one experiment, the temperature will also converge to the stable states because it is analog to start a simulation with a different initial condition. Finally, we note that the eigenvalues of the A matrix in our model are 0.942 and 0.448, both less than 1. This concludes that our DS can track the real physical system and that the system is stable.

5   Conclusion and Future Work

In order to tackle the problem of tracking a physical system with unknown initial conditions, we transformed it into a problem of proving the stability of the system. For a system with an open-loop controller that can enter into a stable state, we showed that a DS starting with a different initial conditions will also reach a similar stable state and track the physical system. Subsequently, we demonstrated this theory using our incubator system fitted with a periodic open-loop controller that resulted in a stable system. By conducting simulations and experiments, we verified our theory and concluded that a DS can track the periodic physical system if the system is stable and only has one stable state.

In the future, we are interested in analyzing the system with a closed-loop controller since this is more robust and general compared to an open-loop system. Also, we want further develop the DS of our incubator system, making it capable of coping with disturbances from opening the lid, etc.

References

1. Bauernhansl, T., Hartleif, S., Felix, T.: The Digital Shadow of production – a concept for the effective and efficient information supply in dynamic industrial environments. Procedia CIRP **72**, 69–74 (2018). https://doi.org/10.1016/j.procir.2018.03.188
2. Blochwitz, T.: Functional mockup interface 2.0: the standard for tool independent exchange of simulation models. In: 9th International Modelica Conference, pp. 173–184. Linköping University Electronic Press, Munich, November 2012. https://doi.org/10.3384/ecp12076173
3. Brecher, C., Buchsbaum, M., Storms, S.: Control from the cloud: edge computing, services and digital shadow for automation technologies. In: 2019 International Conference on Robotics and Automation (ICRA), pp. 9327–9333, May 2019. https://doi.org/10.1109/ICRA.2019.8793488

4. Brembeck, J., Pfeiffer, A., Fleps-Dezasse, M., Otter, M., Wernersson, K., Elmqvist, H.: Nonlinear state estimation with an extended FMI 2.0 co-simulation interface. In: 10th International Modelica Conference, pp. 53–62. Linköping University Electronic Press; Linköpings Universitet, Lund, March 2014. https://doi.org/10.3384/ecp1409653

5. Feng, H., Gomes, C., Sandberg, M., Thule, C., Lausdahl, K., Larsen, P.G.: Developing a physical and digital twin: a process model. In: 3rd International Workshop on Multi-paradigm Modeling for Cyber-Physical Systems, Fukuoka, Japan (2021, to appear)

6. Feng, H., Gomes, C., Thule, C., Lausdahl, K., Iosifidis, A., Larsen, P.G.: Introduction to digital twin engineering. In: 2021 Annual Modeling and Simulation Conference (ANNSIM), pp. 1–12, July 2021. https://doi.org/10.23919/ANNSIM52504.2021.9552135

7. Feng, H., Gomes, C., Thule, C., Lausdahl, K., Sandberg, M., Larsen, P.G.: The incubator case study for digital twin engineering, February 2021. arXiv:2102.10390 [cs, eess]

8. Friedland, B.: Control System Design: An Introduction to State-Space Methods. Courier Corporation (2012)

9. Goebel, R., Sanfelice, R.G., Teel, A.R.: Hybrid Dynamical Systems: Modeling, Stability, and Robustness. Princeton University Press (2012). http://www.u.arizona.edu/~sricardo/index.php?n=Main.Books

10. Gomes, C., Thule, C., Broman, D., Larsen, P.G., Vangheluwe, H.: Co-simulation: a survey. ACM Comput. Surv. **51**(3), 49:1-49:33 (2018). https://doi.org/10.1145/3179993

11. Johansson, M.: Piecewise Linear Control Systems, vol. 284. Springer, Heidelberg (2003). https://doi.org/10.1007/3-540-36801-9

12. Jungers, R.: The Joint Spectral Radius: Theory and Applications, vol. 385. Springer, Heidelberg (2009). https://doi.org/10.1007/978-3-540-95980-9

13. Ladj, A., Wang, Z., Meski, O., Belkadi, F., Ritou, M., Da Cunha, C.: A knowledge-based Digital Shadow for machining industry in a Digital Twin perspective. J. Manuf. Syst. **58**, 168–179 (2021). https://doi.org/10.1016/j.jmsy.2020.07.018

14. Legat, B., Gomes, C., Karalis, P., Jungers, R.M., Navarro-López, E.M., Vangheluwe, H.: Stability of planar switched systems under delayed event detection. In: 2020 59th IEEE Conference on Decision and Control (CDC), pp. 5792–5797, December 2020. https://doi.org/10.1109/CDC42340.2020.9304152

15. Liberzon, D.: Switching in Systems and Control. Springer, Boston (2012). https://doi.org/10.1007/978-1-4612-0017-8

Convergence Properties of Hierarchical Co-simulation Approaches

Irene Hafner[1](✉)(iD) and Niki Popper[1,2](iD)

[1] dwh GmbH, Neustiftgasse 57-59, 1070 Vienna, Austria
{irene.hafner,niki.popper}@dwh.at
[2] Institute of Information Systems Engineering, TU Wien,
Favoritenstraße 9-11, 1040 Vienna, Austria

Abstract. In this paper, local error estimates for hierarchical co-simulation approaches are presented. In hierarchical structures, systems with stronger dependencies on one another, which frequently occur in large-scale cyber-physical systems, may be combined in further co-simulations on one or more lower levels. This allows the selection of individual synchronization times for subsystems in these co-simulations. The estimates presented in this paper show that with this approach, no additional errors compared to traditional co-simulation are to be expected: on the contrary, results from the simulation of a benchmark example show that in case of sensible selection of lower-level couplings, accuracy and stability may even be increased as error propagation slows down.

Keywords: Co-simulation · Hierarchy · Consistency

1 Introduction

Hierarchical structures are no novelty in modeling and simulation in general, confer for example the Discrete Event System Specification (DEVS [19]) or partitioned integration methods [6,7,12,14]. However, hierarchical co-simulation as explained in the following is scarcely found in the literature. Although several frameworks and standards do not prohibit further co-simulations within a co-simulation, and some authors acknowledge the possibility of nested co-simulation [16,17], hierarchical co-simulation has, to the best of our knowledge, not been investigated with regard to error estimates up to now. Compared to hierarchical partitioned multirate schemes, subsystems may still be implemented in individually suitable simulation tools in a hierarchical co-simulation approach.

The idea of the introduction of further co-simulation levels is illustrated in Fig. 1. Such further division and nesting of co-simulations can be motivated by highly diverse time constants or other subsystem properties that require closer interaction between certain subsystems. In a traditional co-simulation approach, this could enforce a rather small macro step and thus, synchronization of all subsystems and consequently high computation time. With the introduction of further levels, more closely dependent subsystems may communicate with a small

A. Cerone et al. (Eds.): SEFM 2021 Workshops, LNCS 13230, pp. 156–172, 2022.
https://doi.org/10.1007/978-3-031-12429-7_12

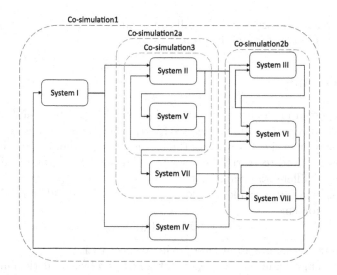

Fig. 1. Schematic depiction of a hierarchical co-simulation approach. Coordination takes place on several levels by one top-level co-simulation that manages the communication between subsystems and further co-simulations. These may again coordinate subsystems and co-simulations on lower levels [9].

macro step on a lower-level co-simulation while exchanging values with all other systems on a distinctly larger upper-level macro step, thus increasing accuracy without drastically slowing down the whole simulation.

A typical application example with these properties would be a manufacturing process where machines have to exchange data rather frequently with logistics while only from time to time transferring their waste heat data to a slow varying, thermal room model. This, in turn, has to be synchronized with an HVAC simulation controlling the room temperature. The latter would not require any communication with machines or logistic devices themselves, let alone evaluation and data exchange at the same, considerably small, time steps. Holistic simulation of urban energy systems likewise intrinsically brings along several different levels of consideration: households, factories, traffic, network, and power plants can each prove complex enough to be addressed by individual co-simulations, which then have to communicate in order to portray the overall system.

In the following, investigations on convergence of the proposed method are presented, starting with estimates on the consistency error for traditional, single-level co-simulation, extending them to hierarchically structured approaches and presenting error studies that illustrate the improvement in accuracy. Investigations on zero-stability and numerical stability, which are essential in addition to consistency to guarantee convergence in case of coupled DAE systems or ODEs with multi-step integration algorithms, are found in [8,9]. There it is shown that stability issues can be tackled by introducing another layer of communication instead of having to decrease the overall communication step size, thus providing an innovative method for stabilization.

2 Consistency

It has been shown in the literature that local error control is a valid method to bound the global co-simulation error (see f.i. [1,4,20]). This justifies investigating the consistency error, i.e. the error of the method in one step, in a co-simulation. For this aim we need to start by calling to mind some background information on numerics of differential equations.

In the following, we consider a uniquely solvable ordinary differential equation initial value problem

$$\dot{\boldsymbol{x}} = f(t, \boldsymbol{x}), \quad \boldsymbol{x}(t_0) = \boldsymbol{x}_0 \tag{1}$$

with Lipschitz continuous right side f with respect to $\boldsymbol{x}$.

For a given approximation $\boldsymbol{x}_{t_n+h}$ of $\boldsymbol{x}$ at time t_n+h by a numerical integration method with step size h, the consistency error is defined as the error of the method in one step and therefore, calculated by

$$\mathcal{E}(t_n, \boldsymbol{x}_n, h) = \boldsymbol{x}(t_n + h) - \boldsymbol{x}_{t_n+h} \tag{2}$$

for given initial values $\boldsymbol{x}(t_n) = \boldsymbol{x}_n$. A method is called consistent if

$$\lim_{h \to 0} \left(\frac{\mathcal{E}(t_n, \boldsymbol{x}_n, h)}{h} \right) = \boldsymbol{0} \tag{3}$$

for every choice of $t_n, \boldsymbol{x}_n$. A method is called consistent of order p if there exists a constant $C > 0$ with

$$\left\| \frac{\mathcal{E}(t_n, \boldsymbol{x}_n, h)}{h} \right\| \leq C \cdot h^p. \tag{4}$$

Since the consistency error is a measure for the *local* error of a method, state values are taken to be exact for all previous points in time.

Remark 1. In case they are not directly needed in the following calculations, the initial values $t_n, \boldsymbol{x}_n$ will be omitted in the notation of $\mathcal{E}$ to simplify the notation.

Important for the error estimates following below are Gronwall's Lemma (Theorem 1) and "the fundamental lemma" (Theorem 2).

Theorem 1. (Gronwall's Lemma *[15]). Let the real function $m(t)$ be continuous in $J := [0, a]$, and let*

$$m(t) \leq \alpha + \beta \int_0^t m(\tau) d\tau \quad \text{in } J \text{ with } \beta > 0$$

then

$$m(t) \leq \alpha e^{\beta t} \quad \text{in } J.$$

Theorem 2. (The "fundamental lemma" [10]). *Supposing that $\boldsymbol{x}(t)$ is a solution of the system of differential Eqs. 1 with f Lipschitz continuous in the second argument with Lipschitz constant L, and $\boldsymbol{v}(t)$ an approximate solution fulfilling*

$$\|\dot{\boldsymbol{v}}(t) - f(t, \boldsymbol{v}(t))\| \leq \epsilon,$$

then, for $t \geq t_0$, we have the error estimate

$$\|\boldsymbol{x}(t) - \boldsymbol{v}(t)\| \leq \|\boldsymbol{x}(t_0) - \boldsymbol{v}(t_0)\| \, e^{L(t-t_0)} + \frac{\epsilon}{L}\left(e^{L(t-t_0)} - 1\right).$$

Remark 2. If $\boldsymbol{v}$ is also an exact solution of $\dot{\boldsymbol{x}} = f(t, \boldsymbol{x})$, from Theorem 2 follows

$$\|\boldsymbol{x}(t) - \boldsymbol{v}(t)\| \leq \|\boldsymbol{x}(t_0) - \boldsymbol{v}(t_0)\| \, e^{L(t-t_0)},$$

which directly implies that in case of the same initial values, $\boldsymbol{v}$ is identical to $\boldsymbol{x}$.

2.1   Consistency in Co-simulation

To investigate consistency in co-simulation, we consider a system of N coupled ODEs given as follows[1]:

$$\dot{\boldsymbol{x}}^i(t) = \boldsymbol{f}^i(\boldsymbol{x}^i, \boldsymbol{u}^i, t), \quad \boldsymbol{x}^i(t_0) = \boldsymbol{x}_0^i \tag{5a}$$

with $i = I, \ldots, N$, $\boldsymbol{x}^i \in \mathbb{R}^{n_x^i}$, $\boldsymbol{u}^i \in \mathbb{R}^{n_u^i}$, and

$$\boldsymbol{u}^i = \boldsymbol{L}^i \boldsymbol{x} = \begin{bmatrix} \boldsymbol{L}^{i,I} & \ldots & \boldsymbol{L}^{i,i-1} & 0 & \boldsymbol{L}^{i,i+1} & \ldots & \boldsymbol{L}^{i,N} \end{bmatrix} \begin{bmatrix} \boldsymbol{x}^I \\ \vdots \\ \boldsymbol{x}^{i-1} \\ \boldsymbol{x}^i \\ \boldsymbol{x}^{i+1} \\ \vdots \\ \boldsymbol{x}^N \end{bmatrix} \tag{5b}$$

with $\boldsymbol{L}^{i,j} \in \mathbb{R}^{n_u^i \times n_x^j}$ $\forall i, j \in \{I, \ldots, N\}$ and the elements of $\boldsymbol{L}^{i,j}$ being equal to zero or one, thus describing the output-input dependencies between the individual subsystems. Thereby, we assume again a unique solution and Lipschitz continuous right-side functions $\boldsymbol{f}^i$ in the first and second argument.

In the following, investigations on convergence of traditional co-simulation analogously as given by Knorr [11][2] are presented and extended on hierarchical

---

[1] Notation with elements of $\mathbb{G} := \{I, II, \ldots\}$ is used to avoid confusion with exponents and allow easy identification of subsystems. In arithmetic operations where elements of $\mathbb{G}$ and $\mathbb{N}$ are mingled, these are to be understood as operations between elements of $\mathbb{N}$ by assigning every element of $\mathbb{G}$ its image under the bijection that uniquely assigns the i-th element of $\mathbb{G}$ the i-th element of $\mathbb{N}$.

[2] The investigations in [11] are restricted to two participating subsystems where the larger micro step size is also taken as macro step size. Following this strategy, we allow an arbitrary number of participating subsystems and macro step size H with the possibility of $H > h_i$ for all subsystem solver step sizes h_i in this work.

approaches in Sect. 2.2. We start by considering the i-th subsystem of (5). In case of a multirate co-simulation, values $\boldsymbol{u}^i$ have to be extrapolated in between two synchronization time steps and will be named $\tilde{\boldsymbol{u}}^i$. Depending on the order q_i of the chosen extrapolation method, $\left\|\boldsymbol{u}^i(t_n + h) - \tilde{\boldsymbol{u}}^i(t_n + h)\right\| \leq Ch^{q_i+1}$ with a constant $C > 0$ holds for a step of size $h > 0$ assuming $\tilde{\boldsymbol{u}}^i(t_n) = \boldsymbol{u}^i(t_n)$. Further, $\boldsymbol{x}^i(t)$ will denote the exact solution of (5a) and $\tilde{\boldsymbol{x}}^i(t)$ the exact solution of

$$\dot{\boldsymbol{x}}^i(t) = \boldsymbol{f}^i(\boldsymbol{x}^i, \tilde{\boldsymbol{u}}^i, t), \quad \boldsymbol{x}^i(t_0) = \boldsymbol{x}_0^i. \tag{6}$$

The approximated solution of (6) at $t_{n,k}$ will be named $\tilde{\boldsymbol{x}}_{n,k}$.

To begin with, we regard the error $\mathcal{E}^i(t_{n,k}, \boldsymbol{x}_{n,k}, h_i)$ of the i-th subsystem in one micro step h_i at $t_{n,k}$, where n is the current macro step and k the current micro step, counted anew for each macro interval. Thus $t_{n+1} := t_{n+1,0} = t_{n,m_i} = t_n + m_i \cdot h_i = t_n + H$ in case of m_i micro steps per macro step, hence m_i denoting the multirate factor of subsystem i in case of fixed, equidistant micro steps which are integer divisors of the (also fixed) macro step size H, which we will assume w.l.o.g.[3] in the following calculations.

Starting with the consistency of the integration of every subsystem for one micro step, we will deduce consistency of the integration of every subsystem for one macro step and further of the co-simulation.

Lemma 1 (Consistency error for one micro step). *Let p_i denote the consistency order of the original method and q_i the order of extrapolation for input values $\boldsymbol{u}^i$. Then*

$$\left\|\frac{\mathcal{E}^i(t_{n,k}, \boldsymbol{x}_{n,k}, h_i)}{h_i}\right\| = \mathcal{O}\left(h_i^{\min\{p_i, q_i+1\}}\right). \tag{7}$$

Proof. Considering exact values at $t_{n,k}$, per definition

$$\left\|\mathcal{E}^i(t_{n,k}, \boldsymbol{x}_{n,k}, h_i)\right\| = \left\|\boldsymbol{x}^i(t_{n,k} + h_i) - \tilde{\boldsymbol{x}}_{n,k+1}^i\right\| = \left\|\boldsymbol{x}^i(t_{n,k+1}) - \tilde{\boldsymbol{x}}_{n,k+1}^i\right\| \tag{8}$$

with the notation described above. Adding and subtracting $\tilde{\boldsymbol{x}}(t_{n,k+1})$ gives

$$\left\|\mathcal{E}^i(t_{n,k}, \boldsymbol{x}_{n,k}, h_i)\right\| \overset{\substack{\text{triangle}\\\text{inequ.}}}{\leq} \left\|\boldsymbol{x}^i(t_{n,k+1}) - \tilde{\boldsymbol{x}}^i(t_{n,k+1})\right\| + \underbrace{\left\|\tilde{\boldsymbol{x}}^i(t_{n,k+1}) - \tilde{\boldsymbol{x}}_{n,k+1}^i\right\|}_{\leq C_{i,1} \cdot h_i^{p_i+1}}. \tag{9}$$

The second term of (9) is the difference of the exact to the approximated solution of the modified system (6) and is therefore bounded by $C_{i,1} \cdot h_i^{p_i+1}$ for a constant $C_{i,1} > 0$ and with p_i being the order of the numerical integration method given for system i.

---

[3] All considerations can be performed analogously for unequally distanced grids with h_i taken as upper bound of all h_{i_j} with $i_j \in \{1, \ldots, m_{i_n}\}$ and m_{i_n} the number of micro steps of subsystem i in the n-th macro step. However, as this would only lead to more complex notation, we will restrict the step sizes as described above for reasons of clarity.

To provide an estimate for the first term in (9), we use the assumption that $\boldsymbol{x}(t)$ and $\tilde{\boldsymbol{x}}(t)$ are the exact solutions of (5a) and (6), respectively, and can therefore be replaced by the integral over their derivatives (since they fulfill conditions like uniqueness, continuity, and differentiability by definition):

$$
\begin{aligned}
\left\| \boldsymbol{x}^i(t_{n,k+1}) - \tilde{\boldsymbol{x}}(t_{n,k+1}) \right\| &= \left\| \int_{t_{n,k}}^{t_{n,k+1}} \left(f^i(\boldsymbol{x}^i, \boldsymbol{u}^i, \tau) - f(\tilde{\boldsymbol{x}}^i, \tilde{\boldsymbol{u}}^i, \tau) \right) d\tau \right\| \\
&\leq \int_{t_{n,k}}^{t_{n,k+1}} \left\| f^i(\boldsymbol{x}^i, \boldsymbol{u}^i, \tau) - f(\tilde{\boldsymbol{x}}^i, \tilde{\boldsymbol{u}}^i, \tau) \right\| d\tau
\end{aligned}
\tag{10}
$$

Adding and subtracting $f(\tilde{\boldsymbol{x}}^i, \boldsymbol{u}^i, \tau)$ gives with the triangle inequality

$$
(10) \leq \int_{t_{n,k}}^{t_{n,k+1}} \left\| f^i(\boldsymbol{x}^i, \boldsymbol{u}^i, \tau) - f(\tilde{\boldsymbol{x}}^i, \boldsymbol{u}^i, \tau) \right\| d\tau + \int_{t_{n,k}}^{t_{n,k+1}} \left\| f^i(\tilde{\boldsymbol{x}}^i, \boldsymbol{u}^i, \tau) - f(\tilde{\boldsymbol{x}}^i, \tilde{\boldsymbol{u}}^i, \tau) \right\| d\tau
$$

$$
\overset{\text{Lipschitz}}{\leq} \int_{t_{n,k}}^{t_{n,k+1}} L_{f^i,x} \left\| \boldsymbol{x}^i - \tilde{\boldsymbol{x}}^i \right\| d\tau + \int_{t_{n,k}}^{t_{n,k+1}} L_{f^i,u} \underbrace{\left\| \boldsymbol{u}^i - \tilde{\boldsymbol{u}}^i \right\|}_{\leq C_{i,2} \cdot h_i^{q_i+1}} d\tau
\tag{11}
$$

$$
\underbrace{\phantom{\int_{t_{n,k}}^{t_{n,k+1}} L_{f^i,u} \left\| \boldsymbol{u}^i - \tilde{\boldsymbol{u}}^i \right\| d\tau}}_{\leq L_{f^i,u} \cdot C_{i,2} \cdot h_i^{q_i+2}}
$$

with Lipschitz constants $L_{f^i,x}$ and $L_{f^i,u}$ of f^i with respect to $\boldsymbol{x}$ and $\boldsymbol{u}$, respectively, and q_i denoting the order of the extrapolation method for the approximation of $\tilde{\boldsymbol{u}}^i$. Declaring $C_{i,3} := L_{f^i,u} \cdot C_{i,2}$ and $\boldsymbol{m}(t) := \left\| \boldsymbol{x}^i(t) - \tilde{\boldsymbol{x}}^i(t) \right\|$, above estimates can be summarized as

$$
\boldsymbol{m}(t_{n,k+1}) \leq \int_{t_{n,k}}^{t_{n,k+1}} L_{f^i,x} \left\| \boldsymbol{x}^i - \tilde{\boldsymbol{x}}^i \right\| d\tau + C_{i,3} \cdot h_i^{q_i+2}.
\tag{12}
$$

Now we can apply the Lemma of Gronwall (Theorem 1) to $\boldsymbol{m}$ with $\alpha = C_{i,3} \cdot h_i^{q_i+2}$ and $\beta = L_{f^i,x}$ and obtain

$$
\boldsymbol{m}(t_{n,k+1}) \leq C_{i,3} \cdot h_i^{q_i+2} \cdot \underbrace{e^{L_{f^i,x} \cdot \overbrace{(t_{n,k+1} - t_{n,k})}^{h_i}}}_{= \sum_{j=0}^{\infty} \frac{(L_{f^i,x} \cdot h_i)^j}{j!}} = \mathcal{O}\left(h_i^{q_i+2} \right)
\tag{13}
$$

and therefore

$$
\left\| \frac{\mathcal{E}^i(t_{n,k}, \boldsymbol{x}_{n,k}, h_i)}{h_i} \right\| \overset{(9),(13)}{\leq} C_{i,1} \cdot h_i^{p_i} + \mathcal{O}\left(h_i^{q_i+1} \right) = \mathcal{O}\left(h_i^{\min\{p_i, q_i+1\}} \right).
\tag{14}
$$

$\square$

This shows that while consistency is maintained in co-simulation, the order may be reduced if the extrapolation order is chosen too low. Constant extrapolation, for example, only maintains the order of integration methods of order one. For higher-order methods, the order is reduced but the method remains consistent (as $\left\| \frac{\mathcal{E}^i(t_{n,k}, \boldsymbol{x}_{n,k}, h_i)}{h_i} \right\|$ still converges to zero, but only linearly). However, higher order extrapolation can also lead to increased stability issues, which is shown for example in [2].

Lemma 2 (Consistency error per subsystem for one macro step). *With the notations above*

$$\left\| \frac{\mathcal{E}^i(t_n, \boldsymbol{x}_n, H)}{H} \right\| = \mathcal{O}\left(H^{\min\{p_i, q_i+1\}} \right). \tag{15}$$

Proof. To extend the considerations for one micro step to one macro step, we will employ the method of "Lady Windermere's Fan", which is shown f.i. in [10,11]. The main idea of this approach is to describe the error of the approximate solution after an interval – in our case, a macro step – by the analytical solutions at every point of a refined mesh – in our case, every micro step – assuming an exact value at the beginning of the considered interval. This is illustrated for a one-dimensional problem in Fig. 2.

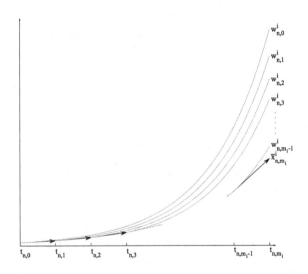

Fig. 2. "Lady Windermere's Fan": exact solutions at every time step of the numerical integration algorithm are used to describe the error of the approximate solution in one macro step (after [11]).

Let $\boldsymbol{w}^i_{n,k}(t), k = 0, \ldots m_i$ denote the exact solution of system (6) but for the initial values $\boldsymbol{w}^i_{n,k}(t_{n,k}) = \tilde{\boldsymbol{x}}_{n,k}$, implying $\boldsymbol{w}^i_{n,0}(t) = \boldsymbol{x}^i(t)\ \forall t > t_n$ since we assume exact values at $t_{n,0}$. Then we can write

$$\left\|\mathcal{E}^i(t_n, \boldsymbol{x}_n, H)\right\| = \left\|\boldsymbol{x}^i(t_{n+1}) - \tilde{\boldsymbol{x}}^i_{n+1}\right\| = \left\|\boldsymbol{x}^i(t_{n,m_i}) - \tilde{\boldsymbol{x}}^i_{n,m_i}\right\| \tag{16}$$

$$\leq \sum_{k=0}^{m_i-1} \left\|\boldsymbol{w}^i_{n,k}(t_{n,m_i}) - \boldsymbol{w}^i_{n,k+1}(t_{n,m_i})\right\|. \tag{17}$$

Since $\boldsymbol{w}^i_{n,k}$ are solutions to the same system with different initial values, we can apply Theorem 2 know that every summand of (17) is bounded by

$$\left\|\boldsymbol{w}^i_{n,k}(t_{n,k+1}) - \boldsymbol{w}^i_{n,k+1}(t_{n,k+1})\right\| \cdot e^{L_{fi,x} \cdot \overbrace{(t_{n,m_i} - t_{n,k+1})}^{(m_i-k-1)h_i}} \tag{18}$$

$$\Rightarrow \left\|\mathcal{E}^i(t_n, \boldsymbol{x}_n, H)\right\| \leq \sum_{k=0}^{m_i-1} \left\|\boldsymbol{w}^i_{n,k}(t_{n,k+1}) - \boldsymbol{w}^i_{n,k+1}(t_{n,k+1})\right\| \cdot e^{L_{fi,x} \cdot (m_i-k-1)h_i}$$

$$\overset{\boldsymbol{w}^i_{n,k}(t_{n,k})=\tilde{\boldsymbol{x}}_{n,k}}{=} \sum_{k=0}^{m_i-1} \left\|\boldsymbol{w}^i_{n,k}(t_{n,k+1}) - \tilde{\boldsymbol{x}}^i_{n,k+1}\right\| \cdot e^{L_{fi,x} \cdot (m_i-k-1)h_i}.$$

As $\left\|\boldsymbol{w}^i_{n,k}(t_{n,k+1}) - \boldsymbol{x}^i_{n,k+1}\right\|$ is the error in one micro step, according to Lemma 1 we can estimate this term with $\mathcal{O}(h_i^{\min\{p_i+1,q_i+2\}})$. Therefore

$$\left\|\mathcal{E}^i(t_n, \boldsymbol{x}_n, H)\right\| \leq \mathcal{O}\left(h_i^{\min\{p_i+1,q_i+2\}}\right) \sum_{k=0}^{m_i-1} e^{L_{fi,x} \cdot (m_i-k-1)h_i} \tag{19}$$

$$\leq \mathcal{O}\left(h_i^{\min\{p_i+1,q_i+2\}}\right) \cdot m_i \cdot e^{L_{fi,x} \cdot (m_i-1)h_i} \tag{20}$$

$$\overset{h_i=H/m_i}{=} \mathcal{O}\left(\left(\frac{H}{m_i}\right)^{\min\{p_i+1,q_i+2\}}\right) \cdot m_i \cdot e^{L_{fi,x} \cdot \frac{m_i-1}{m_i} H} \tag{21}$$

$$= \mathcal{O}\left(H^{\min\{p_i+1,q_i+2\}}\right) \tag{22}$$

$$\Rightarrow (15). \qquad\qquad \square$$

Corollary 1 (Consistency error of co-simulation). *With the notations above, consistency of the co-simulation in one macro step can be determined by*

$$\left\|\frac{\mathcal{E}(t_n, \boldsymbol{x}_n, H)}{H}\right\| = \mathcal{O}\left(H^{\min_{i=I,\ldots,N}\{p_i,q_i+1\}}\right), \tag{23}$$

whereby

$$\min_{i=I,\ldots,N}\{p_i, q_i+1\} := \min_{i=I,\ldots,N}\{\min\{p_i, q_i+1\}\} = \min\{\min_{i=I,\ldots,N}\{p_i\}, \min_{i=I,\ldots,N}\{q_i+1\}\}.$$

Proof. Since $\boldsymbol{x}(t)$ is given as concatenation of all $\boldsymbol{x}^i(t)$, $i = I, \ldots, N$, the approximation of the overall system at a synchronization point t_{n+1} corresponds to the concatenation of the approximations of the states of the N individual subsystems. With this, we can simply infer

$$\|\mathcal{E}(t_n, \boldsymbol{x}_n, H)\| = \|\boldsymbol{x}(t_{n+1}) - \tilde{\boldsymbol{x}}_{n+1}\| \leq \sum_{i=0}^{N} \|\boldsymbol{x}^i(t_{n+1}) - \tilde{\boldsymbol{x}}_{n+1}^i\| \tag{24}$$

$$\leq N \cdot \mathcal{O}\left(H^{\min_{i=1...N}\{p_i+1,q_i+2\}}\right) = \mathcal{O}\left(H^{\min_{i=1...N}\{p_i+1,q_i+2\}}\right) \tag{25}$$

with the estimates from Lemma 2. (23) follows directly by division by H.   □

These estimates show that while overall consistency is maintained in the co-simulation of ODE systems, the convergence order may be reduced in case of lower-order extrapolation of input values. Higher order extrapolation, while enhancing the order of consistency of the coupled method (bounded by the order of the original integration method), can also lead to increased stability issues, as shown f.i. in [1,2]. For DAEs that are only coupled via differential variables, the implicit function theorem (see e.g. [18]) implies that locally, an equivalent ODE system can be found for which above considerations also apply. In case of coupling via algebraic variables, similar estimates (in the sense of dependence on extrapolation orders) are given e.g. in [3,5].

2.2   Consistency in Hierarchical Co-simulation

Now we want to extend above investigations to co-simulation on several levels of hierarchy. From the estimates for traditional co-simulation, which only depend on the error introduced by extrapolation of external input values, we can already expect that this property is not affected by the method used in the respective other subsystems or the time steps and further synchronizations happening there in-between. For detailed estimation, we will first consider the simplest case where hierarchical co-simulation can be applied: Three subsystems of which w.l.o.g. Systems II and III are co-simulated on the lowest level and this co-simulation communicates again on the topmost level with the simulation of System I, as illustrated in Fig. 3.

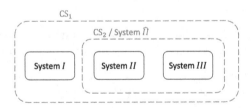

Fig. 3. Illustration of hierarchical co-simulation of three systems on two levels. Co-simulation CS_1 coordinates System I and System $\widehat{II}$, i.e. co-simulation CS_2, which manages the communication between systems II and III.

The co-simulation between Systems II and III will be called CS_2 henceforth, and the corresponding system seen from the perspective of the upper level System $\widehat{II}$. The top-level co-simulation (CS_1) macro step will be denoted H_1

and the second-level co-simulation macro step H_2. For CS_1, we start with the error in one macro step H_1 of System I, for which we obtain from Lemma 2

$$\left\| \frac{\mathcal{E}^I(H_1)}{H_1} \right\| = \mathcal{O}\left(H_1^{\min\{p_I, q_I+1\}} \right). \tag{26}$$

For System $\widehat{II}$, we start by applying Corollary 1 to CS_2, which yields for one step of size H_2

$$\left\| \frac{\mathcal{E}^{\widehat{II}}(H_2)}{H_2} \right\| = \mathcal{O}\left(H_2^{\min\limits_{i=II,III}\{p_i, q_i+1\}} \right). \tag{27}$$

To estimate the error in one macro step H_1, we can repeat the strategy from the proof of Lemma 2 with M_2 describing the quotient of H_1 and H_2 and obtain

$$\left\| \frac{\mathcal{E}^{\widehat{II}}(H_1)}{H_1} \right\| = \mathcal{O}\left(H_1^{\min\limits_{i=II,III}\{p_i, q_i+1\}} \right) \tag{28}$$

and further for the top-level co-simulation CS_1 with (26), (28) and Corollary 1

$$\left\| \frac{\mathcal{E}(H_1)}{H_1} \right\| = \mathcal{O}\left(H_1^{\min\limits_{i=I,II,III}\{p_i, q_i+1\}} \right) \tag{29}$$

and therefore consistency. The order again depends on the extrapolation and consistency orders of all subsystems. This can also be concluded for arbitrary levels of hierarchy and participating subsystems, as Theorem 3 shows.

Theorem 3 (Consistency error of hierarchical co-simulation). *In a hierarchical co-simulation with a total of N participating subsystems, consistency orders p_i, $i = I, \ldots, N$ of their corresponding integration algorithms and extrapolation orders q_i, $i = I, \ldots, N$, the consistency error of the overall co-simulation with macro step H can be estimated as*

$$\left\| \frac{\mathcal{E}(H)}{H} \right\| = \mathcal{O}\left(H^{\min\limits_{i=I,\ldots,N}\{p_i, q_i+1\}} \right). \tag{30}$$

Proof. To begin with, we need to establish comprehensible notation of all considered systems, co-simulations, and step sizes. For this purpose, all participating simulations are depicted in a tree structure, see Fig. 4. We will start from the topmost level, naming the overall co-simulation $S_{1,1}$. Beneath $S_{1,1}$, all further simulations unfold on J levels in total. On every level $j \in 1, \ldots, J+1$ all simulations – be they co-simulations themselves or "leaf" nodes without further branching beneath – are numbered from 1 to K_j. This means that on level j, we find simulations $S_{j,k}$ with $k = 1 \ldots K_j$. While the ordering of these may be arbitrary, this notation is necessary to uniquely identify every co-simulation on every level in a fairly intelligible notation. Nevertheless, to clarify the belonging to the respective co-simulation, the sub-simulations of one node, i.e. all $N_{j,k}$ simulations coordinated by one co-simulation $S_{j,k}$ may be identified by $S_{j,k}^I, S_{j,k}^{II}, \ldots, S_{j,k}^{N_{j,k}}$

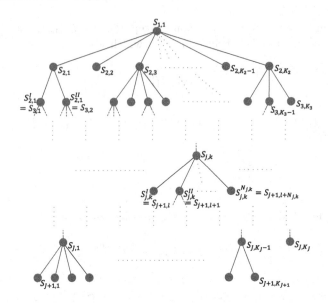

Fig. 4. Illustration of the co-simulation hierarchy in a tree structure.

in addition. This means that the i−th subsimulation of $S_{j,k}$ may be called $S_{j,k}^i$ and equals, using the notation on the next level, $S_{j+1,l}$ for one $l \in \{1, \ldots, K_{j+1}\}$:

$$S_{j,k}^i = S_{j+1,l} \quad \text{for } l = i + \sum_{m=1}^{k-1} N_{j,m} \tag{31}$$

Note that naturally, the sum of all simulations that are co-simulated by simulations on level j equals the number of simulations on level $j + 1$ with the convention that for leaf nodes, $N_{j,k} := 0$.

In analogy to above example with three systems co-simulated on two levels, (30) follows from Lemmata 1, 2 and Corollary 1 when approached bottom-up with induction: On the deepest level $J + 1$, we only have leaf nodes. These systems $S_{J+1,l}, l = 1, \ldots, K_{J+1}$ are integrated with their individual time step $h_{J+1,l}$ and are coordinated by a co-simulation on level J. By considering one of these co-simulations $S_{J,k}$ with macro step size $H_{J,k}$ and its sub-simulations denoted as $S_{J,k}^i, i = I, \ldots, N_{J,k}$, we know from Lemma 2 that for every $S_{J,k}^i$, the error per macro step can be estimated via

$$\left\| \frac{\mathcal{E}_{J,k}^i(H_{J,k})}{H_{J,k}} \right\| = \mathcal{O}\left(H_{J,k}^{\min\{p_{i_{J,k}}, q_{i_{J,k}}+1\}} \right) \tag{32}$$

with $p_{i_{J,k}}$ denoting the consistency order of the integration method of $S_{J,k}^i$ and $q_{i_{J,k}}$ the respective extrapolation order for external input values. With Corollary 1 follows for the consistency order of $S_{J,k}$

$$\left\|\frac{\mathcal{E}_{J,k}(H_{J,k})}{H_{J,k}}\right\| = \mathcal{O}\left(H_{J,k}^{\min\limits_{i=I,\ldots,N_{J,k}}\{p_{i_{J,k}},q_{i_{J,k}}+1\}}\right). \tag{33}$$

For every leaf simulation $S_{J,k}$ on level J with micro step size $h_{J,k}$, we obtain an estimate for the error per micro step with Lemma 1:

$$\left\|\frac{\mathcal{E}_{J,k}(h_{J,k})}{h_{J,k}}\right\| = \mathcal{O}\left(h_{J,k}^{\min\{p_{J,k},q_{J,k}+1\}}\right) \tag{34}$$

As the indexing is unique, we can without confusion with some co-simulation declare $H_{J,k} := h_{J,k}$ and therefore in summary write the estimate for every simulation – cooperative as well as leaf simulation – on level J as

$$\left\|\frac{\mathcal{E}_{J,k}(H_{J,k})}{H_{J,k}}\right\| = \mathcal{O}\left(H_{J,k}^{\min\{p_{J,k},q_{J,k}+1\}}\right) \tag{35}$$

when for co-simulation nodes, we define $p_{J,k} := \min\limits_{i=I,\ldots,N_{J,k}}\{p_{i_{J,k}}\}$ and $q_{J,k} := \min\limits_{i=I,\ldots,N_{J,k}}\{q_{i_{J,k}}\}$.

In the next step, we will assume this estimate for every simulation on a level $j+1$, $j \in \{1,\ldots,J\}$:

$$\left\|\frac{\mathcal{E}_{j+1,k}(H_{j+1,k})}{H_{j+1,k}}\right\| = \mathcal{O}\left(H_{j+1,k}^{\min\{p_{j+1,k},q_{j+1,k}+1\}}\right) \tag{36}$$

again with $H_{j+1,k} := h_{j+1,k}$ if $S_{j+1,k}$ is a leaf node and for co-simulation nodes $S_{j+1,k}$ defining $p_{j+1,k} := \min\limits_{i=I,\ldots,N_{j+1,k}}\{p_{i_{j+1,k}}\}$ and $q_{j+1,k} := \min\limits_{i=I,\ldots,N_{j+1,k}}\{q_{i_{j+1,k}}\}$ (using these definitions recursively in case for an i, the associated simulation $S_{j+1,k}^i$ ($= S_{j+2,l}$ for $l = i + \sum\limits_{m=1}^{k-1} N_{j+1,m}$) is again a co-simulation). Based on that, we consider the simulations on level j. For every leaf node on level j, Lemma 1 can directly be applied:

$$\left\|\frac{\mathcal{E}_{j,k}(h_{j,k})}{h_{j,k}}\right\| = \mathcal{O}\left(h_{j,k}^{\min\{p_{j,k},q_{j,k}+1\}}\right), \tag{37}$$

which with $H_{j,k} := h_{j,k}$ can be written

$$\left\|\frac{\mathcal{E}_{j,k}(H_{j,k})}{H_{i_{j,k}}}\right\| = \mathcal{O}\left(H_{j,k}^{\min\{p_{j,k},q_{j,k}+1\}}\right). \tag{38}$$

For every co-simulation on level j, we can utilize (36) and Corollary 1 to obtain

$$\left\|\frac{\mathcal{E}_{j,k}(H_{j,k})}{H_{j,k}}\right\| = \mathcal{O}\left(H_{j,k}^{\min\limits_{i=I,\ldots,N_{j,k}}\{p_{i_{j,k}},q_{i_{j,k}}+1\}}\right) = \mathcal{O}\left(H_{j,k}^{\min\{p_{j,k},q_{j,k}+1\}}\right) \tag{39}$$

with $p_{j,k} := \min\limits_{i=I,\ldots,N_{j,k}}\{p_{i_{j,k}}\}$ and $q_{j,k} := \min\limits_{i=I,\ldots,N_{j,k}}\{q_{i_{j,k}}\}$ (recursively, if needed). Thus, with (38) we have

$$\left\|\frac{\mathcal{E}_{j,k}(H_{j,k})}{H_{j,k}}\right\| = \mathcal{O}\left(H_{j,k}^{\min\{p_{j,k},q_{j,k}+1\}}\right) \tag{40}$$

for every cooperative and leaf simulation on level j.

This also holds for the topmost level $j = 1$, where only one co-simulation (and, naturally, no leaf node) remains. With $H := H_{1,1}$ and utilizing the fact that in this co-simulation, all $N = \sum_{j=1}^{J} \sum_{k=1}^{K_j} N_{j,k}$ participating leaf simulations and therefore, the consistency and extrapolation orders of every solution algorithm are finally considered, we obtain (30).   □

3   Error Studies on a Coupled Three-Mass Oscillator

In the following, we consider an oscillator with three masses divided by force-displacement decomposition (see [13] for information on the coupling concept), which is illustrated in Fig. 5. With initial values and parameters given according to Table 1, we observe an increase of stiffnesses from left to right, which invites the introduction of another level of hierarchy.

Table 1. Initial values and parameter settings for the benchmark simulation.

x_1	x_2	x_3	$v_1 = v_2 = v_3$	c_1	c_{12}	c_{23}	c_3	d_1	d_{12}	d_{23}	d_3	m_1	m_2	m_3
1 m	2 m	3 m	0 m/s	1E–03 N/m	1E–01 N/m	10 N/m	100 N/m	0.1	0.4	1	2	10 kg	10 kg	10 kg

In a traditional co-simulation, Systems S^I, S^{II} and S^{III} would, in general, all be orchestrated by one algorithm demanding synchronization at the same time step. In a hierarchical approach, Systems S^{II} and S^{III} can be combined in a separate, lower-level co-simulation representing the new system $\widehat{S}^{II}$ that is co-simulated with System S^I on the top-level co-simulation.

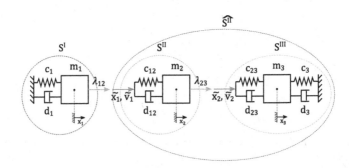

Fig. 5. Illustration of the hierarchical coupling of a three-mass oscillator.

The underlying equations can be interpreted as coupled Dahlquist equations, which invites investigations on stability by this example. Case studies that demonstrate the benefits of a hierarchical versus a traditional co-simulation approach regarding numerical stability are found in [9]. These include detailed

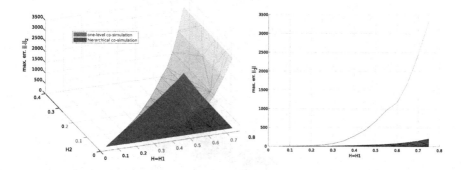

Fig. 6. Error ($\|.\|_2$ of all component errors) for the hierarchical and traditional co-simulation of the test scenario from $t_{start} = 0\,\mathrm{s}$ to $t_{end} = 25\,\mathrm{s}$ depending on macro step sizes.

tables comprising CPU time and errors for the simulation of the test scenario for 100 s, where we see that even for an upper-level macro step size twice as large ($H_1 = 0.2\,\mathrm{s}$) as the one for the traditional co-simulation ($H = 0.1\,\mathrm{s}$), the error can be reduced to less than one seventh if the second-level macro step size is chosen small enough ($H_2 = 0.05\,\mathrm{s}$) while the elapsed computation time is barely increased (from 2.29 s to 2.58 s). Here, on the other hand, we will focus on the impact of varying macro step sizes on both co-simulation levels.

Since the differing stiffnesses result in slower and faster varying subsystems, the step sizes for the individual subsystem solvers are chosen accordingly with $h_I = 0.005\,\mathrm{s}$, $h_{II} = 0.0025\,\mathrm{s}$ and $h_{III} = 0.00125\,\mathrm{s}$. Figure 6 shows the overall error – calculated by $\|.\|_2$ of the maximum errors of all states – depending on the macro step sizes $H = H_1$ for the traditional and upper-level co-simulation in the hierarchical approach, and H_2 for the second-level co-simulation in the hierarchical approach. The duration of all simulations is chosen with 25 s. H_2 ranges from 0.025 s over all multiples that are divisors of H_1 up to $H_1/2$ (for $H_2 = H_1$, the same results as for the traditional approach would be expected).

On the one hand, we immediately observe a faster ascent and more curvature for the error in the traditional approach. In addition, the impact of the choice of H_2 is clearly visible and comes out even more clearly in the separate illustration of the hierarchical approach in Fig. 7.

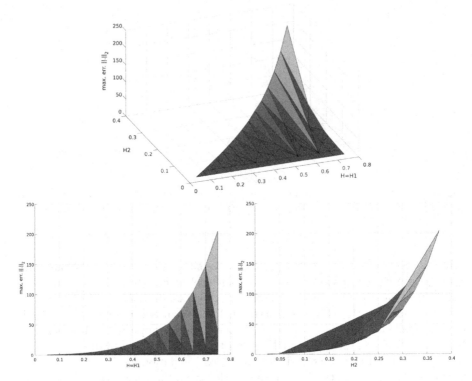

Fig. 7. Error ($\|.\|_2$ of all component errors) for the hierarchical co-simulation of the test scenario from $t_{start} = 0$ s to $t_{end} = 25$ s depending on macro step sizes.

4   Conclusion and Outlook

Above investigations show that consistency is maintained in hierarchical co-simulation, although it may potentially converge with lower order in comparison to the corresponding mono-simulation, depending on the extrapolation of external inputs. Since this is also the case for traditional co-simulation, no further loss of the order of consistency is added by the introduction of further hierarchies. On the contrary, as studies with varying macro step sizes show, error propagation is slowed down and accuracy increased if subsystems with closer dependencies are allowed to communicate more frequently while synchronization intervals with other subsystems can be increased.

Since hierarchical co-simulation is already permitted in certain frameworks and standards for co-simulation, the presented estimates along with investigations on stability in [8,9] provide the assertion that the application of hierarchical methods maintain and may even improve convergence.

Nevertheless, the method offers several aspects for further enhancement. Instead of parallel, non-iterative coupling algorithms with fixed macro step size, zero-order extrapolation and Euler integration methods used in the benchmark example from Sect. 3, strategies that are known to improve stability, performance, or accuracy for traditional co-simulation may be utilized in hierarchical co-simulation as well. Among these, the utilization of sequential, iterative or adaptive orchestration algorithms, different extrapolation orders and higher order and/or multistep subsystem solvers remain a topic for future investigations.

References

1. Arnold, M.: Multi-rate time integration for large scale multibody system models. In: Eberhard, P. (ed.) IUTAM Symposium on Multiscale Problems in Multibody System Contacts. No. 1 in IUTAM Bookseries, pp. 1–10. Springer, Netherlands (2007). https://doi.org/10.1007/978-1-4020-5981-0_1
2. Arnold, M.: Stability of sequential modular time integration methods for coupled multibody system models. J. Comput. Nonlinear Dyn. **5**(3), 031003 (2010). https://doi.org/10.1115/1.4001389
3. Arnold, M., Günther, M.: Preconditioned dynamic iteration for coupled differential-algebraic systems. BIT Numer. Math. **41**(1), 1–25 (2001). https://doi.org/10.1023/A:1021909032551
4. Arnold, M., Hante, S., Köbis, M.A.: Error analysis for co-simulation with force-displacement coupling. PAMM **14**(1), 43–44 (2014). https://doi.org/10.1002/pamm.201410014
5. Busch, M.: Zur effizienten Kopplung von Simulationsprogrammen. Kassel University Press (2012). https://books.google.at/books?id=0qBpXp-f2gQC
6. Esposito, J., Kumar, V.: Efficient dynamic simulation of robotic systems with hierarchy. In: Proceedings 2001 ICRA. IEEE International Conference on Robotics and Automation (Cat. No. 01CH37164), Seoul, South Korea, vol. 3, pp. 2818–2823. IEEE (2001). https://doi.org/10.1109/ROBOT.2001.933049
7. Günther, M., Rentrop, P.: Partitioning and multirate strategies in latent electric circuits. In: Bank, R.E., Gajewski, H., Bulirsch, R., Merten, K. (eds.) Mathematical Modelling and Simulation of Electrical Circuits and Semiconductor Devices, pp. 33–60. ISNM International Series of Numerical Mathematics, Birkhäuser, Basel (1994). https://doi.org/10.1007/978-3-0348-8528-7_3
8. Hafner, I.: Cooperative and multirate simulation: analysis, classification and new hierarchical approaches. Ph.D. thesis, TU Wien, Vienna, Austria (2021)
9. Hafner, I., Popper, N.: Investigation on stability properties of hierarchical co-simulation. SNE Simul. Notes Europe **31**(1), 17–24 (2021). https://doi.org/10.11128/sne.31.tn.10553
10. Hairer, E., Nørsett, S., Wanner, G.: Solving Ordinary Differential Equations I: Nonstiff Problems. Springer Series in Computational Mathematics, vol. 8. Springer, Heidelberg (1993). https://doi.org/10.1007/978-3-540-78862-1
11. Knorr, S.: Multirate-Verfahren in der co-simulation gekoppelter dynamischer Systeme mit Anwendung in der Fahrzeugdynamik. Diploma Thesis, Universität Ulm, Ulm, Germany, October 2002. http://www-num.math.uni-wuppertal.de/fileadmin/mathe/www-num/theses/da_knorr.pdf. Accessed 1 Oct 2021
12. Maten, J.T., Verhoeven, A., Guennouni, A.E., Beelen, T.: Multirate hierarchical time integration for electronic circuits. PAMM **5**(1), 819–820 (2005). https://doi.org/10.1002/pamm.200510381

13. Schweizer, B., Lu, D.: Semi-implicit co-simulation approach for solver coupling. Arch. Appl. Mech. **84**(12), 1739–1769 (2014). https://doi.org/10.1007/s00419-014-0883-5

14. Striebel, M.: Hierarchical mixed multirating for distributed integration of DAE network equations in chip design. Ph.D. thesis, Bergische Universität Wuppertal, April 2006

15. Tahir-Kheli, R.: Ordinary Differential Equations. Springer, Cham (2018). https://doi.org/10.1007/978-3-319-76406-1

16. Thule, C., et al.: Towards reuse of synchronization algorithms in co-simulation frameworks. In: Camara, J., Steffen, M. (eds.) Software Engineering and Formal Methods, pp. 50–66. Springer, Cham (2020). https://doi.org/10.1007/978-3-030-57506-9_5

17. Wang, J., Ma, Z.D., Hulbert, G.M.: A gluing algorithm for distributed simulation of multibody systems. Nonlinear Dyn. **34**(1), 159–188 (2003). https://doi.org/10.1023/B:NODY.0000014558.70434.b0

18. Zeidler, E. (ed.): Springer-Taschenbuch der Mathematik. Springer Fachmedien Wiesbaden, Wiesbaden (2013). http://link.springer.com/10.1007/978-3-8348-2359-5. Accessed 16 Feb 2016

19. Zeigler, B.P.: Object-Oriented Simulation with Hierarchical, Modular Models: Intelligent Agents and Endomorphic Systems. Elsevier Science, Saint Louis (2014). http://qut.eblib.com.au/patron/FullRecord.aspx?p=1876958. Accessed 2 Dec 2019

20. Zhang, H., Liang, S., Song, S., Wang, H.: Truncation error calculation based on Richardson extrapolation for variable-step collaborative simulation. Sci. China Inf. Sci. **54**(6), 1238–1250 (2011). https://doi.org/10.1007/s11432-011-4274-z

Co-simulation-Based Pre-training of a Ship Trajectory Predictor

Motoyasu Kanazawa[1] , Lars Ivar Hatledal[2(✉)] , Guoyuan Li[1(✉)] ,
and Houxiang Zhang[1(✉)]

[1] The Department of Ocean Operations and Civil Engineering,
Faculty of Engineering, Norwegian University of Science and Technology,
Larsgårdsvegen 2, 6009 Ålesund, Norway
{motoyasu.kanazawa,guoyuan.li,hozh}@ntnu.no
[2] The Department of ICT and Natural Sciences, Facultiy of Information Techonology
and Electrical Engineering, Norwegian University of Science and Technology,
Larsgårdsvegen 2, 6009 Ålesund, Norway
laht@ntnu.no

Abstract. A ship trajectory predictor plays a key role in the predictive decision making of intelligent marine transportation. For better prediction performance, the biggest technical challenge is how we incorporate prior knowledge, acquired during the design-stage experiments, into a data-driven predictor if the number of available real-world data is limited. This study proposes a new framework under co-simulation platform Vico for the development of a neural-network-based trajectory predictor with a pre-training phase. Vico enables a simplified vessel model to be constructed by merging a hull model, thruster models, and a controller using a co-simulation standard. Furthermore, it allows virtual scenarios, which describe what will happen during the simulation, to be generated in a flexible way. The fully-connected feedforward neural network is pre-trained with the generated virtual scenarios; then, its weights and biases are finetuned using a limited number of real-world datasets obtained from a target operation. In the case study, we aim to make a 30 s trajectory prediction of real-world zig-zag maneuvers of a 33.9m-length research vessel. Diverse virtual scenarios of zig-zag maneuvers are generated in Vico and used for the pre-training. The pre-trained neural network is further finetuned using a limited number of real-world data of zig-zag maneuvers. The present framework reduced the mean prediction error in the test dataset of the real-world zig-zag maneuvers by 60.8% compared to the neural network without the pre-training phase. This result indicates the validity of virtual scenario generation on the co-simulation platform for the purpose of the pre-training of trajectory predictors.

Keywords: Co-simulation · Trajectory prediction · Informed machine learning

Supported by a grant from NRF, IKTPLUSS project No. 309323 "Remote Control Center for Autonomous Ship Support" in Norway.

A. Cerone et al. (Eds.): SEFM 2021 Workshops, LNCS 13230, pp. 173–188, 2022.
https://doi.org/10.1007/978-3-031-12429-7_13

1   Introduction

Autonomous ship maneuvering hinges on better understanding of ship dynamics. A ship is regarded as a system that consists of many dynamic components, such as a hull model, thruster models, and a controller. In most cases, a simulator of each component is given by its manufacturers and project partners in the format of a black-box model. Therefore, it is of great interest to designers to construct a simplified vessel model easily by connecting black-box sub-models. *Vico*, which is a high-level co-simulation framework [3], enables a simplified vessel model to be constructed by merging a hull model, thruster models, and a controller using the co-simulation standard. Furthermore, it allows virtual scenarios, which describe what will happen during the simulation, to be generated in a flexible way. In the Intelligent Systems Lab [6] at Norwegian University of Science and Technology (NTNU) Ålesund, it has been playing an important role as a cyber testbed of research activity.

In the predictive decision making of autonomous maneuvering, a controller of an autonomous ship makes predictions of its own ship's trajectory based on the current vessel state and future command assumption. Then, a controller can evaluate and manipulate the future command assumption based on the predicted consequences. Hence, accurate motion prediction is the basis of collision avoidance algorithms [5]. According to the definition in [11], the prediction task can be explained as follows. A dataset $\mathcal{D}$ consists of a feature space $\mathcal{X}$ and a marginal probability distribution $P(X)$ where $X = \{x_1, .., x_n\} \in \mathcal{X}$. In the supervised learning, a set of pairs $\{x_i, y_i\}$ of inputs (the current vessel state, commands, and environmental disturbances) $x_i \in \mathcal{X}$ and outputs (predicted trajectories) $y_i \in \mathcal{Y}$ are given in the dataset $\mathcal{D}$. The prediction task of the own ship's trajectory $\mathcal{T}$ is defined as $\mathcal{T} = \{\mathcal{Y}, P(Y|X)\}$ where $Y = \{y_1, ..., y_n\}$. We aim to find an objective predictive function $f = P(Y|X)$ in the development of a predictor.

Ship dynamics is highly nonlinear and complex. In order to comprehend its dynamic characteristics, experiments in the ocean basin and full-scale sea trials are conducted in the design procedure. These data build a dataset $\mathcal{D}_m = \{\mathcal{X}_m, P(X_m)\}$ which is utilized for developing a white-box vessel model f_m using parameter identification algorithm [19]. Its biggest challenge is dataset of a target operation $\mathcal{D}_{\text{target}} = \{\mathcal{X}_{\text{target}}, P(X_{\text{target}})\}$ is not identical to $\mathcal{D}_m$ in most cases. Due to such factors as the shallow water effect and $P(X_{\text{target}}) \neq P(X_m)$, a white-box-model-based prediction can be inaccurate in $\mathcal{D}_{\text{target}}$ [16]. A standard idea of dealing with this problem is to sample $\mathcal{D}_{\text{train}} \subset \mathcal{D}_{\text{target}}$ and develop a black-box model trained using $\mathcal{D}_{\text{train}}$. However, it might be an optimistic expectation that we have $|\mathcal{D}_{\text{train}}|$ that is enough for the training of the black-box model since we would develop a predictor of a ship with less experience in $\mathcal{D}_{\text{target}}$. According to the definition in [18], this challenge is categorized into few-shot learning problem.

This study proposes the framework of co-simulation-based development of a ship trajectory predictor shown in Fig. 1. In a present framework, few-shot learning, which has a limited number of real-world data of a target operation, is informed of prior knowledge by many virtual operations in *Vico*. A simplified vessel model constructed at a small cost in *Vico* and real-world small data of a

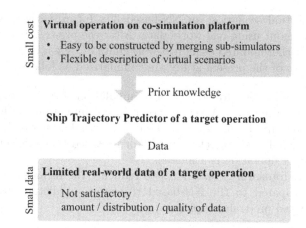

Fig. 1. Co-simulation-based development of a ship trajectory predictor.

target operation compose an accurate ship trajectory predictor only with small cost & data. One should note that we do not re-build a simplified vessel model in *Vico* by employing parameter identification algorithms and $\mathcal{D}_{\text{train}}$ since it may ruin the advantage of using the co-simulation technology, which is an easy construction of a model by merging black-box sub-models. In addition, the present approach is more attractive than a re-construction of a simplified vessel model as a formulation based on physics can not capture highly nonlinear and complex phenomena that lies behind real-world ship dynamics. We use a fully-connected Feedforward Neural Network (FNN) as an architecture of the predictor. The NN is pre-trained using diverse virtual scenarios of a target operation in *Vico* before the NN is trained by a limited number of $\mathcal{D}_{\text{train}}$ real-world data in the $\mathcal{D}_{\text{target}}$. This pre-training informs the NN of prior knowledge f_m that facilitates the training using real-world data of a target operation. A present study is a new approach that is different from any informed-machine-learning-based predictors in previous studies [9, 15–17] in this field. The advantage of the proposed framework is (1) its simple structure of the predictor and (2) the NN can experience diverse virtual scenarios in the pre-training. One should note that a quantitative comparison of prediction performances between predictors is very challenging since the quantitative result could be on the case-by-case basis depending on the fidelity of prior knowledge, real-world dataset of a target operation, and the practical limitation of the implementation of the predictor. Therefore, this study focuses not on a comparison of the performances but on showing the validity of the proposed new framework. In the case study of the present framework, $\mathcal{D}_{\text{target}}$ is real-world data of zig-zag maneuvers of the R/V Gunnerus which is a 33.9m-length research vessel of NTNU. To generate virtual scenarios of zig-zag maneuvers in a pre-training phase, we develop a virtual R/V Gunnerus in *Vico* by merging a hull model provided by SINTEF Ocean, thruster models provided by thruster manufacturers, and a zig-zag controller developed by the authors.

By using the virtual scenarios, an NN-based 30 s future trajectory predictor is pre-trained without using real-world data. After the pre-training, the weights and biases are further updated using real-world data of zig-zag maneuvers $\mathcal{D}_{\text{train}}$. The contributions of the present study are summarized as follows.

- This study introduces a framework of co-simulation-based development of a ship trajectory predictor. A vessel model in co-simulation platform is easily constructed only by merging sub-models. Virtual scenarios facilitate the training of a NN-based trajectory predictor provided that limited real-world data of a target operation is available.
- The proposed framework contributes to reducing the mean prediction error by 60.9% compared to the NN-based trajectory predictor without a co-simulation-based pre-training phase.

2 Related Works

Previous studies aiming at developing accurate ship trajectory predictors are articulated in this section. Trajectory predictors are grouped into two categories; namely, white-box and black-box models. The most concise white-box model is the holonomic [20] and kinematic models [13]. They are widely used in collision avoidance algorithms because of their simple implementation, however, their prediction accuracy is much poorer than that of kinetic dynamic models due to their unrealistic assumptions. Kinetic models are categorized into response models, the Abkowitz model, the Maneuvering Modeling Group (MMG), and vectorial representations. Through experiments in the ocean basin and full-scale sea trials, hydrodynamic parameters of the kinetic models are identified using parameter identification algorithms [19]. The biggest advantage of white-box models is that they require less data to calibrate than their black-box counterparts by virtue of its formulation based on physics. On the other hand, it is a major drawback that tailored experiments take cost & effort. Black-box models exploit a large amount of onboard sensor data using Machine Learning (ML) algorithms. With the rapid development in computational resources and advanced ML algorithms, black-box models are becoming more and more popular recently in this field [7,14].

As we explained in the previous section, a general prediction problem suffers from prediction error due to $P(X_{\text{target}}) \neq P(X_m)$ and a limited number of $\mathcal{D}_{\text{train}} \subset \mathcal{D}_{\text{target}}$. With the aim of introducing prior knowledge f_m to a black-box model trained with $\mathcal{D}_{\text{train}}$, there is a large body of research integrating a mathematical vessel model into a black-box model. In [9], they utilize a reference mathematical vessel model of which dynamic characteristic is similar to the vessel that is subject to the trajectory prediction from the database. The random forest algorithm is trained so that it compensates the error in the model-based predicted acceleration. NN-based error compensation in the model-based acceleration is seen in [15]. Skulstad et al. [16] proposes a multiple-step-ahead trajectory predictor by combining a mathematical vessel model and a Long Short-Term

Memory (LSTM). An LSTM compensates 30 s North and East position errors made by the mathematical vessel model using onboard sensor measurements. Wang et al. [17] develop an NN-based data-driven calibrator that maps trajectory prediction made by a reference vessel model into that of the targeting vessel. In these previous studies, prior knowledge is integrated into the predictor in the form of a mathematical vessel model and the training is conducted using limited real-world data of a target operation. On the other hand, the present study pre-trains an NN-based predictor using diverse virtual scenarios of a target operation. This idea enables the NN to experience the diverse virtual scenarios and acquire prior knowledge before the main training rather than having a complex structure in the predictor with a black-box model and a vessel model.

3 Methodology

The methodology of the present framework is described in this section by taking an example of making trajectory prediction of zig-zag maneuvers of the R/V Gunnerus. It should be noted that the present framework works for any type of operation of any vessels as long as sub-models in *Vico* and real-world data of a target operation $\mathcal{D}_{\text{train}}$ are available.

3.1 Pre-training of a Trajectory Predictor

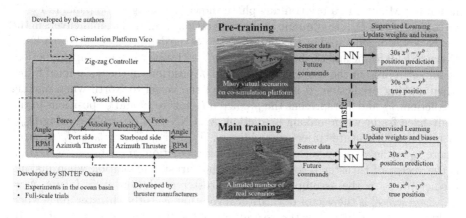

Fig. 2. The framework of co-simulation-based development of a ship trajectory predictor.

If we mix up the augmented and real-world data in one training phase, it may induce a problem of how to balance those two datasets for better prediction accuracy since the fidelity of virtual simulation might not be satisfactory in most cases of ship trajectory prediction. Therefore, the main training using real-world data

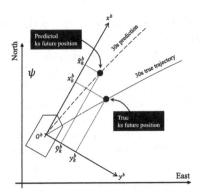

Fig. 3. The diagram of the definition of the North-East-Down frame and the body-fixed frame when making prediction.

is conducted after the pre-training using simulation data is completed. Since trajectory prediction is utilized for the predictive decision making, a vessel needs to have enough time to make a change of course and speed after making prediction. The predictor in this study makes the trajectory prediction for 30 s that is enough for making a change of course and speed of the R/V Gunnerus so that the evaluation of prediction performance is informative in the light of the practical application. Figure 2 shows the overview of the present framework through co-simulation-based pre-training. RPM in Fig. 2 is the abbreviation of the Revolution Per Minute. In a pre-training phase shown in the right top panel in Fig. 2, virtual scenarios $\mathcal{D}_{\mathrm{pre}}$ that mimic a target operation (e.g., zig-zag maneuvers in the case study) are generated in *Vico*. Figure 3 shows the definition of the North-East-Down (NED) frame and the body-fixed $x^b - y^b - z^b$ frame when making prediction. x^b_k and y^b_k represent the true ks future x^b and y^b positions in the body-fixed frame when making prediction. ψ is the heading of the vessel to North when making prediction. $\hat{x}^b_k$ and $\hat{y}^b_k$ represent the predicted ks future x^b and y^b positions in the body-fixed frame when making prediction. An FNN f_v is pre-trained using $\mathcal{D}_{\mathrm{pre}}$. f_v produces 30 s future trajectory prediction $[\hat{x}^b_1, ..., \hat{x}^b_{30}, \hat{y}^b_1, ..., \hat{y}^b_{30}] = f_v(\nu_0, \boldsymbol{n}, \boldsymbol{\delta})$ where ν_0 is the velocity vector when making prediction, $\boldsymbol{n} = [n_0, .., n_k, ..., n_{29}]$ is the vector of thruster revolution, n_0 is the thruster revolution when making prediction, n_k is the assumption of the thruster revolution at ks future, $\boldsymbol{\delta} = [\delta_0, .., \delta_k, ..., \delta_{29}]$ is the vector of thruster revolution, δ_0 is the thruster angle when making prediction, and δ_k is the assumption of the thruster angle at ks future. In the application of this study, the prediction $[\hat{x}^b_1, ..., \hat{x}^b_{30}, \hat{y}^b_1, ..., \hat{y}^b_{30}]$ is used for evaluating the decision making $\boldsymbol{n}$ and $\boldsymbol{\delta}$. Therefore, one should note that $\boldsymbol{n}$ and $\boldsymbol{\delta}$ are given by a controller. As illustrated in the right bottom panel in Fig. 2, the pre-trained NN is transferred to the main training phase. In the main training phase shown in the bottom panel in Fig. 2, the weights and biases of f_v are finetuned using a limited number of real-world data $\mathcal{D}_{\mathrm{train}}$.

3.2   A Virtual Vessel Model in *VICO*

The left panel in Fig. 2 shows the structure of a virtual R/V Gunnerus in *Vico*. It consists of four components; namely, a hull model, a port-side azimuth thruster model, a starboard-side azimuth thruster, and a zig-zag controller. They are packaged into Functional Mock-up Units (FMUs; FMU1.0 for hull and thruster models and FMU2.0 for the controller). The hull model is developed by SIN-TEF Ocean in the SimVal project [2] through experiments in the ocean basin and full-scale sea trials. It is a 6 Degrees Of Freedom (DOF) maneuvering and seakeeping model. Port and starboard-side azimuth thruster models are provided by thruster manufactures. In order to reproduce a target operations in *Vico*, a zig-zag controller is coded in Python by the authors. It provides a pre-defined time series of thruster commands (revolution and angle) to thruster FMUs. An example of the pre-defined time series of thruster angle is shown in Fig. 5. The controller is packaged into a FMU using *PythonFMU* [4].

3.3   An FNN-Based Predictor

Many architectures of ML models have been used for ship trajectory prediction; such as Support Vector Regression (SVR) [7], LSTM [15] and fully-connected FNN [16]. In this study, we use a fully-connected FNN as an architecture of a predictor as it is one of the simplest and well-known ML models that is widely used in the context of transfer learning. The FNN-based predictor consists of the input layer, hidden layers, and output layer. The activation function is the hyperbolic tangent function for the hidden layers and the linear function for the output layer. ν_0, $\boldsymbol{n}$, and $\boldsymbol{\delta}$ are selected as input features through feature selection as explained hereinafter. The output of the predictor is a vector $[\hat{x}_1^b, ..., \hat{x}_{30}^b, \hat{y}_1^b, ..., \hat{y}_{30}^b]$ with a length of 60. The weights and biases of the FNN are updated so that it minimizes the Mean Squared Error (MSE) metric L between the true and predicted position vectors using the Adam [8] optimizer.

$$L = \frac{1}{H} \sum_{k=1}^{H} (\hat{x}_k^b - x_k^b)^2 + (\hat{y}_k^b - y_k^b)^2 \qquad (1)$$

Input features are standardized with their mean and standard deviation in a training dataset in a pre-training phase. The FNN is implemented in Pytorch [12] in Python.

Feature Selection. $[x_1^b, ..., x_{30}^b, y_1^b, ..., y_{30}^b] = f(P, \nu_0, \boldsymbol{n}, \boldsymbol{\delta}) + w$ in theory of ship dynamics where P is a set of hydrodynamic and inertial parameters and w is environmental disturbances caused by wind, wave, and ocean current. In this study, we introduce following assumptions.

- A predictor is trained for a specific loading condition of a specific ship. There-fore, P is ruled out from input features of the NN.

– In most cases, a ship has no accurate measurement of waves and currents. Therefore, environmental forces due to wave and current are not modeled in the predictor.
– In $\mathcal{D}_{\text{target}}$, the effect of wind on the vessel motion is marginal. For the sake of simplicity of the validation study, it is not included in the input features of the NN.

These assumptions yield the predictor $[\hat{x}_1, ..., \hat{x}_{30}, \hat{y}_1, ..., \hat{y}_{30}] = f(\nu_0, \boldsymbol{n}, \boldsymbol{\delta})$. If accurate measurement and mathematical models of environmental disturbances in *Vico* are available, the proposed framework is valid even if the effect of environmental disturbances on ship motion is significant. One might be able to enhance prediction performances by using wind sensor and real-time acceleration data in the future work, however, the scope of this study is not to investigate the best architecture of NN models but to present the validity of present framework.

Hyperparameter Tuning. Hyperparameters are a set of parameters that need to be set prior to the training. It is well known that they have a significant impact on the performance of NNs. In a pre-training phase, the number of hidden layers $nlayers \in [1, 8]$, the number of units in the hidden layers $midunits \in [10, 500]$, learning rate $lr \in [1.0 \times 10^{-4}, 1.0 \times 10^{-1}]$, dropout rate in the input layer $dropin \in [0.0, 1.0]$, and dropout rate in the hidden layer $drophd \in [0.0, 1.0]$ are tuned using hyperparameter tuning framework *optuna* [1] that employs Tree-structured Parzen Estimator as an optimization algorithm. As the range of search of lr is wide, it is searched in the log domain. 50 *optuna* trials are conducted. We check further trials contribute to marginal improvement of the validation loss. In a main training phase, $nlayers$ and $midunits$ are fixed as the NN in the pre-training phase is transfered to the main training phase; then, lr, $dropin$, and $drophd$ are tuned. When a NN is trained without a pre-training phase for comparison purposes apart from the NN trained in pre-training and main training phases, 50 *optuna* trials search an optimal set of hyperparameters $nlayers$, $midunits$, lr, $dropin$ and $drophd$.

4   Case Study

Fig. 4. The starboard view of the R/V Gunnerus employed in the case study [10].

In a case study, we aim to make a 30 s trajectory prediction of real-world zig-zag maneuvers of the R/V Gunnerus. The starboard view of the R/V Gunnerus is shown in Fig. 4.

4.1   Pre-training

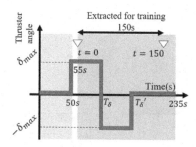

Fig. 5. Pre-defined time series of thruster angle given by a zig-zag controller.

In a pre-training phase, $n_{pre} = 300$ virtual scenarios are generated by merging a hull model, thruster models, and a zig-zag controller of the R/V Gunnerus in *Vico*. A fixed-step algorithm is used and coupling between sub-models are not considered in Vico. The virtual R/V Gunnerus is equipped with two azimuth thrusters in the port and starboard sides. These two thrusters receive the same commands of thruster angle and revolution from a zig-zag controller. Commands are simultaneously applied to thruster models since the difference between command and feedback values of the R/V Gunnerus is very small. The wave, wind, and ocean current are not applied to the virtual vessel. An example of the pre-defined time series of thruster angle is shown in Fig. 5. Each scenario is a 235 s time series. Thruster angle and revolution are set to zero before 50 s. The vessel state is reset to the initial state at 50 s. In order to avoid having impact load due to the reset, a $T_{pre} = 150$ s time series from 55 s ($t = 0$) to 205 s ($t = 150$) is saved in 1 Hz for the experiment with its 30 s future true positions $[x_1^b, ..., x_{30}^b, y_1^b, ..., y_{30}^b]$ and corresponding controller commands at each time step. Thruster angle is δ_{max} until $t = T_\delta$; then, it is changed to $-\delta_{max}$ until $t = T_\delta' = 2T_\delta$ with the maximum change rate. At $t = T_\delta'$, it is turned back to zero with the maximum change rate. Thruster revolution is set to n_{max} from 50 s to 235 s. Each scenario is parameterized by a vector of parameters $S = [\delta_{max}, T_\delta, n_{max}, u_{t=0}]$ where $\delta_{max} \in [-35°, 35°]$, $T_\delta \in [t = 50\,s, t = 75\,s]$, $n_{max} \sim \mathcal{N}(\mu = 130\,\text{RPM}, \sigma = 10\,\text{RPM})$ and the initial surge velocity $u_{t=0} \sim \mathcal{N}(\mu = 4.0\,\text{m/s}, \sigma = 1.0\,\text{m/s})$ are randomly given to each scenario. $\mathcal{N}(\mu, \sigma)$ indicates the Gaussian distribution with the mean value μ and the standard deviation σ. The probability distribution of parameters of S can be assumed based on the general understanding of a target operation $\mathcal{D}_{target}$, however, the discrepancy of the probability distribution of input features in $\mathcal{D}_{pre}$

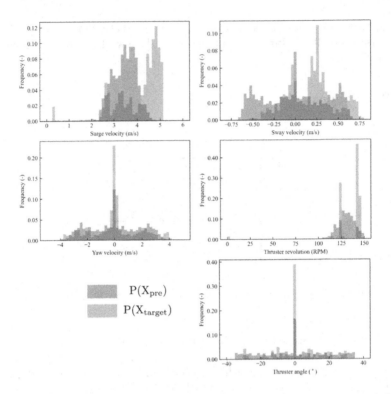

Fig. 6. Probability distributions of input features in the pre-training virtual dataset $P(X_\mathrm{pre})$ and in the target real-world dataset $P(X_\mathrm{target})$.

and $\mathcal{D}_{target}$ is inevitable as shown in Fig. 6. The initial North and East positions, heading, sway velocity, and yaw velocity are set to zero.

$n_\mathrm{pre} = 300$ scenarios are divided into $n_\mathrm{train,pre} = 192$ scenarios in the training dataset, $n_\mathrm{val,pre} = 48$ scenarios in the validation dataset, and $n_\mathrm{test,virtual} = 60$ scenarios in the test dataset. The test dataset is used only for checking the performance of the pre-trained predictor in the pre-training phase as shown in Fig. 9. The NN is trained only by using scenarios in the training dataset. To avoid overfitting the training dataset, the prediction performance of the trained NN in the validation dataset is monitored during the training. If the validation loss does not improve over $n_e = 200$ epochs, the training is automatically stopped; then, the best model is loaded. We checked $n_e > 200$ does not contribute to further improvement of the validation loss.

4.2   Main Training

The zig-zag maneuvers experiments of the R/V Gunnerus are conducted in November 2019 in Trondheim, Norway. Its port-side and starboard-side azimuth thrusters move simultaneously with the same commands of the thruster angle and revolution. Its tunnel thruster is turned off during the experiments. The experiment is a 1600 s time history. We split this time history into $n_{target} = 16$ operations of which length is 100 s. As we need 30 s future positions and commands at each time step in the operation for the training and evaluation purposes, first $T_{target} = 70$ s of each operation is saved as one operation with 30 s true future positions $[x_1^b, ..., x_{30}^b, y_1^b, ..., y_{30}^b]$ at each time step. The R/V Gunnerus is equipped with 13 onboard sensors sampling ship motion in real time. During the experiment, positions in the NED frame (North, East, and heading) and velocities in the body-fixed frame (surge, sway, and yaw speed) are saved in 1 Hz. Equation (2) converts positions in the NED frame into $[x_1^b, ..., x_{30}^b, y_1^b, ..., y_{30}^b]$ in the body-fixed frame:

$$\begin{pmatrix} x_k^b \\ y_k^b \end{pmatrix} = \begin{pmatrix} \cos\psi & \sin\psi \\ -\sin\psi & \cos\psi \end{pmatrix} \begin{pmatrix} N_k - N_0 \\ E_k - E_0 \end{pmatrix} \tag{2}$$

where N_k and E_k are the true ks future North and East positions in the NED frame. N_0 and E_0 are North and East positions when making prediction in the NED frame. At each time step, future command assumptions n and δ are given by the dataset as we examine the prediction performance provided that they are assumed by a controller. As we assume limited real-world data of a target operation are available, we use only $n_{trainval,target} = 12$ operations in the main training and keep the other $n_{test,target} = 4$ operations, that are used only for the evaluation of the prediction performance, untouched in the training process. $n_{trainval,target} = 12$ operations are divided into $n_{train,target} = 9$ operations in the training dataset and $n_{val,target} = 3$ operations in the validation dataset. The NN is trained only by using the training dataset and its performance in the validation dataset is monitored using the validation dataset. If the validation loss does not improve over $n_e = 200$ epochs, the training is automatically stopped; then, the best model is loaded as explained in the previous subsection. By switching the validation dataset four times, four independent NNs are trained (cross-validation). The final prediction to the untouched test dataset is the average of predictions made by these four NNs. In order to examine the contribution of the pre-training, three different strategies of training are investigated as follows.

(A) Without Pre-training. The training of this predictor is conducted without a pre-training phase. Virtual scenarios generated in *Vico* are kept untouched and only limited real-world data of a target operation $n_{train,target}$ is used in the training. It provides a baseline of the comparison study.

(B) Without Finetuning. This predictor is pre-trained with $n_{train,pre}$ virtual scenarios, however, the main training is not performed. The prediction performance of this predictor in $\mathcal{D}_{target}$ reveals the effect of the discrepancy between $\mathcal{D}_{pre}$ and $\mathcal{D}_{target}$.

(C) **Present Study.** This is a predictor that is trained in the manner of the present framework. The main training of this predictor is carried out with real-world data $n_{\text{train,target}}$ after the pre-training with virtual scenarios $n_{\text{train,pre}}$.

4.3   Evaluation Metric

This study introduces an evaluation metric $\overline{S}_{ijk,\text{target}}$ that indicates the mean prediction error over the 30 s prediction horizon in the test dataset of a main training phase.

$$\overline{S}_{ijk,\text{target}} = \frac{1}{n_{\text{test,target}}T_{\text{target}}H} \sum_{i=0}^{n_{\text{test,target}}} \sum_{j=0}^{T_{\text{target}}-1} \sum_{k=1}^{H} S_{ijk,\text{target}} \tag{3}$$

where $S_{ijk,\text{target}}$ is the distance between the predicted and true positions at ks prediction horizon of time step js prediction of ith scenario in the test dataset of a main training phase. In order to examine the prediction performance in a pre-training phase, we use an evaluation metric $\overline{S}_{ijk,\text{pre}}$ as follows.

$$\overline{S}_{ijk,\text{pre}} = \frac{1}{n_{\text{test,pre}}T_{\text{pre}}H} \sum_{i=0}^{n_{\text{test,pre}}} \sum_{j=0}^{T_{\text{pre}}-1} \sum_{k=1}^{H} S_{ijk,\text{pre}} \tag{4}$$

where $S_{ijk,\text{pre}}$ is the distance between the predicted and true positions at ks prediction horizon of time step js prediction of ith scenario in the test dataset in a pre-training phase.

4.4   Results

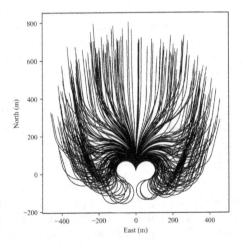

Fig. 7. Trajectories of virtual scenarios in $\mathcal{D}_{\text{pre}}$.

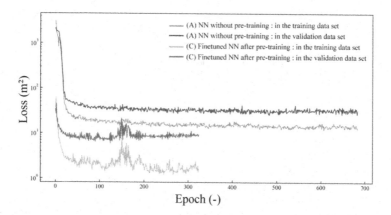

Fig. 8. Histories of the training and validation loss of the first fold of (A) the NN without pre-training and (C) the finetuned NN with pre-training. The training is terminated when the validation loss does not improve over $n_e = 200$ epochs. (Color figure online)

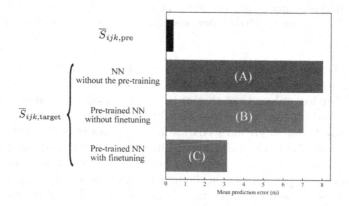

Fig. 9. Mean prediction errors in the test dataset.

Trajectories of $n_{\mathrm{pre}} = 300$ virtual scenarios generated in *Vico* are shown in Fig. 7. It is seen that diverse scenarios are generated thanks to the setting of scenario generation in the pre-training phase. Through the *optuna* hyperparameter optimization in the pre-training phase, $nlayer = 1$ and $midunits = 420$ are selected. Apart from the pre-training phase, (A) without pre-training is trained only by using real-world data $n_{\mathrm{trainval,\ target}}$. $nlayer = 1$ and $midunits = 300$ are selected through *optuna* hyperparameter optimization. After the pre-training phase, the main training is conducted. With a set of optimized hyperparameters, Fig. 8 shows histories of the training and validation losses of the first fold of the cross-validation of (A: blue lines) the NN without the pre-training phase and (C: green lines) the present study with the pre-training phase. The vertical axis displays loss values along a logarithmic scale. It should be noted that the training is

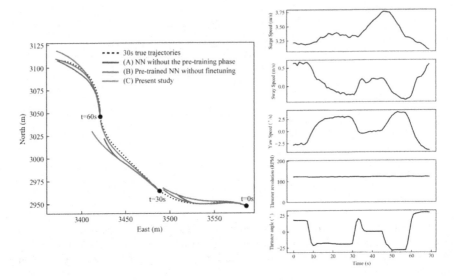

Fig. 10. (left) Snapshots of 30 s prediction at $t = 0$ s, $t = 30$ s, and $t = 60$ s of one scenario in the test dataset (right) Time histories of vessel state and commands.

automatically terminated if the validation loss does not improve over $n_e = 200$ epochs. Since (A) is not pre-trained by virtual scenarios, its training starts with notably higher loss values than that of (C). Accordingly, the validation loss of (A) ends up with higher values than that of (C). This result indicates the pre-training phase based on virtual scenarios facilitates not only the initial stage of the training but also the overall training efficiency in the main training.

Figure 9 shows the mean prediction error in the test dataset in the pre-training phase $\overline{S}_{ijk,\mathrm{pre}}$ and that in the main training phase $\overline{S}_{ijk,\mathrm{target}}$. By comparing the bars in the top and the third from the top in Fig. 9, one can see that the prediction performance of the pre-trained NN deteriorates much in $\mathcal{D}_{\mathrm{target}}$ if it is not finetuned in the main training phase due to the difference between $\mathcal{D}_{\mathrm{pre}}$ and $\mathcal{D}_{\mathrm{target}}$. The bar (A) in the second from the top in Fig. 9 reveals that the training without the pre-training phase produces the largest prediction error in (A), (B), and (C). (C) trained in the present framework with pre-training and main training phases reduces prediction error notably; by 60.8% compared to (A) without pre-training.

The left panel of Fig. 10 shows snapshots of 30 s prediction at $t = 0$ s, $t = 30$ s, and $t = 60$ s of one operation in the test dataset of real-world data in the main training phase. The right panel of Fig. 10 shows time histories of vessel state and commands of the operation. (A) the NN without the pre-training phase deviates significantly from the true trajectories at $t = 0$ s and $t = 30$ s. (B) the NN without finetuning after the pre-training phase succeeded at capturing the trend of the 30 s true trajectory at $t = 0$ s, $t = 30$ s, and $t = 60$ s in the short prediction horizon, however, it ends up with the large prediction error in the distant prediction horizon. (C) present study with the pre-training and main training phases traces the true trajectories more accurately than (A) and (B) at $t = 0\,\mathrm{s}$, $t = 30\,\mathrm{s}$, and $t = 60\,\mathrm{s}$.

5   Conclusion

This study proposed a new framework for co-simulation-based development of a ship trajectory predictor provided that limited real-world dataset of a target operation is available. We integrated prior knowledge, which is virtual scenarios generated by a simplified vessel model in co-simulation platform *Vico*, into training of the neural-network-based trajectory predictor. The neural network is pre-trained using many virtual scenarios generated in *Vico* before it is finetuned using limited real-world dataset of a target operation. In the case study, we employed real-world operations of zig-zag maneuvers of a 33.9m-length research vessel. For pre-training, 300 virtual scenarios of zig-zag maneuvers were generated in *Vico* only by merging sub-models provided by different project partners. The pre-trained neural network was further finetuned using 12 real-world operations of zig-zag maneuvers. The present framework reduces the mean prediction error by 60.8% in the test dataset of real-work operation compared to the neural network without pre-training. Hence, the present framework enables a ship trajectory predictor to be constructed only by using a simplified vessel model in co-simulation platform at a small cost and limited real-world data of a target operation.

References

1. Akiba, T., Sano, S., Yanase, T., Ohta, T., Koyama, M.: Optuna: a next-generation hyperparameter optimization framework. In: KDD 2019: Proceedings of the 25th ACM SIGKDD International Conference on Knowledge Discovery & Data Mining, pp. 2623–2631. Association for Computing Machinery, Anchorage (2019). https://doi.org/10.1145/3292500.3330701

2. Hassani, V., Fathi, D., Ross, A., Sprenger, F., Selvik, Ø., Berg, T.E.: Time domain simulation model for research vessel Gunnerus. In: Proceedings of the International Conference on Offshore Mechanics and Arctic Engineering - OMAE, St. John's, Newfoundland, Canada, vol. 7 (2015). https://doi.org/10.1115/OMAE201541786

3. Hatledal, L.I., Chu, Y., Styve, A., Zhang, H.: Vico: an entity-component-system based co-simulation framework. Simul. Modell. Pract. Theory **108**(September 2020), 102243 (2021). https://doi.org/10.1016/j.simpat.2020.102243

4. Hatledal, L.I., Zhang, H., Collonval, F.: Enabling python driven co-simulation models with PythonFMU. In: 34th International ECMS Conference on Modelling and Simulation (2020). https://doi.org/10.7148/2020-0235

5. Huang, Y., Chen, L., Chen, P., Negenborn, R.R., Gelder, P.H.A.J.M.V.: Ship collision avoidance methods: state-of-the-art. Saf. Sci. **121**(April 2019), 451–473 (2020). https://doi.org/10.1016/j.ssci.2019.09.018

6. Intelligent Systems Lab @ NTNU Aalesund: Intelligent Systems Lab webpage. http://org.ntnu.no/intelligentsystemslab/

7. Kawan, B., Wang, H., Li, G., Chhantyal, K.: Data-driven modeling of ship motion prediction based on support vector regression. In: Proceedings of the 58th Conference on Simulation and Modelling (SIMS 58) Reykjavik, Iceland, 25th–27th September 2017, vol. 138, pp. 350–354 (2017). https://doi.org/10.3384/ecp17138350
8. Kingma, D.P., Ba, J.L.: Adam: a method for stochastic optimization. In: 3rd International Conference on Learning Representations, ICLR 2015 - Conference Track Proceedings, pp. 1–15 (2015)
9. Mei, B., Sun, L., Shi, G.: White-black-box hybrid model identification based on RM-RF for ship maneuvering. IEEE Access **7**, 57691–57705 (2019). https://doi.org/10.1109/ACCESS.2019.2914120
10. Norges teknisk-naturvitenskapelige universitet: Research vessel: R/V Gunnerus. https://www.ntnu.edu/oceans/gunnerus
11. Panigrahi, S., Nanda, A., Swarnkar, T.: A survey on transfer learning. Smart Innov. Syst. Technol. **194**, 781–789 (2021). https://doi.org/10.1007/978-981-15-5971-6_83
12. Paszke, A., et al.: PyTorch: an imperative style, high-performance deep learning library. In: NeurIPS (2019)
13. Perera, L.P.: Navigation vector based ship maneuvering prediction. Ocean Eng. **138**(November 2016), 151–160 (2017). https://doi.org/10.1016/j.oceaneng.2017.04.017
14. Rong, H., Teixeira, A.P., Guedes Soares, C.: Ship trajectory uncertainty prediction based on a Gaussian Process model. Ocean Eng. **182**(December 2018), 499–511 (2019). https://doi.org/10.1016/j.oceaneng.2019.04.024
15. Skulstad, R., Li, G., Fossen, T.I., Vik, B., Zhang, H.: A hybrid approach to motion prediction for ship docking - integration of a neural network model into the ship dynamic model. IEEE Trans. Instr. Meas. **70**, 1–11 (2021). https://doi.org/10.1109/TIM.2020.3018568
16. Skulstad, R., Li, G., Fossen, T.I., Wang, T., Zhang, H.: A co-operative hybrid model for ship motion prediction. Model. Ident. Control: Norwegian Res. Bul. **42**(1), 17–26 (2021). https://doi.org/10.4173/mic.2021.1.2
17. Wang, T., Li, G., Hatledal, L.I., Skulstad, R., Aesoy, V., Zhang, H.: Incorporating approximate dynamics into data-driven calibrator: a representative model for ship maneuvering prediction. IEEE Trans. Ind. Inform. 1 (2021). https://doi.org/10.1109/tii.2021.3088404
18. Wang, Y., Yao, Q., Kwok, J.T., Ni, L.M.: Generalizing from a few examples: a survey on few-shot learning. ACM Comput. Surv. **53**(3), 1–34 (2020). https://doi.org/10.1145/3386252
19. Wang, Z., Zou, Z., Soares, C.G.: Identification of ship manoeuvring motion based on nu-support vector machine. Ocean Eng. **183**(January), 270–281 (2019). https://doi.org/10.1016/j.oceaneng.2019.04.085
20. Zhao, L., Roh, M.I.: COLREGs-compliant multiship collision avoidance based on deep reinforcement learning. Ocean Eng. **191**(May), 106436 (2019). https://doi.org/10.1016/j.oceaneng.2019.106436

Effect of Ship Propulsion Retrofit on Maneuverability Research Based on Co-simulation

Tongtong Wang[1]([✉]) [iD], Lars Ivar Hatledal[2] [iD], Motoyasu Kanazawa[1] [iD],
Guoyuan Li[1] [iD], and Houxiang Zhang[1] [iD]

[1] Department of Ocean Operations and Civil Engineering, Faculty of Engineering,
Norwegian University of Science and Technology (NTNU), Aalesund, Norway
{tongtong.wang,motoyasu.kanazawa,guoyuan.li,hozh}@ntnu.no
[2] Department of ICT and Natural Sciences, Faculty of Information Technology
and Electrical Engineering, Norwegian University of Science and Technology
(NTNU), Aalesund, Norway
laht@ntnu.no

Abstract. Shipping has been dominating the transportation industry
in worldwide trade. During the service life of a vessel, conversions in
mid-life often occur for economic or technical purposes. By replacing
expired components or updating the outdated technology to the latest
operational standards, the service life could be greatly prolonged, and
meanwhile the capability will be enhanced. Bringing ships-in-service to
the latest technology creates the need for advanced methods and tools to
simulate the ship main and auxiliary systems. Co-simulation is emerging as a promising technique in complex marine system modeling. The
Functional Mock-up Interface (FMI) standard enables sub-models representing part of the vessel to be executed individually or as an integrated
part of the overall system. The modularity and re-usability of the sub-
models speed up the simulation cycle and ensure time-cost effectiveness,
which benefits the ship conversion. This paper presents a research related
to the ship propulsion retrofit process based on the co-simulation technique. The ship maneuverability before and after refitting propulsion
units is simulated and analyzed. Through the experiments, propulsion
performance improvements are observed. Eventually, the study supports
that the co-simulation technique to be applied in the maritime field has
an encouraging future.

Keywords: Propulsion retrofit · Ship maneuverability · Co-simulation

1 Introduction

Shipping, as a relatively energy-efficient, environmental-friendly, and sustainable
model of mass transport, is the dominant transportation method for world-wide

This work was supported by a grant from the Research Council of Norway through the
Knowledge-Building Project for industry "Digital Twins for Vessel Life Cycle Service"
(Project no: 270803).

trade. Normally, the life cycle of a ship is estimated to be around 25 years, but the actual age of the short sea fleet, for example, is higher, reaching more than 30–35 years of age for perhaps as much as 40% of the fleet [9]. However, the life cycle of ship systems and major components is much shorter because of the ever faster technological developments. In general, 10–15 years after launching a ship, its main systems are outdated. Upgrading outdated technology in ships to the latest operational standards enhances the capability and prolongs the service life [13]. Furthermore, the international policies fostering the reduction of energy consumption and emissions are always issuing new regulations on energy efficiency and emission reduction [3]. For example, the International Maritime Organization (IMO) has implemented a stricter sulfur content limit–called the IMO 2020 sulfur cap–aiming at improve air quality and protect the environment. Further, IMO has initiated an extensive strategy of the energy efficiency existing ship index for existing ships, which indicate that the energy efficiency of ships should be satisfied during the operation phase. To comply with the new regulations, green technologies are implemented on-board ships [10]. Retrofitting the ships during the operation phase has become a popular choice for the transportation industry [15]. It is possible to upgrade the installed technology with new high-performance machines and significantly improve the system's handling, economic efficiency, as well as emission reduction [8].

Given that modern ships are becoming more complex and integrated, retrofitting them is a complex and intricate engineering task. Optimal performance is relying on all subsystems to work optimally, both individually and aggregated [11,14]. Each subsystem is dedicated to a specific object of the vessel or equipment. Between distributed components, they exchange all relevant ship information, data, or analysis and make coordinated operational decisions. Considering the mutual and multi-disciplinary interaction between subsystems, co-simulation is emerging as a promising technique. Often, it is difficult to describe a truly complex system in a single tool. Instead, people are encouraged to develop models at the partial solution level, such as the dynamic properties check, control strategy design, or energy consumption optimization. It not only dramatically lessens the modeling pressure and promotes efficiency but enables the re-usability of different elements. Furthermore, a branch of components may be generated by different teams or suppliers, each in its own domain and each with its own tools. Using co-simulation, these models can be integrated as black-boxes without revealing the intellectual property of the owner [2]. In addition, considering now the demanding operation of an autonomous vessel, it is better to test ahead in a virtual environment for safety reasons. Co-simulation reduces efforts to conduct pre-training or perform tests by redirecting design attention and reusing the sub-system models. From an efficiency point of view, co-simulation greatly facilitates the ship retrofitting process.

In a co-simulation, different subsystems are modeled separately and composed into a global simulation, where each model is executed independently, sharing information at discrete time points. The Functional Mock-up Interface (FMI) standard is a commonly used standard for co-simulation, and model imple-

Fig. 1. Side view of the research vessel Gunnerus.

menting the FMI is known as a Functional Mock-up Unit (FMU). The FMI enables an FMU exported by one tool to interoperate with a variety of host tools and for host tools to orchestrate interactions between FMUs exported by a variety of other tools [1]. A system can then be modelled as a collection of inter-connected FMUs. Co-simulation thus enables retrofit decisions to be simulated ahead-of-time, cheaply and early in the process.

This study presents the propulsion retrofit process using the co-simulation technique, and the dynamic properties of the retrofitted devices are analyzed and discussed. The research vessel Gunnerus (see Fig. 1), owned and operated by the Norwegian University of Science and Technology (NTNU) serves as the test ship. The simulation fidelity was verified against real ship maneuver in [7] in terms of ship speed, course, and power consumption. Convinced by the high-fidelity resolution of the simulation, further research is conducted with more confidence. As reported in [16], The R/V Gunnrus went through a thruster refit in 2015. The original twin fixed-pitch ducted propellers and rudders were replaced with the Permanent Magnet (PM) rim-drive azimuthing thrusters. The original propellers were 5-bladed, high skew type with a diameter of 2.0 m that rotated in a 19 A type duct profile, and the new azimuthing thrusters incorporates a ring propeller in a tailor-made duct with a diameter of 1.9 m with four blades having a forward skewed shape. Figure 2 shows the propulsion configuration on Gunnerus before and after retrofit, where the left is the origin pitch propeller with ice-fins, and the right is the refitted azimuth thruster provided by Rolls-Royce. The same diesel-electric system supplied the propulsion and maneuvering power before and after the conversion. To document the effect of the change of propulsion system, a simulation test is carried out both before and after retrofitting the

Fig. 2. The propulsion arrangement before and after retrofit.

PM azimuthing thruster in this work. The ship maneuvering capabilities are then verified.

2   Problem Formulation

Thanks to the modularity and flexibility of co-simulation, the effort required to simulate the dynamic properties of the propulsion unit is greatly decreased. In this section, the ship maneuverability and the co-simulation diagram, as well as the FMUs used in this research will be explained.

2.1   Ship Maneuverability

Ship maneuverability is defined as the capability of the craft to carry out specific maneuvers. A maneuvering characteristic can be obtained by changing or keeping a predefined course and speed of the ship in a systematic manner by means of active controls. For most of the surface vessels, these controls are implemented by rudders, propellers and thrusters. The IMO approved standards for ship maneuverability, and the standards specify the type of standard maneuvers and associated criteria. It is always necessary for the vessels to apply these standards, and even some port and flag states adopted some of the IMO standards as their national requirements. To help the vessel prepare for implementation of the standards, prediction of the maneuverability performance in the design stage enables a designer to take appropriate measures in good time to achieve requirements. The prediction could be carried out by using existing data, scaled model test, or numerical simulation [12]. From the practical view, numerical simulation appears an effort-efficient way. Therefore, the ship maneuvering capabilities will be the main concern during simulation experiments.

To examine the course keeping capability of the ship, usually, the turning circle and Kempf's zigzag maneuver are selected. The maneuvers and their characteristic are described as Fig. 3 and Table 1 and 2:

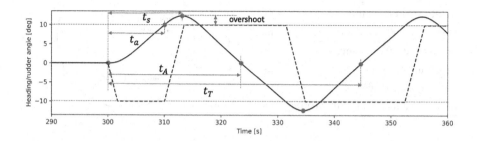

Fig. 3. Schematic of zigzag maneuver and its main characteristics.

Table 1. Zigzag maneuver characteristics.

Characteristic	Reference
Initial turning time t_a	The time from the rudder execution until the heading changes a desired degrees, 10° off the initial course in 10°/10° example
Time to check yaw t_s	The time from the rudder execution until the maximum heading changes
Reach time t_A	The time between the first rudder execution and the instance when the ship's heading is zero
Complete time t_T	The time between the first rudder execution and the instance when the ship's heading is zero after third execution
Overshoot angle	The angle through which the ship continues to turn in the original direction after execution of counter rudder

2.2 Co-simulation Setup

The ship maneuvering simulation is set up as Fig. 4 shows. Each block represents an FMU of which the input and output variables are declared. The experiment is performed in Vico, a generic co-simulation framework based on the Entity-Component-System software architecture that supports the FMI as well as the System Structure and Parameterization (SSP) standards [5]. The user may manipulate the wind, waves, and ocean currents to mimic environmental conditions. An overview of FMUs applied in the maneuvering simulation is presented. All the FMUs, except the *VesselModel* and *PMAzimuth*, are developed by the authors using PythonFMU [6].

1. VesselModel
The vessel model reflects the vessel's hydrodynamic properties, such as the

Table 2. Turning circle maneuver and its main characteristics.

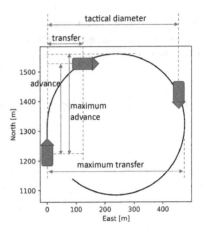

Characteristic	Unit
Steady turning radius	m
Transfer at 90° heading	m
Advance at 90° heading	m
Maximum transfer	m
Maximum advance	m
Tactical diameter at 180° heading	m

mass, resistance, and cross-flow drag, as well as restoring forces. It is a 6° of freedom (DOF) time-domain simulation model developed by MARINTEK's vessel simulator (VeSim) [4]. Summing all the external forces acting on the ship, the dynamic equations of vessel motions are then solved. It can be implemented in sea-keeping and maneuvering problems for marine vessels subjected to waves, wind, and currents based on a unified nonlinear model Eq. 1.

$$(M_{RB} + M_A)\dot{\nu} + C(\nu)\nu + D(\nu) + g(\eta) + \int_0^t h(t - \tau)\nu(\tau)\,d\tau = q \qquad (1)$$

where the 6DOF ship velocity state is expressed as the vector $\nu = [u, v, w, p, q, r]'$ referred to the coordinate shown in Fig. 5. The $[u, v, w]$ are the linear velocity along x_b, y_b, z_b directions, and $[p, q, r]$ are the angular velocities rotating around three directions. $M_{RB} \in \mathbb{R}^{6\times6}$ is the rigid body mass, and $M_A \in \mathbb{R}^{6\times6}$ is the added mass. $C(\nu) = C_{RB}(\nu) + C_A(\nu) \in \mathbb{R}^{6\times6} \times \mathbb{R}^6$ is describing the generalized coriolis-centripetal forces. $D(\nu) \in \mathbb{R}^{6\times6} \times \mathbb{R}^6$ is a vector of damping forces and moments. $g(\eta) \in \mathbb{R}^6$ is a vector of gravitational/buoyancy forces and moments. And $h(\tau)$ refers to impulse response functions calculated by SINTEF OCEAN's potential theory. $q \in \mathbb{R}^6$ is the external forces and moments acting on the ship. The model itself is fully coupled and it can be used for simulation and prediction of coupled vehicle motion.

2. **PID controller**

The PID controller is created to generate shaft speed and rudder angle commands according to Eq. 2. In the control law, the $k_{\{\cdot\}}$ is the parameter enabling tuning, and the predefined approach speed u_d as well as the ship

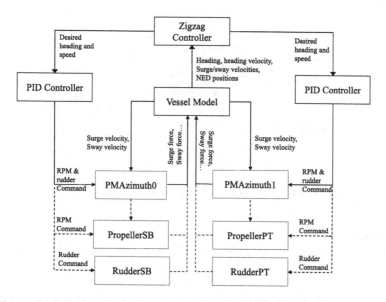

Fig. 4. Diagram showing the relationship of the engaged ship components.

heading ψ_d are issued by the *ZigzagController*.

$$RPM = k_{pu}(u - u_d) + k_{iu}\int_0^t (u - u_d)dt + k_{du}\frac{d}{dt}(u - u_d)$$

$$\delta = k_{ppsi}(\psi - \psi_d) + k_{ipsi}\int_0^t (\psi - \psi_d)dt + k_{dpsi}\frac{d}{dt}(\psi - \psi_d)$$

(2)

3. **Zigzag controller**
   It is a logistic solver without numerical computation. Given the current ship speed and heading, it can tell to which side the rudder should turn and deliver the command saturation to the connected *PID controller*.
4. **PMAzimuth**
   It is a hydrodynamic model of the azimuth thruster without actuator, implemented by the manufacturer Kongsberg Maritime using VeSim. Feeding a specific RPM and angle command, vessel speed, as well as the loss factor into the model, it produces a 3DOF force on heave, surge, and sway directions.
5. **Propeller**
   Both the propeller and rudder are generic models parametrized to R/V Gunnerus. The surge force related to the propeller is calculated with:

$$\tau_p = f(n, u)$$

(3)

where n is the propeller shaft speed (r/min), and u is the vessel's surge velocity. Note that the sway force and yaw moment due to propeller are neglected as they have smaller magnitudes compared to those of hull and rudder components.

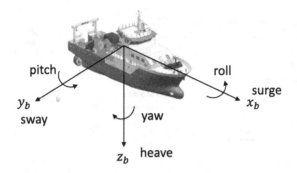

Fig. 5. The ship body coordinate and motion in 6DOF.

6. Rudder

The rudder is modelled according to [17]. It can be expressed as:

$$\tau_r = g(u, v, r, n, \delta, \theta) \tag{4}$$

where u, v, r are the velocities in surge, sway, and yaw directions respectively. And δ is the rudder angle. θ refers to the hull-rudder interaction coefficients.

3   Experiment Results

Experiments are implemented with the designed co-simulation diagram in Vico. The detailed experimental scenarios and the corresponding ship maneuverability, with either pitch propeller or PMAzimuth thruster installed, are presented in this section.

3.1   Simulation Scenarios

Ship maneuvering experiments with a different set of propulsion units are implemented. It is also worth noticing that the ship maneuverability could be affected by water depth, environmental forces, ship speed and hydrodynamic derivatives. To ensure the results comparable, identical settings except only the propulsion units are employed. The ship is assuming cruising on calm and deep water without external environmental disturbances. Eight maneuver test scenarios are defined as Table 3 shows, aiming to investigate the propulsion performance under different execution angles and speeds.

A $10° - 10°$ zigzag test means that the rudder and azimuth angles are given a command of $\pm10°$, and when the ship heading change reaches $10°$ the rudder/azimuth reverse to the opposite side. The $10°$ in turning circle refers to the constant rudder/azimuth angle. As a key parameter, the ship surge speed is given as the steady velocity before the zig-zag/turning circle maneuvers are initiated. During the process, $300\,s$ are saved first to warm up and drive the ship to the pre-defined speed, and $300\,s$ are arranged for operations. The simulation time step is set to $0.05\,s$.

3.2 Results Analysis

In this section, the main maneuver characteristics of the ship before and after conversion will be observed and discussed.

Zigzag Maneuver. Zigzag trajectories for the ship using both the pitch propellers and azimuth thrusters are simulated. Three selected test results are presented and compared in Fig. 6, 7 and 8. Naturally, differences in turning velocities are observed from these figures. A more noticeable yaw velocity distinction between the pitch propeller and azimuth arises during 10° turn command. The

Table 3. Maneuver experiment cases implemented in Vico.

Maneuver	Execution	Speed
Zig-zag	10° − 10°	low
		High
	20° − 20°	Low
		High
Turning circle	10°	Low
		High
	20°	Low
		High

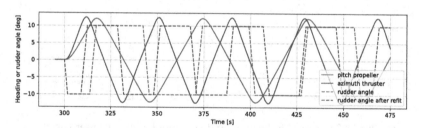

(a) 10°/10° zigzag ship heading and command angle at higher speed.

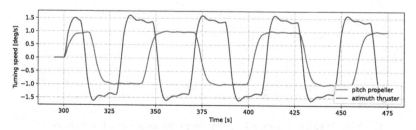

(b) Ship turning velocities under propeller or azimuth actuation.

Fig. 6. 10°/10° zigzag properties at higher speed.

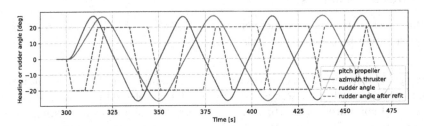

(a) 20°/20° zigzag ship heading and command angle at higher speed.

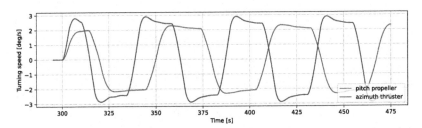

(b) Ship turning velocities under propeller or azimuth actuation.

Fig. 7. 20°/20° zigzag properties at higher speed.

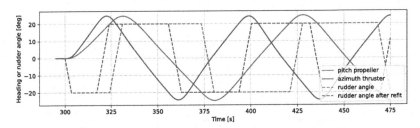

(a) 20°/20° zigzag ship heading and command angle at lower speed.

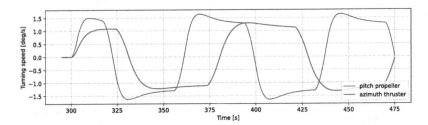

(b) Ship turning velocities under propeller or azimuth actuation.

Fig. 8. 20°/20° zigzag properties at lower speed.

Table 4. The zigzag characteristics for the ship before and after propulsion unit retrofit.

Characteristics		10°/10°			10°/10°			20°/20°			20°/20°		
		pr	azi	gain [%]	pr	azi	gain [%]	pr	azi	gain [%]	pr	azi	gain [%]
Approach speed	[m/s]	4.7	4.78	–	2.4	2.5	–	4.7	4.73	–	2.45	2.47	–
t_a	[s]	13.7	8.9	35	24.25	15.6	35.7	8.95	6.45	27.9	14.9	10.4	30.2
t_s	[s]	3.6	2.9	19.4	5	3.6	28	10.8	8.25	23.6	16.3	12.2	25.2
t_A	[s]	31.9	21.3	33.2	54.4	35.25	35.2	34.5	26.25	23.9	55.6	40.7	26.8
t_T	[s]	60.3	40.75	32.4	103.2	67	35.1	64.15	50.3	21.6	103.8	78.75	24.1
First overshoot angle	[°]	2.17	2.37	−9.2	1.57	1.6	−1.9	6.64	6.7	−0.9	4.6	4.7	−2.1
Second overshoot angle	[°]	2.2	2.42	−10	1.58	1.61	−1.9	6.87	6.68	2.8	4.8	4.36	9.2
Average overshoot angle	[°]	2.2	2.42	−10	1.576	1.6	−1.5	7.01	6.7	4.4	4.92	4.35	11.6

statistic results are summerized in Table 4. It could be observed that the measured key time parameters in the azimuth group are effectively decreased. This conclusion reveals that the ship with azimuth installed reaches the desired course within a shorter time, and it responds more quickly to the given command.

Meanwhile, it is observed in Fig. 6 that the rudder rate of both systems are similar, as they reverse from port-side to starboard in a similar amount of time. Although it takes longer time for the ship using conventional rudders to drive itself to the target course, it does not necessarily generate a larger overshoot angle. Instead, their average overshoot angles are related to the execution command and maneuver speed as indicated in Table 4. If a smaller angle command is given to the azimuth, it would even lead to a slightly larger average overshoot angle compared to the conventional rudder, even with a lower or higher forward speed. With an increasing angle command, the azimuth thrusters are observed to perform outstandingly.

Turning Circle. The turning circle maneuver experiments are conducted under the resembling co-simulation structure (Fig. 4) but replacing the *Zigzag controller* with *Turning controller*. The execution angle and speed are distinguished into two categories: 10° and 20°, higher and lower approach speed, respectively.

The statistical maneuver results are presented in Table 5. Among the four cases, two of them are selected to visualize the differences (See Fig. 9 and 10).

Table 5. The turning characteristics for the ship before and after propulsion unit retrofit.

Characteristics		10°			10°			20°			20°		
		pr	azi	gain [%]	pr	azi	gain [%]	pr	azi	gain [%]	pr	azi	gain [%]
Approach speed	[m/s]	4.7	4.8	–	2.4	2.5	–	4.7	4.7	–	2.4	2.5	–
Steady turning radius	[m]	237.5	185.5	21.9	237.8	186.3	21.6	90.4	91.3	−1	93.2	92.9	0.32
Maximum transfer	[m]	476.2	370.6	22.2	476.4	371.9	21.9	190.7	184.8	3.1	195.1	187.7	3.8
Maximum advance	[m]	266.7	200.5	24.8	265.5	200.3	24.5	127.8	108.9	14.8	128.3	109.6	14.6
Transfer	[m]	227.5	173.7	23.6	227.2	173.9	23.4	88.1	82.2	6.7	88.8	82.5	7.1
Advance	[m]	266.4	200.1	24.9	265.2	199.9	24.6	127	108.1	14.9	127.4	108.7	14.7
Tactical diameter	[m]	475.9	370.2	22.2	476.1	371.5	22.0	189.96	184.1	3.1	194.2	186.9	3.8

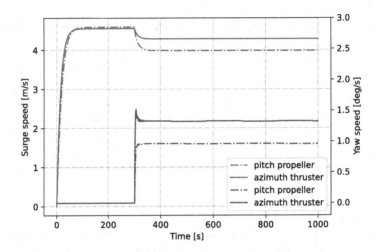

(a) The ship's surge and yaw speed when circling at 10° with a fast speed.

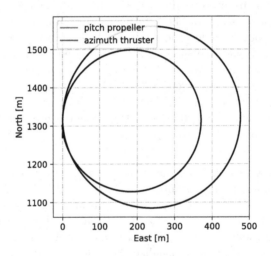

(b) Comparison of propeller and azimuth actuated ship trajectories.

Fig. 9. 10° turning circle properties at higher speed.

The ship equipped with either the conventional pitch propellers and rudders, or azimuths, are approaching at similar speeds before execution. From Fig. 9a, a drop of surge speed is observed when the rudder is instantiated, and the drop of pitch propeller is more obvious compared to that of the azimuth. Meanwhile, a larger turning velocity is offered by the azimuth. The out-performance in response velocities is expected to lead to a narrowed turning radius which is verified in Fig. 9b.

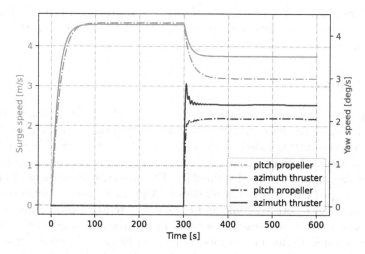

(a) The ship's surge and yaw speed when circling at 20° with a fast speed.

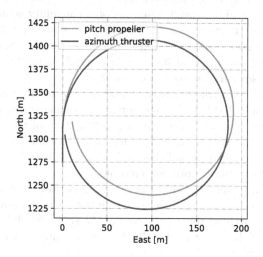

(b) Comparison of propeller and azimuth actuated ship trajectories.

Fig. 10. 20° turning circle properties at higher speed.

Moreover, the statistical results show that the angle command affects the propulsion performance more than the approach speed. Comparing Fig. 9 and Fig. 10, the ship exhibits similar speeds before operation. However, the percentage of decreased surge velocity with 20° rudder angle is higher than that with 10° counter angle. For the azimuth thruster, it drops about 6% in 10° and 19% in 20°. For the propeller, the values are 15% and 32%. When the rudder angle is given 20°, it not necessarily generates a large turning radius, as the propulsion moments could produce a higher yaw rate compared to 10°. This finding leads to a compromise in overall turning performance. Therefore, it is understandable that the steady radius reduction at 20° command is smaller than that of the lower command.

4   Conclusion

The continuously improving knowledge and availability of high-performance machines and drives have created the need for advanced methods and tools to facilitate retrofitting existing ships. Usually, the retrofit is driven by environmental and/or technical reasons, such as to comply with new energy regulations or to upgrade outdated technology. Either way, it is beneficial to ensure fast refitting procedures by allowing easier integration of new components. Co-simulation reduces both the time and the costs of refitting procedures, extending the operative life of a vessel in service. In this research, the authors utilized the co-simulation techniques to model the ship maneuver process before and after propulsion conversion and evaluate the impact of new devices on the ship maneuverability, aiming to support decisions on measures to meet operational standards. By comparing the zigzag and turning circle maneuver characteristics in the present work, an improved course keeping capability is observed after refitting advanced permanent magnet driven azimuth thrusters on the ship. This practice supports that co-simulation enables time cost-effective redesign and fast virtual tests by taking advantage of its modularity and flexibility, and emerges as a promising technology in the maritime industry.

However, it should be clarified that the quality of the simulation model may vary, and the tests conducted in order to compare the maneuvering performance of the two systems, and are not necessarily a good measure of the daily maneuvering capabilities of the vessel. Agreeing with this situation, the experiments performed through co-simulation will be qualitatively informative so that the comparative conclusions drawn upon are credible.

In the present study, only the ship maneuvering performance investigation is within scope, but in many cases, energy consumption is the major concern. Therefore, further research on the energy cost of the ship with different propulsion sets installed will be implemented by taking advantage of co-simulation technology in the future.

References

1. Broman, et al.: Determinate composition of FMUS for co-simulation. In: 2013 Proceedings of the International Conference on Embedded Software (EMSOFT), pp. 1–12. IEEE (2013). https://doi.org/10.1109/EMSOFT.2013.6658580
2. Gomes, C., Thule, C., Broman, D., Larsen, P.G., Vangheluwe, H.: Co-simulation: a survey. ACM Comput. Sur. (CSUR) 51(3), 1–33 (2018). https://doi.org/10.1145/3179993
3. Halff, A., Younes, L., Boersma, T.: The likely implications of the new IMO standards on the shipping industry. Energy Policy 126, 277–286 (2019). https://doi.org/10.1016/j.enpol.2018.11.033

4. Hassani, V., Ross, A., Selvik, Ø., Fathi, D., Sprenger, F., Berg, T.E.: Time domain simulation model for research vessel gunnerus. In: International Conference on Offshore Mechanics and Arctic Engineering, vol. 56550, p. V007T06A013. American Society of Mechanical Engineers (2015). https://doi.org/10.1115/OMAE2015-41786

5. Hatledal, L.I., Chu, Y., Styve, A., Zhang, H.: Vico: an entity-component-system based co-simulation framework. Simul. Model. Pract. Theory **108**, 102243 (2021). https://doi.org/10.1016/j.simpat.2020.102243

6. Hatledal, L.I., Collonval, F., Zhang, H.: Enabling python driven co-simulation models with PythonFMU. In: Proceedings of the 34th International ECMS-Conference on Modelling and Simulation-ECMS 2020. ECMS European Council for Modelling and Simulation (2020). https://doi.org/10.7148/2020-0235

7. Hatledal, L.I., Skulstad, R., Li, G., Styve, A., Zhang, H.: Co-simulation as a fundamental technology for twin ships (2020). https://doi.org/10.4173/mic.2020.4.2

8. Hou, H., Krajewski, M., Ilter, Y.K., Day, S., Atlar, M., Shi, W.: An experimental investigation of the impact of retrofitting an underwater stern foil on the resistance and motion. Ocean Eng. **205**, 107290 (2020). https://doi.org/10.1016/j.oceaneng.2020.107290

9. Koenig, P., Nalchajian, D., Hootman, J.: Ship service life and naval force structure. Nav. Eng. J. **121**(1), 69–77 (2009). https://doi.org/10.1111/j.1559-3584.2009.01141.x

10. Li, K., Wu, M., Gu, X., Yuen, K., Xiao, Y.: Determinants of ship operators' options for compliance with IMO 2020. Transp. Res. Part D: Transp. Environ. **86**, 102459 (2020). https://doi.org/10.1016/j.trd.2020.102459

11. Ling-Chin, J., Roskilly, A.: Investigating a conventional and retrofit power plant on-board a roll-on/roll-off cargo ship from a sustainability perspective-a life cycle assessment case study. Energy Convers. Manage. **117**, 305–318 (2016). https://doi.org/10.1016/j.enconman.2016.03.032

12. Liu, J., Hekkenberg, R., Rotteveel, E., Hopman, H.: Literature review on evaluation and prediction methods of inland vessel manoeuvrability. Ocean Eng. **106**, 458–471 (2015). https://doi.org/10.1016/j.oceaneng.2015.07.021

13. Liu, L., Yang, D.Y., Frangopol, D.M.: Ship service life extension considering ship condition and remaining design life. Mar. Struct. **78**, 102940 (2021). https://doi.org/10.1016/j.marstruc.2021.102940

14. Mauro, F., La Monaca, U., la Monaca, S., Marinò, A., Bucci, V.: Hybrid-electric propulsion for the retrofit of a slow-tourism passenger ship. In: 2020 International Symposium on Power Electronics, Electrical Drives, Automation and Motion (SPEEDAM), pp. 419–424. IEEE (2020). https://doi.org/10.1109/SPEEDAM48782.2020.9161920

15. Peri, D.: Robust design optimization for the refit of a cargo ship using real seagoing data. Ocean Eng. **123**, 103–115 (2016). https://doi.org/10.1016/j.oceaneng.2016.06.029

16. Steen, S., Selvik, Ø., Hassani, V.: Experience with rim-driven azimuthing thrusters on the research ship Gunnerus. In: Proceedings of High-Performance Marine Vessels (2016)

17. Yasukawa, H., Sakuno, R.: Application of the MMG method for the prediction of steady sailing condition and course stability of a ship under external disturbances. J. Mar. Sci. Technol. **25**(1), 196–220 (2019). https://doi.org/10.1007/s00773-019-00641-4

Co-simulation of a Model Predictive Control System for Automotive Applications

Cinzia Bernardeschi[1] , Pierpaolo Dini[1] , Andrea Domenici[1(✉)] ,
Ayoub Mouhagir[2] , Maurizio Palmieri[1] , Sergio Saponara[1] ,
Tanguy Sassolas[2] , and Lilia Zaourar[2]

[1] Department of Information Engineering, University of Pisa, Pisa, Italy
andrea.domenici@unipi.it
[2] Université Paris-Saclay, CEA, List, 91120 Palaiseau, France

Abstract. Designing a Model Predictive Control system requires an accurate analysis of the interplay among three main components: the plant, the control algorithm, and the processor where the algorithm is executed. A main objective of this analysis is determining if the controller running on the chosen hardware meets the time requirements and response time of the plant. The constraints, in turn, should be met with a satisfactory tradeoff between algorithm complexity and processor performance. To carry out these analyses for an autonomous vehicle control, this paper proposes to leverage parallel co-simulation between the plant, the model predictive controller and the processor.

Keywords: Model predictive control · Co-simulation · Autonomous vehicles

1 Introduction

Control algorithms based on *model predictive control* (MPC) are increasingly being employed in embedded systems with high-performance requirements and stringent constraints, such as automotive applications. MPC relies on the availability of a mathematical model of the controlled plant, used at each sampling period to evaluate a prediction of the plant's future behaviour over a given timespan (the *prediction horizon*) and choose optimal values for the control variables, to be applied at the next sampling period [13].

Designing an MPC system for embedded applications requires an accurate analysis of the interplay among three main components: the plant, the control

A. Domenici—This work has been partially supported by the European Processor Initiative (EPI) project, which has received funding from the European Union's Horizon 2020 research and innovation program under Grant Agreement № 826647, and by the Italian Ministry of Education and Research (MIUR) in the framework of the CrossLab project (Department of Excellence).

A. Cerone et al. (Eds.): SEFM 2021 Workshops, LNCS 13230, pp. 204–220, 2022.
https://doi.org/10.1007/978-3-031-12429-7_15

algorithm, and the processor where the algorithm is executed. With the model-based design approach, the analysis exploits the results of the simulations. In particular, a detailed simulation of the processing architecture executing the control software is needed, given the role of processor performance in meeting real-time constraints. This is a typical situation where co-simulation provides substantial support to developers who need to model subsystems from different areas of expertise: algorithms, processor architecture and plant physics.

This work presents an approach to enable the analysis of MPC systems through co-simulation. To this end, an open source library for MPC algorithm, GRAMPC [10], has been extended with the implementation of a standard interface for co-simulation, FMI (Functional Mock-up Interface) [5]. The MPC algorithm contains a *prediction* model of the plant that is distinct from the *actual* model. This allows the analysis of the controller under a variation of the actual model parameters. The approach also encompasses MPC performance analysis, thanks to the use of the VPSim [7] virtual prototyping tool for complex electronic Systems-on-Chip (SoC) from the SESAM framework [26], which supports FMI co-simulation.

The application of the proposed approach is shown in a case study from autonomous vehicle control, where the plant, the model predictive controller, and the processor are simulated in parallel. The plant is an autonomous car that must reach a destination along a road with a given geometry, avoiding obstacles. The vehicle is simulated with a standard kinematic model implemented in C with the GRAMPC framework, adapting an example from the GRAMPC distribution. The processor architecture is an ARMv8 multi-core processor, simulated in the VPSim framework. The metrics used to analyse different co-simulation runs are the difference of the actual trajectory from the reference one and the execution time of the GRAMPC algorithm, that should be less than the co-simulation step.

The paper is organised as follows: Sect. 2 introduces a selection of related works; background on MPC and the GRAMPC library is briefly reported in Sect. 3; Sect. 4 illustrates the proposed approach for multi-model simulation of automotive systems, while Sect. 5 shows the application to a case study from autonomous driving; finally, Sect. 6 contains conclusions and further work.

2   Related Work

Among the many works available to readers looking for an extensive background on wheeled vehicles dynamics, we may cite [14]. More specifically, Yurtsever et al. [28] provide a survey on recent work about autonomous driving. Also the literature on model predictive control offers many fundamental texts, e.g. [13].

In model-driven development, co-simulation [11] can be applied in the analysis of complex cyber-physical systems that integrate a high-level control algorithm with pre-existing closed implementations of lower-level plant dynamics.

Lee et al. [17] report on the co-simulation of an MPC-controlled heating, ventilation and air-conditioning plant, using an ad-hoc Python-based infrastructure to connect a building simulator with a Matlab controller. Similar ad-hoc solutions have been proposed in several works, e.g., von Wissel et al. [27], who use

Simulink S-functions to connect a powertrain model developed on the Siemens LMS Amesim simulator with an MPC controller developed in the Honeywell OnRAMP environment. Using S-functions to couple a Simulink model to different simulators is a common technique, used, e.g., in [4], where a Simulink model of a human heart was coupled to an executable formal model of a pacemaker.

The Functional Mockup Interface 2.0 [5] is a *de facto* standard for co-simulation, and INTO-CPS [16] is an integrated tool chain for model-based design based on FMI.

An FMI infrastructure based on the TISC co-simulation platform was presented by Gräber et al. [12]. In their work, FMUs simulate the plant, an optimizer, and a system estimator. The plant can be modeled with different tools, the optimizer is built with the MUSCOD-II software package using a direct multiple shooting method. A vapor compression cycle is discussed as an application example. An FMI-based infrastructure was used by Ceusters et al. [6], who generate an FMU from a Modelica simulator of multi-energy systems, and use it to communicate with a Python-based environment that models two alternative controllers, one based on MPC and one on reinforcement learning. Another FMI-based framework for co-simulation of human-machine interfaces was presented in [21]. Co-simulation has been paired with formal methods to validate and verify control systems of various kinds [2,3], including robot vehicles [9,20].

In the automotive field, the interaction between multi-physics modelling/simulation environments and embedded software development environments has been addressed by many works. Recently, the eFMI (FMI for embedded systems) standard has been proposed as a result of the EMPHYSIS (Embedded systems with physical models in the production code software) project [19].

3   Model Predictive Control and the GRAMPC Framework

This section introduces a very succint description of the concept of model predictive control and of the GRAMPC framework [10] for the simulation of MPC systems.

3.1   Model Predictive Control

A model predictive control system iteratively solves an Optimal Control Problem (OCP) of the following form [10], where $t \in [0, T]$ is the MPC-internal time coordinate and T is the prediction horizon:

$$\min_{u} J(u, x_k) = V(x(T)) + \int_0^T l(x, u, \tau)d\tau \qquad (1)$$

$$M\dot{x} = f(x, u, t_k + \tau) \qquad (2)$$

$$x(0) = x_k \qquad (3)$$
$$x(\tau) \in [x_{\min}, x_{\max}] \qquad (4)$$
$$x(T) \in \Omega_\beta \qquad (5)$$
$$u(\tau) \in [u_{\min}, u_{\max}], \qquad (6)$$

where (1) is the cost functional, which depends on the time evolution of the control variables' vector u and of the sampled state variables' vector x_k. The first term $(V(x(T)))$ of the cost functional represents the *terminal* cost associated with the final state at the end of the prediction horizon, while the second term represents the *integral* cost computed over the whole trajectory over the prediction horizon. The system dynamics are expressed by (2), where the mass matrix M defines the inertial properties of the system, and $t_k = t_0 + k\Delta t$, with $0 < \Delta t < T$, is the k-th sampling instant.

The state of the system at the beginning of the k-th control interval is given by (3). The remaining relations express constraints on the state and the control inputs. In particular, Ω_β is the set of states such that the terminal cost is less than or equal to β. This constraint is typically used to ensure stability.

The controller computes the trajectory of control variables that minimizes (1), and its first segment of simulated length Δt is applied as a plant input during the actual (real-time) control period $[t_k, t_{k+1})$.

A common form for the cost functional uses quadratic norms of the form

$$V(x) = \|x - x_{\text{des}}\|_P^2 \qquad (7)$$
$$l(x, u) = \|x - x_{\text{des}}\|_Q^2 + \|u - u_{\text{des}}\|_R^2, \qquad (8)$$

where $(x_{\text{des}}, u_{\text{des}})$ is the desired set-point and the norms are weighted by the positive (semi-)definite matrices P, Q and R, defined according to the application.

3.2   The GRAMPC Framework

The GRAMPC (Gradient-Based MPC) framework supports simulation of nonlinear systems under MPC by providing a highly configurable optimization algorithm. Users must supply a model of the plant to be simulated, by coding a set of C-language functions implementing well-defined, yet flexible interfaces. Users also set options and parameters to customize the optimization and simulation algorithms. In particular, a user may choose one of a set of available solvers for the optimizer.

The optimization algorithm implements an augmented Lagrangian method, based on the gradient-descent paradigm, also exploited in other adaptive [8] and learning [15] control techniques for modern mechatronic systems. Such an algorithm, at each iteration, requires the evaluation of the plant's dynamics (for the prediction), of the cost functional, of their partial derivatives w.r.t. state and control variables, and of the constraint relations. As mentioned above, all these computations are specified by the user with C functions. For example, function ffct computes the plant dynamics, Vfct and lfct compute the terminal and integral cost, respectively, dldx computes the gradient of the integral cost w.r.t. the state variables, and hfct checks the inequality constraints.

The implementation of the optimization and simulation algorithm maintains a structure (grampc) that contains all the information of the problem at hand (including the current state of the plant, the values of the command variables, the time). A function named grampc_run takes the grampc structure as an input and executes a step of the MPC algorithm, updating grampc.

In order to execute a simulation, a user writes a source file with the functions modeling the system (ffct etc.) and a file with a main program where parameters and options are initialized. Then, a loop starts, invoking grampc_run at each iteration. The resulting values of control and state variables can be further processed and printed out.

Figure 1 summarises how a simulation is built on top of the GRAMPC framework. The framework is composed of a library providing the implementations of the core MPC functions, and the declarations of the interfaces to be implemented by the user, who provides the problem formulation in two source files, one with the prediction plant model and the other with the initialization and the main loop. It may be observed that, in the GRAMPC framework, it is not possible to simulate the controlled plant and the controller separately, since the controlled plant model coincides with the prediction plant model used in the controller.

4   Proposed Approach

This work is based on the idea of embedding a GRAMPC model into an FMU written in C where the time advancing function fmi2DoStep invokes the grampc_run function. The control values updated in grampc are forwarded as FMI output variables and the current state stored in grampc is overwritten by the FMI input variables as shown in Fig. 2. This is achieved by exploiting a preexistent FMU generator such as [18] or [22] to create the basic FMU structure that should be compiled together with the whole GRAMPC library source files and the two GRAMPC files of the system at hand.

The file with the model of the plant does not require changes, while the file with the implementation of the algorithm requires some minor changes: the initialization of the GRAMPC parameters should be wrapped into a function that will be invoked by the FMU initialization function fmi2SetupExperiment, while the code for executing the MPC algorithm should be wrapped in a function that will be invoked by the fmi2DoStep function. Finally, the values stored in

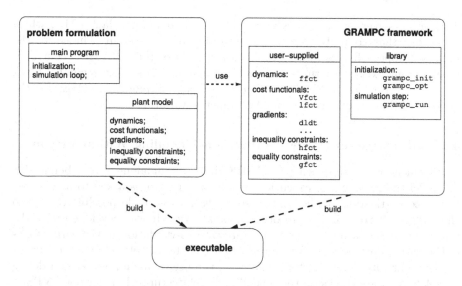

Fig. 1. GRAMPC library schema.

the `grampc` structure should be linked to the buffers where the FMU variables are stored.

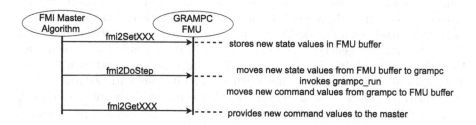

Fig. 2. Interaction between Master and GRAMPC.

4.1   Advantages of GRAMPC as an FMU

The benefits deriving from the proposed approach are related to the general advantages of an FMI based co-simulation, i.e., the possibility of easily coupling the GRAMPC controller with other tools such as Simulink or OpenModelica for the controlled plant component. In particular, the GRAMPC FMU may require a simple model for the predicted trajectory, while a more complex and accurate model can be created for the controlled plant in another FMU, using tools that fit the problem domain. Moreover, the time required to run a co-simulation can be easily reduced by exploiting a simple numerical integration solver (e.g., a Euler solver) in the GRAMPC FMU while a more accurate and computation demanding solver (e.g., Runge-Kutta) is only used in the plant FMU, which is the one that computes the actual evolution of the system. Thanks to the proposed approach it is also possible to run tests of the GRAMPC controller by exploiting existing features such as the Simulink white noise generator block for sensor errors, or the INTO-CPS Design Space Exploration (DSE) for the analysis of the behaviour with small parametric variations.

4.2   Advantages of Hardware Platform Modelling Within VPSim

Complementary to separating GRAMPC into a standalone FMU, being able to model the `grampc_run` execution on a real hardware is a key to assess performance bottlenecks of the control strategy. This is made possible thanks to the VPSim [7] SoC virtual prototyping capabilities. It was developed with the purpose of accelerating the software/hardware co-validation in the early stages of the design development. VPSim makes it easy to model and emulate various hardware architectures. At the same time, the user can simply test and debug complete software stacks on these emulated architectures. Furthermore, VPSim is distinguished by its ability to host third-party subsystems using many standard and non-standard interfaces. In particular, it fully supports the FMI standard [24]. Therefore, it can interface easily with other modelling tools and simulators within an FMI-based co-simulation. From the user view, FMI in VPSim is exposed as a proxy component that must be connected to a compatible hardware communication interface, such as CAN bus or I2C slave. In addition, VPSim proposes a user-friendly method for automatic generation of the virtual platform FMU, based on a high-level description of the hardware/software platform. Figure 3 shows the general architecture of an FMI co-simulation involving an FMU with a GRAMPC model executed on a processor emulated with VPSim. The deployment of GRAMPC on a simulated architecture with VPSim enables (i) evaluating the behavior of `grampc_run` on the target hardware architecture, (ii) identifying the best hardware support for the control code, and (iii) devising software improvement strategies such as parallel implementation.

5   Case Study

The specific case study concerns the autonomous driving of a vehicle modelled for simplicity by kinematics laws through the GRAMPC library. The problem addressed is to follow a sinusoidal trajectory that follows the carriageway, avoiding some obstacles (modelled as circular areas). Figure 4 shows a possible trajectory of a car avoiding an obstacle (red circle).

The case study is taken from an example available in the GRAMPC distribution. Obviously, in a realistic case, the MPC control algorithm, which is a low-level control, will have to be integrated with the vision and decision system for the waypoints to be reached at each iteration. This information is assumed to be given as input to the system statically, at the beginning of the simulation, and the vision system will not be considered.

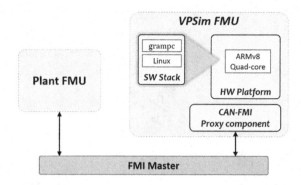

Fig. 3. The architecture for a co-simulation with VPSim FMU.

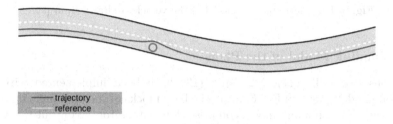

Fig. 4. Example of a trajectory.

5.1   Vehicle Model

The model used for the case study is the kinematic bicycle model of the vehicle shown in Fig. 5, adapted from [23]. This model, commonly used in the field of MPC, approximates a four-wheel vehicle by replacing the two wheels of each axle with one wheel on the longitudinal axis.

The kinematic behaviour of the model is described by the following equations, where the control command inputs are the acceleration a (m/s^2) and the front wheel steering angle δ (rad).

$$\dot{x} = V\cos(\psi + \beta(\delta)) \tag{9}$$

$$\dot{y} = V\sin(\psi + \beta(\delta)) \tag{10}$$

$$\dot{V} = a \tag{11}$$

$$\dot{\psi} = \frac{V}{l_r + l_f}\cos\left(\beta(\delta)\right)\tan\left(\delta\right) \tag{12}$$

Angle β is the slip angle at the centre of gravity G and it is described by Eq. (13), where l_r and l_f are the distances from G of the rear and front wheel, respectively.

$$\beta(\delta) = \arctan\left(\tan(\delta)\frac{l_r}{l_r + l_f}\right) \tag{13}$$

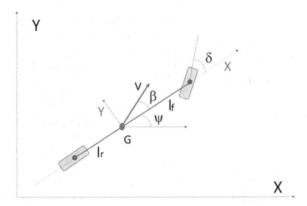

Fig. 5. Kinematic bicycle model of the vehicle, redrawn from [23].

5.2 Simulink Model of the Plant

Equations (9), (10), (11), (12), and (13) have been implemented with the Simulink model shown in Fig. 6, using the base blocks of the Simulink trigonometry library. The Simulink model comprises four integrators to output the actual values of the variables. The initial state of these integrators corresponds to the initial values of the plant's state variables.

The Simulink environment generates an FMU whose model parameters (such as l_r, l_f, and the initial state) can be set in the INTO-CPS co-simulation environment. The Simulink environment also chooses the ode45 variable step size and the default parameters for the explicit Runge-Kutta integrator.

5.3 Vehicle and Controller in GRAMPC

The model of the vehicle in GRAMPC is shown in Listing 1.1 and matches the equations shown in Sect. 5.1, using the notation of C. The same model is used to solve the optimisation problem and for executing a self-contained simulation in the framework. Listing 1.2 shows the four optimisation constraints:

```
double beta = ATAN(param[18]*TAN(u[0])/(param[18]+
param[19]));
out[0] = COS(x[2]+beta)*x[3];
out[1] = SIN(x[2]+beta)*x[3];
out[2] = x[3]*COS(beta)*TAN(u[0])/(param[18]+param
[19]);
out[3] = u[1];
```

Listing 1.1. Implementation of the vehicle model in GRAMPC.

– out[0] on line 1 represents the constraint for obstacle avoidance and is represented by a circle of radius 1 located at (50, −0.2) in the XY plane.

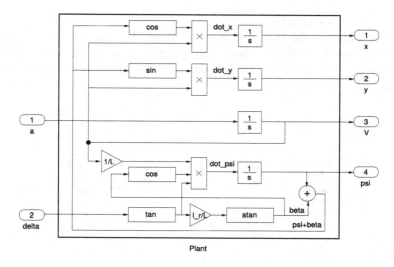

Fig. 6. Simulink model of the plant. Parameter L equals $l_r + l_f$.

- out[1] on line 2 and out[2] on line 3 are the constraints representing the edges of the road, represented by two sinusoids,
- out[3] on line 4 represents the speed limit that the vehicle must respect.

```
1   out[0] = (2 - POW2(-50 + x[0]) - POW2(( 0.2 + x[1]))
    );
2   out[1] = -x[1] + 4*SIN(2 * pi * 0.01 * x[0])-1.5;
3   out[2] =  x[1] - 4*SIN(2 * pi * 0.01 * x[0])-4.5;
4   out[3] = x[3] - 40;
```

Listing 1.2. Definition of the constraints in GRAMPC.

```
1  State* tick(State* st) {
2    grampc->sol->xnext[0] = (typeRNum)st->x;
3    grampc->sol->xnext[1] = (typeRNum)st->y;
4    grampc->sol->xnext[3] = (typeRNum)st->V;
5    grampc->sol->xnext[2] = (typeRNum)st->psi;
6    grampc_setparam_real_vector(grampc,"x0",grampc->sol->
       xnext);
7    grampc_run(grampc);
8    t = t + grampc->param->dt;
9    st->a = grampc->sol->unext[1];
10   st->delta = grampc->sol->unext[0];
11 }
```

Listing 1.3. Algorithm evolution in GRAMPC.

Listing 1.3 shows the custom function tick, called by the master every co-simulation step through the fmi2Dostep function: lines 3–7 move the values received from the controlled plant to grampc; lines 9–10 invoke the execution of grampc and increase the time variable by dt; lines 12–13 save the newly generated commands. The co-simulation architecture is shown in Fig. 7, with the INTO-CPS Co-simulation Orchestration Engine (COE) playing the role of the FMI master algorithm.

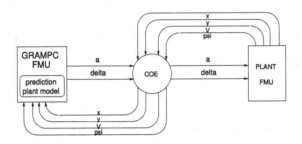

Fig. 7. Co-simulation architecture of the case study.

5.4   Hardware Platform with VPSim

VPSim can simulate a large variety of architectures using both its integrated models and external model providers such as QEMU [1], ARM fast models, or open virtual platforms. In the context of this paper, as shown in Fig. 3, VPSim emulates a quad-core ARMv8 64-bit processor architecture using QEMU. Each core has private L1 & L2 caches. All the cores share four slices of LLC banks, which are connected to the NoC and peripheral devices. The platform runs a Linux OS which executes the GRAMPC algorithm. A CAN controller model provides FMI interfaces and makes it possible to receive and transmit control I/O data to and from the grampc_run application that uses the SocketCAN API [25] to retrieve them, as would be the case on real hardware.

It must be stated that a real-time OS could be supported by the proposed methodology. It would be required for industrial development and validation to ensure the periodic scheduling of the grampc_run function. Yet, using a standard Linux - executing a single application triggered by CAN events - is relevant for the exploration of the control strategy while accounting for potential execution performance bottlenecks. Indeed, if grampc_run executes in less time than the period of CAN messages, it is then periodically executed. Otherwise, it will fail to process all incoming messages and meet its deadlines, as would be the case when considering an RTOS.

5.5   Results

The GRAMPC framework uses a single model as the prediction model, needed for MPC optimization, and as the controlled plant model. While this choice

Table 1. Parameter values.

Parameter	Value	
δ_{min}	-0.5	rad
δ_{max}	$+0.5$	rad
a_{min}	-11.2	m/s^2
a_{max}	$+5.34$	m/s^2
Front track l_f	1.670	m
Rear track l_r	1.394	m
Time horizon	1	s
MPC solver	Euler	

is often convenient, it may be the case that an embedded application must use a prediction model simplified with respect to the controlled plant model. In such cases, co-simulation makes it possible to use two distinct models. As a preliminary step towards the co-simulation of distinct models, in this work it has been checked that co-simulation does not introduce significant deviations from the case of GRAMPC simulation with a single model. In order to verify the consistency between the two simulation methods, the same mathematical vehicle model has been implemented in C for the prediction model, and in Simulink for the plant model. The results of co-simulations are consistent with the results of self-contained simulations in GRAMPC, producing a difference less than 1 mm.

This section reports results of co-simulations in case of the decoupling of the two models, making the co-simulated system more realistic with respect to the GRAMPC self-contained one. The main parameters of the analyzed scenarios are shown in Table 1.

Nominal Co-simulation Results. The vehicle starts at the position (0,0) and must follow the sinusoidal trajectory avoiding the obstacle at (50, -0.2). Figure 8 shows a run with a fixed step size of 0.001 s and an end time of 20 s.

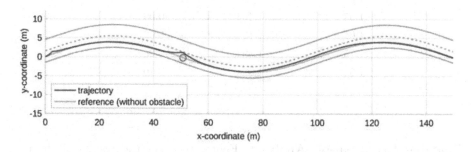

Fig. 8. Results of a co-simulation run.

The maximum computation time of the GRAMPC algorithm is less than a millisecond and the simulation time is 50 s on an Intel® Core™ i7-7700 CPU @ 3.60 GHz × 8. The choice of the Euler solver inside the MPC model, together with the GRAMPC setup, leads to a computation time less than the co-simulation step-size, which guarantees a realtime-like throughput. Notoriously, the Euler solver is computationally less demanding than other solvers.

As shown in Fig. 8, the vehicle follows the trajectory avoiding the obstacle. The mean error of the actual trajectory against the reference trajectory without the obstacle is 0.08 m (first row in Table 2) and the maximum absolute error is 0.76 m, both evaluated excluding the area around the obstacle. With respect to the limits imposed on the constrained optimisation problem, as far as the obstacle avoidance section is concerned, it was verified that the trajectory calculated by GRAMPC is such that the centre of mass of the vehicle completely avoids the obstacle. This translates into verifying that the distance between the trajectory and the centre of the obstacle is always greater than the radius of the circle that formally defines the obstacle region itself. The closest distance between the obstacle and the centre of mass of the vehicle is 0.4 m.

Response to Physical Parameter Variation. By exploiting the decoupling of the model used within GRAMPC and the model used for plant in Simulink it is possible to run robustness tests against small variations of the physical parameters of the vehicle under analysis. Table 2 shows the four different scenarios where the parameters l_f and l_r of the Simulink model have a ±5% deviation with respect to the nominal values used in the first experiment, reported in the first row of the table. This variation of the parameters emulates a reasonable measurement error. It may be stated that the GRAMPC algorithm is robust as the mean error is scarcely affected by physical changes. Moreover, the maximum absolute error is not affected by physical variations and therefore it is not shown.

Table 2. Different parameters.

Front track l_l (m)	Rear track l_r (m)	Error (m)
1.67	1.394	0.08
1.67	1.464	0.09
1.67	1.324	0.09
1.75	1.394	0.08
1.59	1.394	0.10

Enabling Perturbation Analysis. Figure 10 shows the results of the co-simulation with a perturbation in the y and ψ values produced by the plant FMU and consumed by the GRAMPC algorithm. The perturbation has been

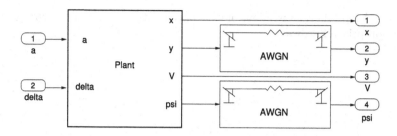

Fig. 9. Simulink model of the plant with AWGN.

implemented with the Simulink AWGN block, which has been applied to the y coordinate with a variance of 0.1 (10 cm of measurement error) and to angle ψ with a variance of 0.01 (1° of measurement error), obtaining the model in Fig. 9. As shown in Fig. 10, the vehicle is still capable of avoiding the obstacle but the error has increased to 0.240 m and the maximum absolute error has increased to 1.58 m. The framework can be used for perturbation analysis by considering more cases.

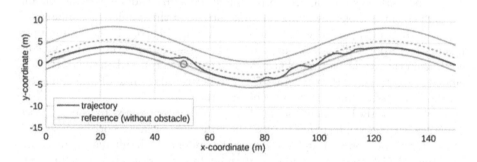

Fig. 10. Results of a co-simulation run with AWGN on sensors.

Co-simulation with VPSim. As a benefit of the proposed approach, it is possible to generate the FMU that executes the grampc_run algorithm on a specific hardware platform. In the following, we show the nominal co-simulation in case of GRAMPC executed on top of an ARMv8 quad-core processor emulated with VPSim. With respect to Fig. 7, the FMU generated with VPSim can replace the GRAMPC FMU. The results, shown in Fig. 11, present a mean error of 0.262 m, while the vehicle is still able to avoid the obstacle. This increase in the mean error is consistent with the fact that the average execution time of grampc_run (2.4 ms) is longer than the expected co-simulation stepsize (1 ms). From this point on, several improvement strategies to the grampc_run implementation could be sought by the designer such as parallelizing the algorithm, changing the prediction window or the optimization solver, or even choosing more appropriate hardware.

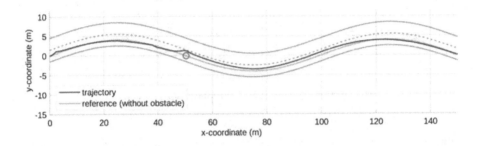

Fig. 11. Results of a co-simulation run with VPSim.

6   Conclusions

This paper has shown an approach to enable the analysis of an MPC algorithm through co-simulations involving relevant aspects of three different domains: (i) physical laws, defining the evolution of the system, (ii) control algorithm, optimising the response of the system, and (iii) processor architecture imposing constraints on the execution time. The proposed approach is based on the strategy of embedding into an FMU a GRAMPC controller running on a VPSim-simulated processor to assess the performance of the processor.

The case study has shown some of the possibilities opened by the proposed approach, such as the response to physical parameter variations or the evaluation of the impact of processor architecture on the system. Each different analysis can be extended with knowledge and tools deriving from the respective domain. Users expert in parallel computation could improve the architecture performances by optimising the code, users expert in fault tolerance could decrease the impact of faulty sensors by applying, for example, redundancy in the plant model. Finally, experts in MPC could be interested in finding the best trade-off between accuracy and performance. Working all together on the same artefacts, and combining effort in different fields would lead towards the implementation of optimal and robust systems.

As further work, in order to get better results in the parallelisation of MPC, it could be also interesting to investigate the usage of a quite different structure of the control algorithm allowing a more effective parallelism.

References

1. Bellard, F.: QEMU, a fast and portable dynamic translator. In: 2005 USENIX Annual Technical Conference (USENIX ATC 05). USENIX Association, Anaheim, CA (2005)

2. Bernardeschi, C., et al.: Cross-level co-simulation and verification of an automatic transmission control on embedded processor. In: Cleophas, L., Massink, M. (eds.) Software Engineering and Formal Methods. SEFM 2020 Collocated Workshops. LNCS, vol. 12524, pp. 263–279. Springer, Cham (2021). https://doi.org/10.1007/978-3-030-67220-1_20

3. Bernardeschi, C., Dini, P., Domenici, A., Palmieri, M., Saponara, S.: Formal verification and co-simulation in the design of a synchronous motor control algorithm. Energies **13**(16), 4057 (2020). https://doi.org/10.3390/en13164057

4. Bernardeschi, C., Domenici, A., Masci, P.: A PVS-simulink integrated environment for model-based analysis of cyber-physical systems. IEEE Trans. Soft. Eng. **44**(6), 512–533 (2018). https://doi.org/10.1109/TSE.2017.2694423

5. Blochwitz, T., et al.: Functional mockup interface 2.0: the standard for tool independent exchange of simulation models. In: Proceedings of the 9th International MODELICA Conference, pp. 173–184, no. 76 in Linköping Electronic Conference Proceedings, Linköping University Electronic Press (2012). https://doi.org/10.3384/ecp12076173

6. Ceusters, G., et al.: Model-predictive control and reinforcement learning in multi-energy system case studies. In: eprint 2104.09785 (2021)

7. Charif, A., Busnot, G., Mameesh, R.H., Sassolas, T., Ventroux, N.: Fast virtual prototyping for embedded computing systems design and exploration. In: Chillet, D. (ed.) Proceedings of the Rapid Simulation and Performance Evaluation: Methods and Tools, RAPIDO 2019, Valencia, 2019, pp. 3:1–3:8. ACM (2019). https://doi.org/10.1145/3300189.3300192

8. Dini, P., Saponara, S.: Design of adaptive controller exploiting learning concepts applied to a BLDC-based drive system. Energies **13**(10), 2512 (2020). https://doi.org/10.3390/en13102512

9. Domenici, A., Fagiolini, A., Palmieri, M.: Integrated simulation and formal verification of a simple autonomous vehicle. In: Cerone, A., Roveri, M. (eds.) Software Engineering and Formal Methods. LNCS, vol. 10729, pp. 300–314. Springer Int. Publishing, Cham (2018). https://doi.org/10.1007/978-3-319-74781-1_21

10. Englert, T., Völz, A., Mesmer, F., Rhein, S., Graiche, K.: A software framework for embedded nonlinear model predictive control using a gradient-based augmented lagrangian approach (GRAMPC). Optim. Eng. **20**, 769–809 (2019). https://doi.org/10.1007/s11081-018-9417-2

11. Gomes, C., Thule, C., Broman, D., Larsen, P.G., Vangheluwe, H.: Co-simulation: a survey. ACM Comput. Surv. (CSUR) **51**(3), 1–33 (2018). https://doi.org/10.1145/3179993

12. Gräber, M., Kirches, C., Scharff, D., Tegethoff, W.: Using functional mock-up units for nonlinear model predictive control. In: 9th International MODELICA Conference vol. 076, pp. 781–790 (2012). https://doi.org/10.3384/ecp12076781

13. Grüne, L., Pannek, J.: Nonlinear Model Predictive Control - Theory and Algorithms. Springer, London (2017). https://doi.org/10.1007/978-0-85729-501-9

14. Guiggiani, M.: The Science of Vehicle Dynamics - Handling, Braking, and Ride of Road and Race Cars. Springer, Dordrecht (2018). https://doi.org/10.1007/978-94-017-8533-4

15. He, W., Gao, H., Zhou, C., Yang, C., Li, Z.: Reinforcement learning control of a flexible two-link manipulator: an experimental investigation. IEEE Trans. Syst. Man Cybern. Syst. **51**(12), 7326–7336 (2020). https://doi.org/10.1109/TSMC.2020.2975232

16. Larsen, P.G., et al.: Integrated tool chain for model-based design of cyber-physical systems: the INTO-CPS project. In: Modelling, Analysis, and Control of Complex CPS (CPS Data), 2016 2nd International Workshop on, pp. 1–6. IEEE (2016)

17. Lee, D., Lim, M.C., Negash, L., Choi, H.L.: EPPY based building co-simulation for model predictive control of HVAC optimization. In: 2018 18th International Conference on Control, Automation and Systems (ICCAS), pp. 1051–1055 (2018)

18. Legaard, C., Tola, D., Schranz, T., Macedo, H., Larsen, P.: A universal mechanism for implementing functional mock-up units. In: Proceedings of the 11th International Conference on Simulation and Modeling Methodologies, Technologies and Applications - SIMULTECH, pp. 121–129. SciTePress (2021)

19. Lenord, O., et al.: eFMI: an open standard for physical models in embedded software. In: Proceedings of the 14th International Modelica Conference 2021, pp. 57–71. Modelica Association (2021). https://doi.org/10.3384/ecp2118157

20. Palmieri, M., Bernardeschi, C., Masci, P.: Co-simulation of semi-autonomous systems: the Line Follower Robot case study. In: Cerone, A., Roveri, M. (eds.) Software Engineering and Formal Methods. LNCS, vol. 10729, pp. 423–437. Springer, Cham (2018). https://doi.org/10.1007/978-3-319-74781-1_29

21. Palmieri, M., Bernardeschi, C., Masci, P.: A framework for FMI-based co-simulation of human-machine interfaces. Softw. Syst. Model. **19**(3), 601–623 (2020). https://doi.org/10.1007/s10270-019-00754-9

22. Palmieri, M., Macedo, H.D.: Automatic generation of functional mock-up units from formal specifications. In: Camara, J., Steffen, M. (eds.) Software Engineering and Formal Methods, pp. 27–33. Springer, Cham (2020). https://doi.org/10.1007/978-3-030-57506-9_3

23. Polack, P., Altché, F., Novel, B., de La Fortelle, A.: The kinematic bicycle model: a consistent model for planning feasible trajectories for autonomous vehicles? In: 2017 IEEE Intelligent Vehicles Symposium (IV), pp. 812–818 (2017). https://doi.org/10.1109/IVS.2017.7995816

24. Saidi, S.E., Charif, A., Sassolas, T., Guay, P.G., Souza, H., Ventroux, N.: Fast virtual prototyping of cyber-physical systems using SystemC and FMI: ADAS use case. In: Proceedings of the 30th International Workshop on Rapid System Prototyping (RSP 2019), pp. 43–49. Association for Computing Machinery, USA (2019). https://doi.org/10.1145/3339985.3358488

25. SocketCAN (2020). https://www.kernel.org/doc/html/latest/networking/can.html

26. Ventroux, N., et al.: SESAM: an MPSoC simulation environment for dynamic application processing. In: 2010 10th IEEE International Conference on Computer and Information Technology, pp. 1880–1886 (2010). https://doi.org/10.1109/CIT.2010.322

27. Von Wissel, D., Talon, V., Thomas, V., Grangier, B., Lansky, L., Uchanski, M.: Linking model predictive control (MPC) and system simulation tools to support automotive system architecture choices. In: 8th European Congress on Embedded Real Time Software and Systems (ERTS 2016), France (2016)

28. Yurtsever, E., Lambert, J., Carballo, A., Takeda, K.: A survey of autonomous driving: common practices and emerging technologies. IEEE Access **8**, 58443–58469 (2020). https://doi.org/10.1109/ACCESS.2020.2983149

Running Large-Scale and Hybrid Real-Time Aircraft Simulations in an HLA Framework

Jean-Baptiste Chaudron$^{(\boxtimes)}$ (ID), Aleksandar Joksimović (ID), Pierre Siron (ID), Rob Vingerhoeds (ID), and Xavier Carbonneau (ID)

ISAE-SUPAERO, Université de Toulouse, Toulouse, France
{jean-baptiste.chaudron,aleksandar.joksimovic,pierre.siron,
rob.vingerhoeds,xavier.carbonneau}@isae-supaero.fr

Abstract. Aircraft systems and associated technologies have been constantly evolving over the years with an increasing number of components and actors interacting together in a global chain. Given this ever-growing complexity, the use of simulation in the design and validation phases of emerging systems is essential and allows to significantly accelerate some prototyping phases, especially in cross-domain concepts. This article describes analysis and experiments done at ISAE-SUPAERO for running large scale and hybrid real-time simulations for aircraft using the HLA standard (High Level Architecture).

Keywords: Aircraft simulation · Distributed systems · Co-simulation · HLA

1 Introduction

The complexity of aircraft systems in their environment makes it crucial to analyse and understand their characteristics and behavior early in the design phases. Modeling and simulation (M&S) play an important role in aircraft systems development, where simulation models are used for prototyping and evaluating systems in order to identify potential problems or even misconceptions and allow for decisions at all development stages. Aircraft sub-systems simulation targets cross-domain, multi-scale, multi-physics and multi-fidelity models. The high increase of computer processing power enables many possibilities such as high fidelity simulations. Moreover, emerging computer network technologies, especially the rise of Ethernet technology from low [21] to high speeds [14], have led to distributed computing, where different computers are interacting over a communication network to achieve a common goal, thus providing more power and efficiency to solve very complex problems. However, such configurations come with a cost and designers of distributed applications have had to face several problems such as handling the heterogeneity of the hardware and software distributed components. The efforts to tackle these interoperability issues have led to the development of multiple standards such as Common Object Request

A. Cerone et al. (Eds.): SEFM 2021 Workshops, LNCS 13230, pp. 221–237, 2022.
https://doi.org/10.1007/978-3-031-12429-7_16

Broker Architecture (CORBA) [26], Data Distribution Service (DDS) [26], Functional Mock-up Interface (FMI) [22] or the non-standardised approach proposed in Robotic Operating System (ROS) [10,11,27]. Such middleware standards provide the following advantages:

- *Interoperability*: defining a common and shared framework for the connected systems to interact together and exchange data,
- *Modularity*: easing the operation of adding or replacing entities, and
- *Distribution*: allowing the processing entities to be located on different computers on a communication network.

For simulation purposes, the specialised High-Level Architecture (HLA) standard [31–33] has been developed to provide a shared and documented framework for interoperability and reusability of heterogeneous distributed simulation. Although interoperability and reusability remain essential, both high performance and availability have also to be considered to fulfill the requirements of the simulation. For many years now, we have been using our implementation of the HLA standard (called CERTI) to develop and maintain hybrid real-time aircraft simulations and allow testing of real hardware components within the simulation loop. This article presents our framework using the HLA standard for aircraft simulations, its architecture and particularities, and an open-source implementation. An aero-propulsive aircraft/engine test scenario for distributed simulation from this open source implementation is presented as illustration. The remainder of this paper is structured as follows:

- Section 2 presents an overview of our HLA simulation framework.
- Section 3 continues with an outline of the application architecture and its particularities.
- Section 4 describes the engine models for the case-study.
- Section 5 discusses the experiments and the obtained results.
- Finally, the paper is wrapped up with some conclusions and perspectives for future work.

2   An HLA Based Framework

2.1   CERTI

In distributed systems, "middleware" is the term used to describe a software layer implementing the services and framework (defined in a standard) to connect the distributed entities. For HLA, the middleware is referred to as the Run-Time Infrastructure (RTI) and implements the HLA Interface Specification part of the standard [31]. Thus, it provides a set of software services to all the connected HLA-compliant applications called "HLA federates". As depicted in Fig. 1, the RTI can be seen as the intermediary layer allowing the global execution of the distributed simulation and hiding the complexity of the underlying distributed system. In reality, each computers embed a local component handling the services implementation.

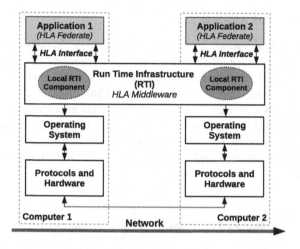

Fig. 1. HLA/RTI as middleware for distributed simulations.

For years, ISAE-SUPAERO has been developing and maintaining an RTI called CERTI. This project started in 1996 [30] and is available as open-source software [25]. CERTI is recognizable through its original architecture of communicating processes implemented in C++. Each HLA-compliant simulation (called a "federate") is connected to a local RTI Ambassador (RTIA) process and linked to a library (libRTI) providing the HLA interface services. A global shared process called "RTI Gateway" (RTIG) acts as a software-based router and connects all available RTIAs. Each federate process interacts locally with its RTIA via a Unix-domain socket and the RTIA processes exchange messages over the network, in particular with the RTIG process via Transmission Control Protocol (TCP) sockets. This architecture is depicted in Fig. 2.

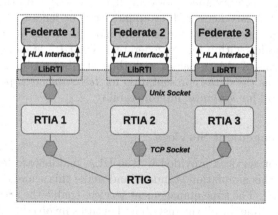

Fig. 2. CERTI architecture.

2.2   Federates and Models

When building a simulation, the studied system and its environment can be considered as a set of components with various interactions between them. From CERTI point of view, we must distinguish the model for each simulation component and the federate process (a C++ code) that integrates and implements one or more models (see Fig. 3). One way to design and to build a distributed simulation (i.e. a HLA federation) is to associate a federate only one component model. A federate may also encompass several models for the sake of simplicity of the overall architecture or to enable a performance optimization of the global simulation [12].

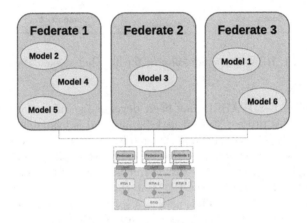

Fig. 3. Illustration of three federates with different models.

The interoperability between the individual models within one federate belongs to the federate designer. However, it is necessary to describe the HLA interfaces of the federate i.e. the data updated by the federate model(s) to the RTI and the data consumed by the federate model(s) from the RTI (data provided by other models within remote federates). The HLA standard requires this description in terms of object classes with their attributes plus some additional information such as the data representation for *technical* interoperability or the unit and referential for *semantic* interoperability.

2.3   Time Management Services

Time management services provided by the HLA middleware form one of the main benefits of this simulation standard and make this standard special compared to others [15]. The time management services enable the access to a consistent global logical time for all the distributed entities involved in the simulation. A time-stamp is assigned to each message and the RTI must ensure that all the messages are delivered with respect to time-stamp order. Therefore no message is allowed to be delivered to a federate in its past (from the logicial time scale point of view). This allows to have a deterministic and reproducible simulation which respects all the computing dependencies expressed with a logical time ordering. Different algorithms can be used to implement time management services and provide this consistent logical baseline. The most used algorithm is the so-called Null Message Algorithm (NMA) from Chandy and Misra [4], implemented in CERTI. The approach is based on a contract for each federate called "looka-head". Each federate undertakes not to send simulation messages with a logical timestamp less than its local time plus its lookahead. This algorithm is suitable for real-time and we have proposed some formal analysis [5] as well a new optimized version, called Null Message Prime (NMP) [9]. In addition, we have shown that these algorithms can be particularly useful when solving distributed differential equations such as in a complex aircraft distributed simulation [8].

3   SMARTIES Project

3.1   Background

In the context of the Research Platform for Embedded Systems Engineering Project (PRISE), we developed a first version of an aircraft flight simulation [7,16], called SDSE (French acronym for Distributed Simulation of Embedded Systems). However, although relevant and functional, this application had two disadvantages:

1. The aerodynamic model of the aircraft (i.e. the definition of its aerodynamic coefficients) as well as the propulsive system model were basic and their realisation could not be validated against real-life data nor any publication.
2. The part concerning the autopilot and flight control systems were integrated in a HLA simulator, but not integrated on a real dedicated on-board system (i.e. not a simulated system). Therefore, this HLA simulation was not a real Hardware-In-the-Loop (HIL) simulation.

Based on these findings, a new project, called Simulation Modules for Aircraft Real-TIme Embedded Systems (SMARTIES), was initiated. In order, to address the first limitation of SDSE, two well-documented flight dynamics models with available open source references have been implemented:

– A high-fidelity model for a Boeing 747-100 as described in National Aeronautics and Space Administration (NASA) references [17,18].
– A high-fidelity model for a F16 as described in Air Force Institute of Technology and NASA references [20,24].

The information provided in these references is very detailed and makes it possible to faithfully reproduce behaviour of these two aircraft and thus allow for realistic simulation on the platform.

The second limitation was addressed by the design and implementation of a full avionics bay connected to the HLA simulator entities and which are running autopilots, fly-by-wire and flight control systems on dedicated on-board computers including the fault tolerance redundancy handling [6].

3.2   Architecture Overview

The SMARTIES architecture, inherited from SDSE, is composed of 10 HLA federates, each representing a specific part or system of the aircraft and its environment (see Fig. 4).

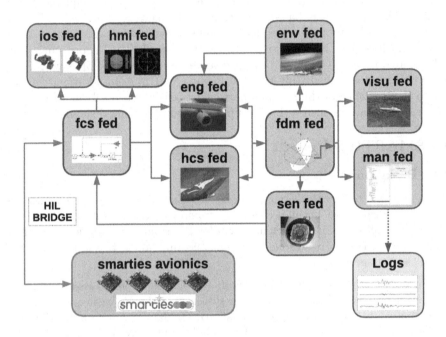

Fig. 4. SMARTIES architecture illustration.

- The Inputs/Outputs federate (*ios fed*) acquires the pilot's inputs in the cockpit from the multiple interfaces such as the side sticks, throttle, etc. The elevator, aileron, and rudder axes can then be commanded through this interface with a processing via the flight control laws implemented in the fcs and avionic systems.
- The Human-Machine Interfaces federate (*hmi fed*) is used to display essential flight information in the cockpit interfaces such as Primary Flight Display (PFD), Navigation Display (ND) or Electronic Centralized Aircraft Monitor (ECAM).
- The Flight Control System federate (*fcs fed*) is in charge of the aircraft control and implements two types of functions:
  1. the autopilot functions such as speed and altitude hold control systems;
  2. the low level flight control functions (i.e. control and stability augmentation systems).

The implemented control laws are a combination of the control laws documented in the reference documents as well as improvements realised in-house. These flight control laws can run either on a simulation entity (within a federate) or on the specialized SMARTIES avionics bay [6].

- The Engine federate (*eng fed*) simulates either 4 high-bypass turbofan engines for the B747 aicraft or 1 low-bypass turbofan engine for the F16 aircraft. The engine characteristics change with the atmospheric conditions as well as the aircraft flight Mach number. This federate receives the required throttle commands from the flight control system federate. A new model for this federate is presented in Sect. 4.
- The Hydraulic Control Surfaces federate (*hcs fed*) implements all the hydraulic actuators behaviors whose deflections change the aerodynamic forces and influence the aircraft motion. The different control surfaces are described in previously mentioned references for each aircraft and modeled either by first or second-order filters with position and rate saturation to enforce realism.
- The Flight Dynamics Model federate (*fdm fed*) represents the core of the simulation and computes the equations of motion. The aircraft trajectory and altitude will evolve with respect the action of aerodynamic, gravity and propulsion forces. The aerodynamic coefficients are implemented in the form of interpolation tables with multiple entries as described in the references.
- The Environment federate (*env fed*) reproduces the weather and climate conditions. It models the US Standard Atmosphere 1976 [23] and depending on the altitude it calculates the corresponding atmospheric variables such as temperature, pressure or air density. It also integrates different types of wind models, such as wind shears and gusts, as well as turbulence models like the ones from Dryden [13,35] and Von Karman [34].
- The Sensors federate (*sen fed*) simulates the different aircraft sensors such as Inertial Measurements Units (IMU) or Global Positioning System (GPS). Each sensor having its own dynamics, we used first or second-order low-pass Butterworth filters [3] augmented with delay, bias, drift, and noise phenomena in order to augment realism.
- The Visualization federate (*visu fed*) is in charge of the 3D virtual environment display and The Manager federate (*man fed*) enables configuration features for the simulation entities along with logs.

3.3 Open-Source Version

We decided to release part of our work for the F16 aircraft simulation as open-source software. In order to ease the usage and deployment of such application for an end-user, the native SMARTIES architecture was simplified. As described in Sect. 2.2, some models can be aggregated within one federate as long as the local scheduling of the entities is guaranteed. Figure 5 illustrates the architecture of our open-source version: 5 models (eng, hcs, fdm, env and sen) are combined within one federate (also named *fdm fed*). Also, the ios and hmi models are combined into a cockpit federate (i.e. ***ckp fed***). The complex avionics bay composed of several redundant units and corresponding network entities has been replaced to allow users to run the fcs code on an embedded device (having an ethernet interface) of their choice. In the remainder of this paper, the open-source application is considered (the code is available at https://github.com/ISAE-PRISE/smarties_f16).

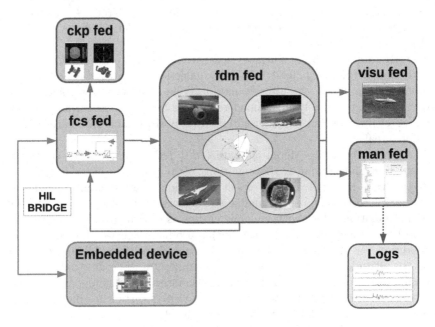

Fig. 5. Open-source SMARTIES architecture illustration.

4   Towards a New F16 Engine Model

4.1   Design Tool

A dedicated software called PROOSIS [2] was employed to create new models for the F16 engine. PROOSIS is an object-oriented software, primarily dedicated to gas turbine engine cycle modelling. It enables the user to both size a new engine architecture and/or to explore engine system design space by evaluating the performance of a pre-sized model through its three-layer structure:

1. Schematic: component-level modelling and whole system architecture assembly,
2. Partition: boundary value definition and initialisation of non-linear terms of the equation system that describes the constructed architecture, and

3. Experiment: engine cycle design and/or off-design simulations in steady state, transient, optimisation scenarios, etc.

Using this tool, the engine cycles were sized for a single operating point corresponding to a typical cruise condition of the F16 aircraft. The boundary conditions chosen to control the model are two engine fuel flows (in core and the afterburner), directly linked to the throttle command of the coupled flight simulator.

4.2   New Engine Models

A new F16 engine schematic (illustrated in Fig. 6) representing an afterburning two-spool low bypass-ratio F110 turbofan architecture was assembled based on the information provided in [1,19].

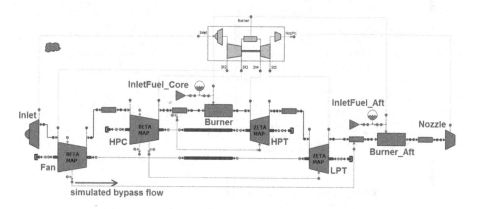

Fig. 6. F110-GE-129 engine architecture (PROOSIS schematic).

The fan was implemented using a compressor component, with the bypass mass flow modelled as bleed at the compressor inlet, re-injected into the main flow right after the last turbine stage. This was done in order to avoid certain inherent difficulties arising from the default library fan model being adapted to single stage components used in civil applications, as opposed to the compact three-stage fan employed in the current engine. The high pressure turbine cooling was modelled as a simple offtake from the high pressure duct re-injected into the high-pressure turbine, since no information concerning this feature was found in the literature. The post-combustion was modelled with an additional burner component situated between the low pressure turbine and the exhaust nozzle. Since no relevant data was found in the public domain, the exhaust nozzle component does not include variable cross-section feature typical for fighter aeroplane engines. As this drawback may only influence the value of the calculated performance, and has no further influence on the framework development at hand, the engine is assumed to have a fixed nozzle, which will be corrected in the future developments if additional data is acquired. Robustness of the resulting cycle model was verified with parametric steady state simulations for the following operating envelope: altitude (from 2000 m to 4500 m), Mach number (from 0.3 to 0.9) and net thrust (from 65000 N to 104750 N).

4.3   HLA/PROOSIS Coupling

There are two possibilities to link the PROOSIS tool with a HLA environment. PROOSIS (based on C++) allows the access (via an API) to certain features, internal variables and functions. Therefore, it is possible to implement an internal HLA federate such as depicted in Fig. 7.

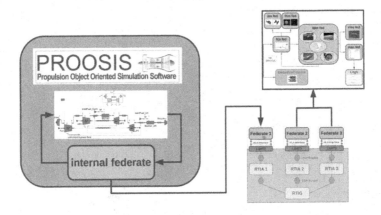

Fig. 7. Illustration of a PROOSIS internal federate.

However, this solution is difficult to handle and requires a lot of knowledge on the internal behavior of the tool. In order to implement a consistent global HLA simulation, it is important to understand the parameters impacting the frequency behavior of models components, their interactions and identify correlations between these models and implemented solvers within PROOSIS. This work is ongoing and not suitable for the open-source version. We therefore generated appropriate look-up interpolation tables (see Fig. 8) that have been integrated into the federate code to provide the necessary engine performance parameters to the simulator.

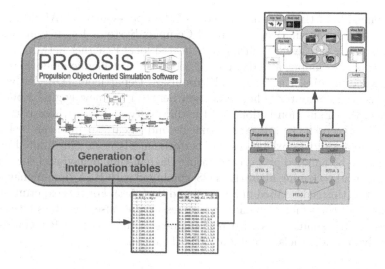

Fig. 8. Illustration of PROOSIS engine performance table generation.

5 Experiments and Results

5.1 Testing Environment

Our experiment takes place at the ISAE-SUPAERO facilities as illustrated in Fig. 9.

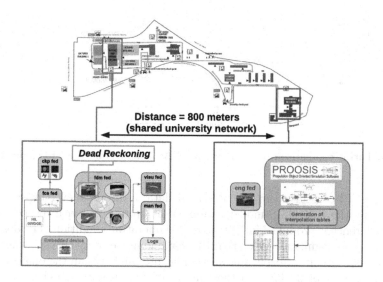

Fig. 9. Testing environment illustration between DISC and DAEP.

On one side of the campus, the core of the open-source SMARTIES application is running in the Department of Complex Systems Engineering (DISC) building. These entities are sharing a dedicated local Ethernet network. On the other side of the campus at a distance of 800m, the new engine model is running in the Aerodynamics, Energetic and Propulsion Department (DAEP) building on a dedicated computer. The connection between the new engine model and the rest of the simulation is done via the university network, whose bandwidth is shared between all the university students and researchers and can be limited for the simulation data exchanges (i.e. performance issues). This limited performance has to be handled properly to ensure the real-time behavior of the simulation using dead reckoning algorithms [28] combined with HLA time-management services [15] (this is described in Sect. 5.3 and Sect. 5.4). For this experiment, the distribution of the simulation entities is necessary because it enables future possibilities of HIL simulations with real engines running at DAEP facilities and aircraft dynamics models running at DISC facilities.

5.2   Engine Models Comparison

Figure 10 shows the resulting curves comparing the thrust value produced by the new model versus the original values from the NASA reference documents.

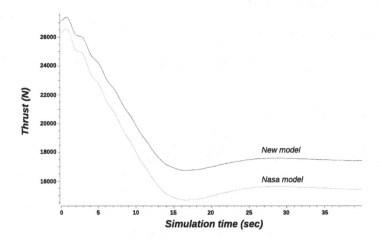

Fig. 10. Thrust comparison between original NASA model and the PROOSIS model.

As it can be observed in the presented data, a discrepancy (thrust overestimation by the PROOSIS model) stabilised at around 7% with respect to the NASA reference is produced for the simulated operating condition. The observed difference is not significant given that the maximum operating thrust (order of 75 kN for cruise without afterburning) is significantly higher than the one currently simulated (order of 20 kN), so the validity of the current model is not compromised. In addition, the NASA model used as a baseline is not a full engine cycle model in itself, but a look-up table providing engine thrust as a function of flight conditions and throttle setting. As such, we have no way of knowing the underlying assumptions concerning either the variable nozzle or any other engine component. It is nevertheless of interest to pinpoint the following drawbacks of the current PROOSIS model, which will inform future improvements:

1. The current engine thermodynamic cycle was sized only for one operating point at 75 kN thrust and in steady state, based on scarce data assembled from several different references. With sufficient reference data, the inherent limitations of single-point cycle sizing method can be overcome in future developments by employing the so-called *multi-point* cycle sizing process [29]. This more complex method is available in PROOSIS and can ensure a broader operating envelope for the sized engine.
2. Variable nozzle cross-section (typical feature to regulate military jet engine performance over its operating envelope) was not taken into account in the developed model.

5.3   Real-Time Performance Results

During the execution, each federate computation and communication is scheduled by time management principles (see Sect. 2.3). A suitable deployment of these techniques ensures a consistent behaviour on the simulation logical time reference. The progress in the logical time is then correlated to an hardware clock to ensure the respect of real-time constraints for federate executions.

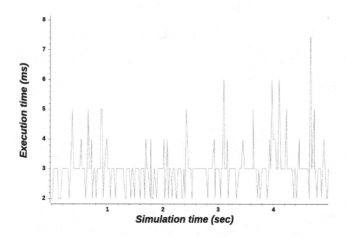

Fig. 11. Fdm federate step times with 10 Hz remote engine federate.

In a nominal mode, the frequency of the simulation 100 Hz which means that a federate must receive, process and send data every 10 ms (deadline). Due to our large scale distributed testing environment, we had to analyse the real-time behavior of the simulation according to the remote engine federate. Our experiments have shown that this federate has to run with a frequency 10 Hz to ensure proper real-time execution. Higher frequencies are causing the 100 Hz federates to wait for the remote federate and violating their 10 ms deadlines. Figure 11 shows the execution steps measurements from the fdm federate with the remote engine federate running 10 Hz; it can be seen that the real-time behavior is correct.

5.4   Dead-Reckoning Performance Results

The real-time performance of the entire distributed simulation is guaranteed using 10 Hz remote engine federate. However, the fdm federate still need relevant thrust inputs 100 Hz. To address this, dead reckoning algorithms [28] have been implemented into the fdm federate in combination with its HLA time management handling. A shadow engine model is integrated into the federate and uses dead reckoning to predict and mimic 100 Hz behavior based on the incoming data received from the remote federate 10 Hz. Two types of dead reckoning techniques have been used: one is only considering the rate of change of the remote value (noted DR1) and one considering the acceleration of change as well (noted DR2). The formulas 1 and 2 are given below with P the value (here the predicted or received thrust value), V is the rate of change of this received value and A its acceleration of change. These predications compared to the remote model are displayed in Fig. 12.

$$P_{DR1} = P_{10Hz} + V_{10Hz} * \Delta t \tag{1}$$

$$P_{DR2} = P_{10Hz} + V_{10Hz} * \Delta t + \frac{1}{2} * (A_{10Hz} * \Delta t)^2 \tag{2}$$

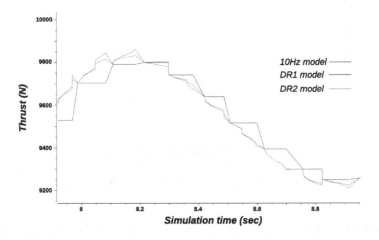

Fig. 12. Thrust comparison 10 Hz model and dead reckoning estimations.

6 Conclusion

This article focuses on the use of an HLA framework to develop, run and extend large-scale, hybrid and real-time simulations for aircraft systems. The approach building on HLA technology was presented, along with a dedicated implementation of a high-accuracy aircraft model from the public domain. The existing engine model was extended using a new dedicated design tool and the new model has been validated with previous model from the literature. A distributed large scale run-time execution experiment was then performed over a university network. Proper handling of the simulation could be confirmed, some enhancements to cover timing issues have been put in place. We have released an open-source package and therefore, our application with its implementation details are accessible, can be used, reproduced and extended. For future work, an extension of the engine model validation is foreseen, along with addressing the mentioned open points, and the overall large-scale simulation scenario will be further enhanced so to allow larger scale simulations to take place. In particular, we are targeting to pursue our effort towards new HIL simulations which will integrate real engines into the simulation loop. We are also working on the integration of multiple aircraft in the simulation loop using different existing simulators such as X-Plane (see https://github.com/ISAE-PRISE/xplane_hla).

Acknowledgements. This work has been realized in the context of the PRISE project (Research Platform for Embedded Systems Engineering) which tends to provide a platform for study, evaluation and validation of new embedded system concepts and architectures to our students and researchers. In addition, this project has inherited from the research work carried out in the scope of the AEGIS research chair, established between SAFRAN Group and ISAE-SUPAERO in 2016 with aim to promote and help develop multi-disciplinary preliminary design and optimisation methods applied to highly integrated propulsive systems.

References

1. Development of the F110-GE-100 Engine, Turbo Expo: Power for Land, Sea, and Air, vol. Volume 2: Aircraft Engine; Marine; Microturbines and Small Turbomachinery (1984). https://doi.org/10.1115/84-GT-132
2. Alexiou, A.: Introduction to gas turbine modelling with PROOSIS, 4th edn. Technical report Empresarios Agrupados Internacional (2020)
3. Butterworth, T.: On the theory of filter amplifiers. Exp. Wirel. Eng. J. **7**, 536–541 (1930)
4. Chandy, K., Misra, J.: Distributed simulation: a case study in design and verification of distributed programs. IEEE Trans. Softw. Eng. **5**(5), 440–452 (1979). https://doi.org/10.1109/TSE.1979.230182
5. Chaudron, J.B., Noulard, E., Siron, P.: Design and model-checking techniques applied to real-time RTI time management. In: Spring Simulation Multiconference-SpringSim 2011. Boston, United-States (2011)
6. Chaudron, J.B., Saussié, D.: Towards the design of a distributed aircraft flight control system connected to simulation components. In: 12e Conférence Internationale

de Modélisation. Optimisation et Simulation (MOSIM 2018), pp. 1–8. Toulouse, France (2018)

7. Chaudron, J.B., Saussié, D., Siron, P., Adelantado, M.: Real-time distributed simulations in an HLA framework: application to aircraft simulation. SIMULATION **90**(6), 627–643 (2014). https://doi.org/10.1177/0037549714531054

8. Chaudron, J.B., Saussié, D., Siron, P., Adelantado, M.: How to solve ODEs in real-time HLA distributed simulation. In: Simulation Innovation Workshop (SIW), pp. 1–12. Orlando, US (2016)

9. Chaudron, J.B., Siron, P., Adelantado, M.: Analysis and optimization of time-management services in CERTI 4.0. In: 2018 Fall Simulation Innovation Workshop (SIW), pp. 1–10. Orlando, US (2018)

10. Cousins, S.: Exponential growth of ROS. Robot. Autom. Mag. IEEE **18**, 19–20 (2011). https://doi.org/10.1109/MRA.2010.940147

11. Cousins, S., Gerkey, B., Conley, K., Garage, W.: Sharing software with ROS. IEEE Robot. Autom. Mag. **17**(2), 12–14 (2010). https://doi.org/10.1109/MRA.2010.936956

12. Deschamps, H.: Scheduling of a cyber-physical system simulation. Ph.D. thesis, ISAE-SUPAERO, Université de Toulouse, France (2019)

13. Dryden, H.L., Abbott, I.H.: The design of low-turbulence wind tunnels. Technical report Report 940, National Advisory Committee For Aeronautics, Washington, D.C., USA (1949)

14. D'Ambrosia, J., Law, D., Nowell, M.: 40 gigabit ethernet and 100 gigabit ethernet technology overview (2010)

15. Fujimoto, R.M.: Time management in the high level architecture. Simulation **71**(6), 388–400 (1998)

16. Gervais, C., Chaudron, J.B., Siron, P., Leconte, R., Saussié, D.: Real-time distributed aircraft simulation through HLA. In: 16th IEEE/ACM International Symposium on Distributed Simulation and Real Time Applications DS-RT, pp. 251–254. Dublin, Ireland (2012). https://doi.org/10.1109/DS-RT.2012.45

17. Hanke, C., Nordwall, D.R.: The simulation of a jumbo jet transport aircraft - volume i: mathematical model. Technical report, Prepared by Boeing Company Wishita Division, Kansas for. National Aeronautics and Space Administration, Ames Research Center, Moffet Flield California (1970)

18. Hanke, C., Nordwall, D.R.: The simulation of a jumbo jet transport aircraft - volume ii: modeling data. Technical report, Prepared by Boeing Company Wishita Division, Kansas for. National Aeronautics and Space Administration, Ames Research Center, Moffet Flield California (1970)

19. Holzman, J.K., Webb, L.D., Burcham, F.W.: Flight and static exhaust flow properties of an F110-GE-129 engine in an F-16XL airplane during acoustic tests (1996)

20. Marchand, M.A.: Pitch rate flight control for the F-16 aicraft to improve air-to-air combat. Technical report AFIT/GGC/EE/77-7, Air Force Institute of Technology (1977)

21. Metcalfe, R.M., Boggs, D.R.: Ethernet: distributed packet switching for local computer networks. Commun. ACM **19**, 395–404 (1976)

22. Modelica association project FMI: functional mock-up interface for model exchange and co-simulation. MODELISAR consortium (2020)

23. National oceanic and atmospheric administration, national aeronautics and space administration, US air force: U.S. standard atmosphere (1976). https://doi.org/10.1016/0032-0633(92)90203-Z

24. Nguyen, L.T., Ogburn, M.E., Gilbert, W.P., Kibler, K., Brown, P.W., Deal, P.: Simulator study of stall/post-stall characteristics of a fighter airplane with relaxed longitudinal static stability. Technical report 1538, NASA (1979)
25. Noulard, E., Rousselot, J.Y., Siron, P.: CERTI, an open source RTI, why and how. In: Spring Simulation Interoperability Workshop, pp. 1–11. San Diego, United-States (2009)
26. Object management group: minimum CORBA specification. OMG available specification, version 1.0, formal/02-08-01 (2002)
27. Quigley, M., et al.: ROS: an open-source robot operating system. In: ICRA Workshop on Open Source Software (2009)
28. Ryan, P.: Performance of dead reckoning algorithms across technology eras (2018)
29. Schutte, J.S.: Simultaneous multi-design point approach to gas turbine on-design cycle analysis for aircraft engines. Ph.D. thesis, Georgia Institute of Technology (2009)
30. Siron, P.: Design and implementation of a HLA RTI prototype at ONERA. In: Fall Simulation Interoperability Workshop (1998)
31. The Institute of Electrical and Electronics Engineers (IEEE) Computer Society: IEEE Standard for Modeling and Simulation (M&S) High Level Architecture (HLA) - Federate Interface Specification. Simulation Interoperability Standards Committee (2010)
32. The Institute of Electrical and Electronics Engineers (IEEE) Computer Society: IEEE Standard for Modeling and Simulation (M&S) High Level Architecture (HLA) - Framework and Rules. Simulation Interoperability Standards Committee (2010)
33. The Institute of Electrical and Electronics Engineers (IEEE) Computer Society: IEEE Standard for Modeling and Simulation (M&S) High Level Architecture (HLA) - Object Model Template (OMT) Specification. Simulation Interoperability Standards Committee (2010)
34. Von Kármán, T.: Progress in the statistical theory of turbulence. Natl. Acad. Sci. J. **34**, 530–539 (1948)
35. Yeager, J.C.: Implementation and testing of turbulence models for the F18-HARV simulation. Technical report NASA/CR-1998-206937, NASA Langley Research Center, Hampton, Virginia, USA (1998)

Comparison Between the HUBCAP and DIGITBrain Platforms for Model-Based Design and Evaluation of Digital Twins

Prasad Talasila[2]([✉]) [iD], Daniel-Cristian Crăciunean[1] [iD], Pirvu Bogdan-Constantin[1] [iD], Peter Gorm Larsen[2] [iD], Constantin Zamfirescu[1] [iD], and Alea Scovill[3] [iD]

[1] Lucian Blaga University of Sibiu, Sibiu, Romania
{daniel.craciunean,bogdan.pirvu,
constantin.zamfirescu}@ulbsibiu.ro
[2] DIGIT, Aarhus University, Aarhus, Denmark
{prasad.talasila,pgl}@ece.au.dk
[3] Agrointelli ApS, Aarhus, Denmark
als@agrointelli.com

Abstract. Digital twin technology is an essential approach to managing the life-cycle of industrial products. Among the many approaches used to manage digital twins, co-simulation has proven to be a reliable one. There have been multiple attempts to create collaborative and sustainable platforms for management of digital twins. This paper compares two such platforms, namely the HUBCAP and the DIGITbrain. Both these platforms have been and continue to be used among a stable group of researchers and industrial product manufacturers of digital twin technologies. This comparison of the HUBCAP and the DIGITbrain platforms is illustrated with an example use case of industrial factory to be used for manufacturing of agricultural robots.

Keyword: Digital twins · Industrial products · Model-Based Design · Co-simulation · HUBCAP · DIGITbrain · Agricultural robots · CPS

1 Introduction

In order to optimize the use of industrial products such as manufacturing equipment, it is paramount to consider digital technologies. Here digital models, digital shadows and digital twins play an important role for the Cyber-Physical Systems (CPSs) and the manufacturing processes. Unfortunately, many of the software tools have rather high costs for getting started. This means that there is a risk that many Small and Medium

The work presented here is partially supported by the HUBCAP and the DIGITbrain projects. The HUBCAP and the DIGITbrain projects are funded by the European Commission's Horizon 2020 Programme under Grant Agreements 872698 and 952071 respectively. The Poul Due Jensen Foundation that has funded subsequent work on taking this forward towards the engineering of digital twins.

A. Cerone et al. (Eds.): SEFM 2021 Workshops, LNCS 13230, pp. 238–244, 2022.
https://doi.org/10.1007/978-3-031-12429-7_17

Enterprises (SMEs) do not have enough financial power to join the digital transition. There is also a need for close collaboration between industrial product manufacturers, product users and CPS software providers so that reuse of CPS models and tools is possible. Thus, it is essential to establish sustainable and affordable collaboration platforms that lower the barriers to the use of digital twin technologies among SMEs.

In this paper we focus on and compare the platforms targeted by two H2020 Innovation Actions (IA) called HUBCAP [1–3] and DIGITbrain [4]. The focus of the HUBCAP platform is on Model-Based Design (MBD) for CPSs for any application domain. The HUBCAP collaboration platform contains interesting innovation called a sandbox enabling users to try MBD tools and models directly from an internet browser without having to install anything locally. The focus of the DIGITbrain platform is on evolving Digital Twins (DTs) for Manufacturing as a Service in a platform essentially enabling DTs from cradle to grave for industrial products. The main innovation here is a clear distinction between data, models and algorithms of a DT and in addition a high level of configurability on the execution environment.

We are going to take a closer look at the capabilities and differences of the HUBCAP and the DIGITBrain platforms. A case study of digital twin for industrial factory of agricultural robots has been used to illustrate the importance of collaborative MBD, publication, reuse, and evaluation of digital twins using the HUBCAP and the DIGITbrain platforms.

An important point to note is that the paper contains a review of the conceptual abstractions and implementations of two ongoing projects. Thus, the description provided in this paper is limited to capabilities implemented while writing this paper. The rest of this paper is organized as follows: Sect. 2 presents an overview of the HUBCAP and the DIGITBrain platforms, Sect. 3 compares these two digital twin platforms, and Sect. 4 deals with the Agrointelli's factory simulation case study.

2 The HUBCAP and the DIGITbrain Platforms

The HUBCAP and the DIGITbrain platforms are aimed at sustaining the ecosystem of CPSs for industrial users. The general approach taken is to help the users create, exchange and reuse DTs for industrial products (IPs). Both the projects have a marketplace for publishing, reuse of DT among manufacturers, consultants, software vendors and users.

Both the projects treat DT models and software as assets on the platform that can be reused. In the case of HUBCAP, the emphasis is on creation of models in general using the available software authoring tools (such models can for example be used in a DT context). The collaborative development of DT models is the primary advantage of the HUBCAP platform [2, 3]. The HUBCAP platform contains HUBCAP Sandbox Middleware (HSM) which is a formally proven, secure platform for management of collaboration among users based on Sandbox concept [5, 6].

The DIGITbrain platform enables creation of DTs for IPs using reusable models, algorithms/tools and data sources. The DIGITbrain platform, which is under development can host reusable data (D), models (M), algorithms (A), and Model-Algorithm (MA) pair assets [7, 8]. The DTs are composed from these reusable assets and are evaluated on distributed computing infrastructure.

3   Comparison of the HUBCAP and the DIGITbrain Platforms

In this section, we provide a comparison of the workflows involved in creation and execution of a digital twin on both the HUBCAP and DIGITbrain platforms.

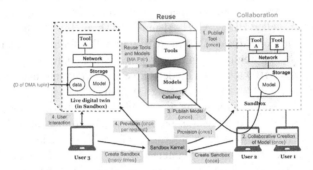

Fig. 1. Workflow in the HUBCAP project leading to execution of reusable digital twins.

Figure 1 shows a typical workflow in the HUBCAP platform if it is used in a DT context. The HSM is responsible for managing the workflow in the creation and execution of reusable DTs. The HSM manages collaboration among platform users in secure sandbox environments. The sandbox kernel, a part of HSM, is responsible for on-demand provisioning of a sandbox for users. Software providers can use a live sandbox to package their tool/software and publish the same into the catalog. The owner of a tool can also update and version their software. This is *step-1* in the workflow.

Each sandbox has a shared storage using which files can be exchanged in a sandbox. These usually are model, data, or log files that may be needed for collaboration within a sandbox. In *step-2*, user(s) typically develop a model, download the same to their computers before publishing it to the catalog (*step-3*). Model owners can also upload new versions of their models. In the HUBCAP project, the collaborative development (i.e., authoring) of tools and models is very easy. The publishing of a model or a tool need to be only once.

In *step-4*, users of a digital twin model can request the sandbox kernel to provision a sandbox in which model(s) and tool(s) needed for creation of a DT are available. Users can then upload required data to use the selected digital twin. Thus, even the HSM supports Data-Model-Algorithm (DMA) abstraction of a DT. Each sandbox has a unified web interface via which participants of a sandbox can collaborate. Thus, real-time collaboration between users is possible in both authoring and usage phases of digital twins. The provision of all sandboxes happens on a single server thus limiting the number of concurrent sandboxes that can be run.

Figure 2(a) shows a typical workflow involved in the DIGITbrain platform (which itself is based on the CloudBroker platform [9]). The DIGITbrain platform allows for publishing of tool/software/algorithm package. The *first step* is the publication of model-algorithm pair on the CloudBroker platform. At present, the authoring of models and tools is outside the main part of the DIGITbrain platform.

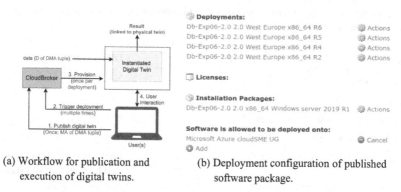

(a) Workflow for publication and execution of digital twins.

(b) Deployment configuration of published software package.

Fig. 2. The use of DIGITbrain project for evaluation of digital twins.

One advantage of the DIGITbrain platform is the ability of users to use the published software package and configure the on-demand deployment of a DT. Figure 2(b) shows different levels of configuring the deployment of a software package. The installation packages allow for the selection of the base computing environment (operating system, library dependencies, graphics hardware etc.) and installation of the software package on the deployment environment. The deployment environment specifies the hardware computing infrastructure onto which the software is configured to be deployed. In *step-2*, users can trigger automated deployment of selected software via a web interface.

The selected deployment environment is usually available among competing cloud service providers. In the present case, the on-demand deployment happens on the Microsoft Azure - a public cloud service provider. It is also possible to select different kinds of hardware configurations and save them as deployment configurations. The DT users can select one of the possible deployments. In *step-3*, the CloudBroker platform provisions the software on selected computing infrastructure and makes the DT ready to use. Users can then import models, data or both into the live DT.

Table 1 summarizes the relative advantages of each implementation platforms explained above. An interesting side effect of both the implementations is a ready to use a DT published on the platforms. Users have the option to just publish the tools, or tools-models or tools-models-data. Publication of model, data within a packaged tool available on the platform might help users check the setup of tool with an example model-data pair before using their own data-model-algorithm tuple.

Table 1. Comparison on the relative advantages of implementing digital twins on the HUBCAP and the DIGITbrain platforms.

Workflow	HUBCAP	DIGITbrain
Create new models	Yes	No
Collaboration between users	Yes	No
Publish and reuse models	Yes	No
Publish tools/software	Yes	Yes
Ease of publishing models and tools	Easy	Involves multiple steps each of which require significant domain knowledge
Reuse of models by customers	Yes	Yes
Deployment of software and model for instantiation of digital twins	Manual	Automated
Specification of deployment infrastructure	Not possible	Configurable in the publication phase
Execution environment	Single server	On scalable cloud resources
Export results to user	Manual	Supports both manual and automated
Reuse of evaluation environment	Yes, but manual deployment in each scenario	Nearly automated, click of a button

4 Agrointelli Case Study

Agrointelli (AGI) is a Danish SME which produces customizable autonomous field robots that can perform various operations in a farmer's field (e.g., light tillage, seeding, weeding, spraying, etc.). As AGI increases their sales, it is expected that there will be spare part orders from existing customers which will cause interruptions in the production process. For this, AGI needs to optimize their production process while considering spare part orders from farmers and/or dealers. Thus, creating a production model that can be scaled up, in terms of detail and potential integration of real-time data (e.g., resource, process etc.) for realistic decision-making is critical for its future growth of AGI.

In this case study, a co-simulation model of factory production process has been created using the CPPS-SimGen tool. The INTO-CPS toolchain has been used as the tool for executing the factory co-simulation model [11]. Thus, the factory DT is used to simulate the behaviour of the manufacturing process for a period of several months. The factory manufacturing process is co-simulated at maximum production capacity. The result of the co-simulation is taken over by a Decision Support System (DSS) which outputs new production schedule. The decision process to update production schedule begins with the arrival of a new order for a set of broken parts. This new order will have to be included in the current production plan and completed as soon as possible without disturbing the ongoing production plan. By simulating the manufacturing process and using a DSS the estimated delivery date for the part orders is generated.

The factory co-simulation model has been created using the CPPS-SimGen tool running in a HUBCAP sandbox [10]. The factory model is then downloaded onto the user's computer. The model is published onto the HUBCAP models repository. This factory co-simulation model can be executed using the INTO-CPS toolchain, in a new sandbox on the HUBCAP platform. The required parts order data need to be uploaded onto a running sandbox. A new instance of factory DT can be created by using factory model and the INTO-CPS tool available in the HUBCAP catalogue.

In case of the DIGITbrain project, authoring of the factory model is not possible. In the DIGITbrain parlance, the factory co-simulation model is the DIGITbrain model and the INTO-CPS toolchain is the DIGITbrain algorithm. This model-algorithm (MA) pair is then published as a software package onto the CloudBroker platform. The published software package is configured with four deployment configurations and coupled with one commercial cloud service provider for on-demand execution of the factory DT. The required parts order data is directly uploaded into the cloud computer on which the factory DT is being executed. In this way, the Agrointelli use case has been used to demonstrate the capabilities of the HUBCAP and the DIGITbrain platforms.

5 Concluding Remarks

There is a timely need for collaborative and sustainable development platforms for DT stakeholders. The HUBCAP and DIGITbrain project platform are two prototype demonstrations of DT platforms. The HUBCAP platform is geared towards collaborative development and publication of models and tools that can be used for DTs. The DIGITbrain platform enables industrial product manufacturers, users, software providers and domain experts to collaborate on the publication and use of DTs. The comparative advantages of these two platforms have been demonstrated using an industrial case study. Even though there are overlaps in some of the features of these DT platforms, both are complementary in nature. Thus, the HUBCAP and the DIGITbrain platforms are important collaborative efforts to sustain the adoption of DT technologies among SMEs.

Acknowledgements. We would also like to express our thanks to the anonymous reviewers.

References

1. Chapurlat, V., Nastov, B.: Deploying MBSE in SME context: revisiting and equipping digital mock-up. In: 2020 IEEE International Symposium on Systems Engineering (ISSE), pp. 1–8 (2020). https://doi.org/10.1109/ISSE49799.2020.9272230
2. Larsen, P.G., et al.: A cloud-based collaboration platform for model-based design of cyber-physical systems. In: Proceedings of the 10th International Conference on Simulation and Modeling Methodologies, Technologies and Applications, INSTICC, vol. 1, pp. 263–270. SIMULTECH (2020). https://doi.org/10.5220/0009892802630270
3. HUBCAP Project. https://www.hubcap.eu/. Accessed 07 Oct 2021
4. DIGITbrain Project. https://digitbrain.eu/. Accessed 07 Oct 2021

5. Kulik, T., et al.: Extending the formal security analysis of the HUBCAP sandbox. In: Macedo, H.D., Thule, C., Pierce, K. (eds.) Proceedings of the 19th International Overture Workshop, 1 October 2021, p. 36 (2021). https://arxiv.org/abs/2110.09371

6. Kulik, T., Macedo, H.D., Talasila, P., Larsen, P.G.: Modelling the HUBCAP sandbox architecture in VDM: a study in security. In: Fitzgerald, J., Oda, T., Macedo, H.D. (eds.) Proceedings of the 18th International Overture Workshop, 7 December 2020, p. 20 (2020). https://arxiv.org/pdf/2101.07261.pdf

7. Sandberg, M., et al.: Deliverable 5.1 Models for Digital Twins, DIGITBrain Project, 30 June 2021

8. Pena, S., et al.: Deliverable 7.1 Minimum Viable Digital Product Brain, DIGITBrain Project, 30 June 2021

9. Kiss, T.: A cloud/HPC platform and marketplace for manufacturing SMEs. In: 11th International Workshop on Science Gateways, IWSG 2019 (2019)

10. Functional Mock-up Interface for Model Exchange and Co-Simulation, Document version: 2.0.1, 2nd October (2019). https://fmi-standard.org/. Accessed 21 Oct 2021

11. The INTO-CPS Tool Chain User Manual, Version: 1.3, September 2019. https://into-cps-association.readthedocs.io/en/latest/. Accessed 28 November 2021

OpenCERT 2021 - 10th International Workshop on Open Community approaches to Education, Research and Technology

OpenCERT 2021 Organizers' Message

The 10th International Workshop on Open Community approaches to Education, Research and Technology (OpenCERT 2021) aimed at promoting the use of Open Community approaches in Education and Research while also exploiting them to achieve wide diffusion and proper assessment of new, innovative Technology.

The workshop received five full paper submissions, which were reviewed for quality, correctness, originality, and relevance. Each submission was posted on GitHub and reviewed by at least four Program Committee (PC) members. This first phase of the review process was carried out as an interactive, open discussion between the authors and the reviewers. Authors had then the chance to revise their works and submit the revised versions. A final review process, based on the revised papers, was carried out by the same reviewers of the first phase using the EasyChair system and was concluded by a closed discussion among the PC members. Four contributions were accepted for presentation at the workshop. Three of these are included in this volume: the tool paper "DrPython–WEB: a tool to help teaching well-written Python programs" by Battistini, Isaia, Sterbini, and Temperini, the survey paper "Formal Methods Communities of Practice: A Survey of Personal Experience" by Bowen and Breuer, and the learning experience paper "Learning from Mistakes in an Open Source Software Course" by Zhangeldinov.

The workshop program also featured two keynote talks: "A Life-long Learning Education Passport Powered by Blockchain Technology and Verifiable Digital Credentials: The BlockAdemiC Project" by Ioannis Stamelos, Aristotle University of Thessaloniki, Greece, and "Open Source Discovery, Adoption, and Use: An Informal Perspective" by Tony Wasserman, Carnegie Mellon University, USA. The contributions of the keynote speakers are also included in this volume.

We would like to thank the keynote speakers for accepting our invitations and for their very stimulating talks. We are also grateful to the Program Committee members, for their enthusiasm and effort in actively participating in the open review process and for their extensive work in providing constructive feedback and collaborating with the authors during the revision process, and the authors, for their valuable contributions and for wisely exploiting the collaboration with the reviewers to improve their works. Finally, we would like to thank all workshop attendees for their active participation in discussions and for the feedback they provided to the authors.

March 2022

Antonio Cerone
Marco Temperini

Organization

Program Committee Chairs

Antonio Cerone — Nazarbayev University, Kazakhstan
Marco Temperini — Sapienza University of Rome, Italy

Steering Committee

Luís Soares Barbosa — University of Minho and UNU-EGOV, Portugal
Peter T. Breuer — Hecusys LLC, USA
Antonio Cerone — Nazarbayev University, Kazakhstan

Program Committee

Roberto Bagnara — BUGSENG and University of Parma, Italy
Luís Soares Barbosa — University of Minho and UNU-EGOV, Portugal
Leonor Barroca — Open University, UK
Peter T. Breuer — Hecusys LLC, USA
Nicola Capuano — University of Basilicata, Italy
Antonio Cerone — Nazarbayev University, Kazakhstan
Stefano De Paoli — Abertay University, UK
Tania Di Mascio — University of L'Aquila, Italy
Elsa Estevez — Universidad Nacional del Sur, Argentina
Rosella Gennari — Free University of Bozen-Bolzano, Italy
Ralf Klamma — RWTH Aachen University, Germany
Padmanabhan Krishnan — Oracle Labs, Australia
Matteo Lombardi — Griffith University, Australia
Andreas Meiszner — Scio, Portugal, and University of Liverpool, UK
John Noll — University of East London, UK, and Lero - The Irish Software Research Centre, Ireland
Kyparissia Papanikolaou — School of Pedagogical and Technological Education (ASPETE), Greece
Donatella Persico — Institute for Educational Technologies (CNR-ITD), Italy
Alexander K. Petrenko — Institute for System Programming of the Russian Academy of Sciences (ISP RAS), Russia

A Life-Long Learning Education Passport Powered by Blockchain Technology and Verifiable Digital Credentials: The BlockAdemiC Project

Sofia Terzi[1,2]([⊠])(iD), Stamelos Ioannis[1], Konstantinos Votis[2], and Thrasyvoulos Tsiatsos[1]

[1] Aristotle University of Thessaloniki, Thessaloniki, Greece
{sofiaterzi,stamelos,tsiatsos}@csd.auth.gr
[2] Center for Research and Technology Hellas, Thessaloniki, Greece
{sofiaterzi,kvotis}@iti.gr

Abstract. The academic and business domains are in constant transformation due to the technological advances. In this context, proof of knowledge, skills and training are becoming crucial for the students and employees. In a competitive environment like that, forgery of diplomas and certificates is a frequent problem that has not been faced properly yet. Additionally, the lack of a formal representation and acknowledgment of informal and life-long learning outcomes costs to all educational stakeholders and employee seekers. By using open source blockchain technology as a game changer throughout the educational process, security, privacy, integrity and immutability of the data related to diplomas, certificates and skills acquired by learners can be guaranteed. The result of building a system supported by blockchain for life-long learning, a positive impact in trust and transparency of the education achievements, institutions, students and companies is achieved.

Keywords: Education · Blockchain · Verifiable credentials · Anti-forgery · Digital identity

1 Introduction

The business world is facing a surge in demand for technical skills as it becomes more digitized and it finds it difficult to hire qualified and trained personnel [1]. In this environment, proof of skills, training and education are important in order to get hired by companies. This proof is mainly provided in forms of diplomas and certificates which include information for the candidate's educational and knowledge background. Unfortunately, the forgery of diplomas and certificates is a well-known problem in the academic sector leading universities and other institutions to even stop issuing them in paper [2]. At the same time, there are many informal ways of being educated and trained apart from universities. Life-long learning

A. Cerone et al. (Eds.): SEFM 2021 Workshops, LNCS 13230, pp. 249–263, 2022.
https://doi.org/10.1007/978-3-031-12429-7_18

(LLL) and digital education courses help to acquire microcredentials - which are a representation of the smaller units of knowledge that are not included in a standard diploma or degree - and are not currently represented in national qualification frameworks [3] and thus they lack a formal proof, though they represent skills that are needed for certain job positions. Massive Open Online Courses (MOOCS) for example have evolved from single-course into multi-course with associated degrees, such as the XSeries degree by Coursera and the Nanodegree by Udacity, accredited with microcredentials to prove the skills and training [4] someone has achieved after successfully completing them.

In parallel, the academic world is also being further digitized than just in the administrative tasks. Especially after the pandemic of Covid-19 in many EU countries the educational processes are taking place either fully online or in hybrid mode both in physical and digital presence. This comes with the opportunity to capture learners' interactions through the learning management systems (LMS) and monitor their learning outcomes in order to assign fine-grained badges and microcredentials to them. These microcredentials can then be combined with or accompany the student's diploma to demonstrate the upskill achieved. When speaking about diplomas and microcredentials representation in this paper, we are always referring to their digital format, unless noted otherwise.

Since proof of education and skills is often used to apply for subsequent studies to other national or international education institutions and for job vacancies at companies as well, it is crucial to eliminate any chances of this proof being counterfeit. Blockchain (BC) technology can be used to sustain availability of such credentials while preventing the forgery of the digital records that are stored on it. This is possible because of BC's principal characteristics which include a decentralized append-only database type called ledger, with its records - called transactions - being timestamped, digitally signed, added to blocks connected as a merkle-tree forming the ledger and written only after a consensus among the participating servers - which are called nodes - has been achieved [5]. Taking advantage of open source BC technologies and their features, the BlockAdemiC (BCA) project will create a digital cybersecure system (platform) for certifying and verifying educational activities, diplomas, certificates and skills in higher education (HE) and LLL domains. This system is supporting a cryptographically secured wallet that cannot be tampered, for safe storing all the digital credentials (DC) related to a person's education and training, forming a digital education passport (EDP). BCA differentiates from all other projects that support storing DCs because it goes a step further to record, apart from diplomas and certificates, the educational activities covering special skills - such as scientific and technical - and general skills. Thus, by adding a lower level of student activity recording, it achieves to capture specific actions that lead to acquiring these special and general skills, and stores them immutably on the BC ledger, making them integral part of the EDP. The rest of this paper is separated in three subsections describing a) the BCA potential and the problems it attempts to solve, b) what is blockchain and how it helps to solve the described problem, c) expected impact after piloting the platform, and d) conclusions and future work.

2 The BlockAdemiC Project

2.1 BlockAdemiC Overview

BCA as a project provides a decentralized immutable lifelong cybersecure EDP. To this direction it adopts the European Credit Transfer and Accumulation System (ECTS) for HE institutions' (HEI) students to represent learning outcomes and make academic qualifications recognized across EU countries [6]. The platform that has been developed will be applied at the Aristotle University of Thessaloniki, the largest university in Greece, and at least three more educational institutions in Greece and abroad. Students registered through the BCA will hold a student's digital wallet, that will be used as the EDP, to store their DCs such as diplomas and certificates. By utilizing BC technology HEIs and LLL institutions (LLLI) will issue digital tokens to store soft and hard skills as DCs. A token is an educational credit that reflects in the qualifications that learners have acquired after completing a learning activity.

Qualifications according to the Greek National Qualifications Framework [7] take the form of learning outcomes classified into predefined levels. Learning outcomes - for example what the person knows, understands and can do after completing a learning process - are categorized into knowledge, skills and abilities. A wallet in BC technology is a permanent digital personal storage and in BCA can be held on-chain or off-chain. The difference between the two types is that when stored on-chain this wallet is decentralized hosted on the BC network, when stored off-chain it is hosted on the user's personal computer storage. BCA's goal as a project goes beyond previous BC based education platforms such as Blockcerts and BCdiploma that only store the diplomas [8,9] or EduCTX that uses as a learning unit only the results of written exams to calculate the total ECTS earned by the students [10]. What differentiates BCA is its ambition to capture the total effort of the overall activities that students have on LMS and that can be expressed in ECTS and qualifications. This perspective makes the learning measurement unit more precise than just a single exam result.

Special care has been taken to protect the BCA users' personal and sensitive information by choosing open source technologies supporting the platform to form a private BC network where only registered users can participate. This also differentiates BCA from other education platforms that use public BCs, such as the Ethereum network [11,12], not only in terms of privacy, but also in terms of lowering the costs for storing information and operating the network, having better response times and storage speed, while maintaining security, decentralization, immutability and transparency. This way, information exchange processes between the participating institutions is being reinforced and secured at the same time by the BC technology. The same applies for all the other participants on this network, namely students, alumni and companies offering job opportunities. All of the participants are able to exchange information only after being authenticated and authorized. After research, four main roles - called actors - have been identified at BCA's design stages to be the end users of the system, and their interactions have been defined. After defining the roles, their interactions formed the use case scenarios that would be supported by the system. The actors are

- The educational institutions, which include the HEIs and the LLLIs
- The teachers that are part of the educational institutions
- The students and alumni of the educational institutions
- The companies that are offering job opportunities and are interested in trained personnel with special skills and knowledge

The actions that each of these actors can perform are the following

- Educational Institutions and Teachers
  - HE Institutions
    * Register for an institution account for authentication and authorization
    * Issue verifiable DCs (diplomas, degrees, certificates)
    * Verify DCs issued by any educational institution
  - LLL Institutions
    * Register for an institution account for authentication and authorization
    * Issue verifiable DCs (certificates)
    * Verify DCs issued by any educational institution
  - Teachers
    * Register for a personal teacher's account for authentication and authorization
    * Issue verifiable DCs (microcredentials)
- Students
  - Register for a personal student or alumni account for authentication and authorization
  - Hold a life-long EDP
  - Participate to activities and courses in order to earn DCs (diplomas, degrees, certificates, microcredentials)
  - Present their EDP to HEIs and LLIs for subsequent studies or to employers looking for personnel
- Companies/Employers
  - Register for a company account for authentication and authorization
  - Announce job vacancies and define additional special skills needed to apply
  - Search with specific criteria for alumni in order to offer job opportunities
  - Verify DCs issued by any educational institution

The specific use case scenarios covered by the common interactions between the above mentioned actors are explained next.

The first scenario BCA covers is the need for issuing DCs like diplomas and certificates and capturing microcredentials all along the educational process. Furthermore, by storing the verifiable DCs on the immutable BC ledger, BCA prohibits tampering of all kinds of DC tokens contained in the EDP addressing the fraud of diplomas and certificates forgery. Furthermore, BCA confronts not only the forgery of diplomas and microcredentials, but also makes them auto-verifiable by other users of the BC network using complicate cryptographic techniques covered in detail at Sect. 2.3. The actors interacting in this use case are HEIs, teachers and students. Figure 1 demonstrates the high level steps for storing a token as a verifiable DC on the BC representing a microcredential or a diploma. BC hosts the student's wallet on-chain in this case, which serves as the storage for the DCs which formulate as explained before the EDP. The wallet, and by that the EDP, is accessible by the students and they can have an overview of the complete list of DCs they have acquired throughout their learning journey.

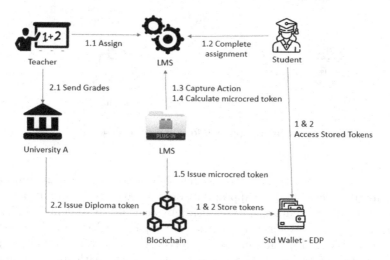

Fig. 1. Issue tokens as verifiable digital credentials

The second scenario BCA covers is the students' need for subsequent studies at other HEIs than the one they received their degree and therefor the assistance of students' mobility between EU countries, or even worldwide. Students' mobility is a priority in EU, and special programmes, like the Erasmus+ [13], are dedicated to support individual applicants and organizations such as universities, training centers and companies to take part in education and training in Europe, apply for grants and facilitate the complete mobility process. The actors interacting in this use case are the students and the HEIs. What is worth noticing for Fig. 2 which presents the high-level steps for this scenario, is that the HEI that issued the diploma as a DC and digitally signed it - named university A - in the previous scenario takes no part in this information exchange and DC verification. That happens because BC eliminates the need for verifying DCs by a third-party other than the ones that are involved in this specific transaction [14]. Additionally, in this use case instead of a university acting as an actor for auto-verifying the integrity of the diploma, could be a LLL training center with the only condition to have registered for an account and participate in the BCA network.

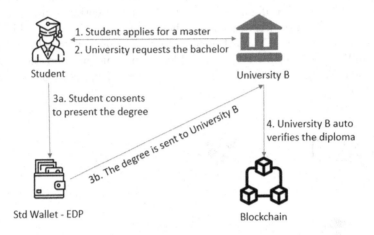

Fig. 2. Applying for a master course

The third scenario BCA covers is the need of companies searching for job candidates in order to find the appropriate ones based on their training and skills. Moreover, it allows companies to publish job vacancies so students fulfilling the requirements can apply for a them. After a company registers and creates their profile, it is assigned an account which serves as an identity for accessing the BCA network. After that, they are able to issue queries with specific criteria to receive results from a pool of students and alumni holding diplomas, certificates and skills and who have consent to be included in this list. That use case will facilitate the need for connecting the education institutions students and alumni and the labor market, easing the process called school-to-work transition [15,16]. Figure 3, demonstrates the high-level steps when a student or alumni applies for a job vacancy and the auto-verification of the DCs that are needed and presented for this position. As explained before, the verification of the DCs presented does not require the involvement of the issuer as argued before, disengaging the necessity for third-party verifiers, speeding up the whole process. In this particular scenario, the forming of a subset of DCs on the fly from the EDP - rather than sending just one DC for proving the candidate fulfills the specific skills and education - inserts a level of flexibility per use case and is a paradigm of how an EDP can serve as a life-long secure education passport in different circumstances and occasions.

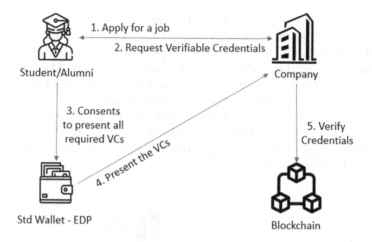

Fig. 3. Applying for a job

2.2 Innovation and Objectives

The project is innovative to its grounds due to the fact that it uses BC decentralized technology for security, confidentiality and combines it with learning analytics, as follows:

- It ensures the authenticity and makes impossible to falsify the learning outcomes, as well as the acquired knowledge, abilities and skills through the registration of the tokens and the relevant diplomas/certificates on the ledger. This is achieved by distributing copies of data to all BC nodes hosted by the participating organizations and institutions. The distribution of data enhances also the speed and the consensus (agreement) and avoids the single point failure in the education network.
- It ensures access to the system only to authorized users and allows the definition of classified rights. For example, only certified instructors/training providers are able to record students' results, while students/trainees are in control of sharing with third parties, such as companies, only the data they choose to share.
- It automates the process of certification of learning outcomes. For this purpose, SCs have been developed, which are code (software) that is executed automatically as long as certain predefined conditions are met.
- It enhances the participation of students/learners in the process of consent and verification of learning activities. Suitable decentralized blockchain-based mobile decentralized applications (DApp) has been developed for this purpose.
- It ensures interoperability with the support of international standards and their implementation on the BCA platform to lay the foundations for integration with other systems of similar educational applications in either the national or international educational environment.

- It assesses the learning activities with additional metrics other than learning outcomes. Such metrics aim at assessing the skills necessary for the student to enroll for a particular course or general skills such as communication, collaboration, understanding, taking initiatives, problem solving. Those metrics are usually collected through calibrated psychometric tests and in BCA through user actions like contributing to collaborative activities.
- It increases usability and end user friendliness through the adoption of gamification techniques, interfacing with existing digital training platforms and connecting to social media, mainly professional as Linkedin, and academic as ResearchGate and academia.eu (Fig. 4).

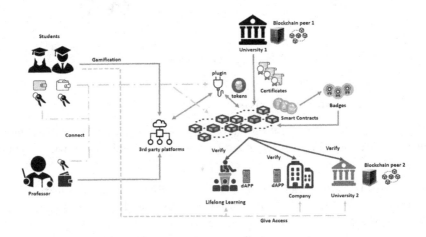

Fig. 4. BlockAdemiC overview

BCA successful implementation is divided in four separate and equally important objectives as described below:

- **Objective 1:** To create a complete system of distributed security with open source technologies, which will extend and ensure the certification and verification of educational activities, skills and qualifications in HE and LLL. BCA will provide an individual, tamper-proof and therefore inviolable educational passport, based on BC technology. In this EDP, both the diplomas and the relevant educational records will be stored in the form of convertible digital educational credits (tokens), adopting appropriate mechanisms of learning analytics. The tokens will be linked to the formal education system (in the case of HEI), and the labor market through the adoption of corresponding interoperability mechanisms that will allow the recognition of educational 'parchments' at a pan-European and global level. In this way it is expected to facilitate the mobility of students, but also professionals, by opening new prospects for professional and academic development, including personalized learning.

- **Objective 2:** To design and develop the appropriate framework and mechanisms of learning analytics in order to automatically link learning metrics with the assessment of (technical) knowledge (hard skills) and horizontal skills (soft skills) of learners. The proposed framework of learning analytics will directly link the individual learning activities of the trainees with the relevant knowledge and horizontal skills through a credit system compatible (where possible) with ECTS. At the same time, this framework will allow learners - in collaboration with trainers - to choose their learning path (personalized training), adapted to their learning goals, knowledge and skills ensuring the interoperability of learning tools. The learning analytics mechanisms that have been developed are integrated into a gamification framework in order to make the overall solution more attractive to the learners and to give them additional incentives to adopt it.
- **Objective 3:** To develop a technological platform of distributed digital security for the registration of educational activities, degrees and certifications on a BC ledger forming an EDP. The platform supports storing, presentation and verification of diplomas and other kinds of tokens as well as the ability to share student VCs with stakeholders and companies. BC solutions improve verification processes and eliminate false training credits claims that were never obtained or that have undergone forgery. At the same time, they enable transactions such as the transfer and convertibility of educational units, and the recognition of learning activities.
- **Objective 4:** To pilot and validate the operation of the proposed technological solution in real conditions. The integrated system, as well as its individual functions will be piloted in 2 broader scenarios: in a postgraduate program of AUTh and in at least 3 more collaborating educational institutions in Greece and abroad, the network of the partner Web2Learn. The results of the pilot application will be evaluated for the purpose of (a) the performance of the proposed system in relation to the requirements analysis and project objectives, (b) the recording and implementing of individual improvements in the system, and (c) the evaluation of possibilities for exploiting the results beyond the framework of the project.

2.3   BlockAdemiC Technological Background

As argued before, BCA is a project that utilizes open source BC technology in order to create a secure, decentralized and tamper-proof EDP for storing DCs. The reason behind using the BC technology for the system is mainly to prevent forgery of diplomas in HE and LLL domains and to create a LLL EDP. BC can increase transparency and credibility with its immutable storage, the ledger. Additionally, the immutable transactions that are being added on the ledger create permanent records, with an associated permanent link, called anchor. This anchor, along with the digital signature for each transaction makes auto-verification of the records possible. In more detail, when a transaction such as a diploma is stored on the ledger for a graduate student, the digital signature of the institution issuing this diploma is stored along with the transaction. Digital signatures have the characteristic of being self-verified, because they contain the public key of the signer, in this case the HEI's or the LLLI's [17].

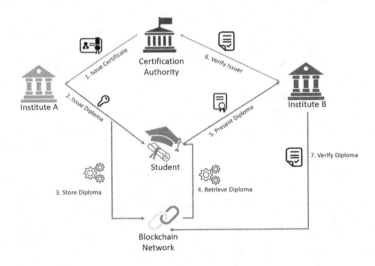

Fig. 5. Auto-verification of a diploma's validity

The verification of the diploma based on the HEI's digital signature that is stored along with the diploma, follows certain steps that are implied by the Public Key Infrastructure specification framework, utilizing the X.509 HEI's digital certificate issued by a trusted Certification Authority (CA) [18]. First, institute A is issued a certificate by a trusted CA and holds a private and a public key. Then, institution A issues a diploma as a DC for a student and digitally signs it with its private key. The DC contains, apart from the related information and the holder of the diploma, the digital signature and the public key of the issuer, in this case institution A. When the holder of the diploma presents it to institution B to apply for a master course for example, institution B receives the DC that contains information about the issuer and retrieves the issuer's public key. With this information alone, institution B is able to verify the validity of institution's A certificate and the integrity of the information regarding the diploma record on the BC which institution A signed before. The high-steps of this process are displayed in Fig. 5.

What is worth noting regarding the BCA implementation is that institution A and institution B have no direct interaction or information exchanged between them. Instead, the PKI infrastructure and the BC network supporting the solution allows the auto-verification of the diploma's origins and integrity directly from the BC solution. The only difference in BCA regarding a traditional PKI implementation, is that the CA is running on the BC network as a service and it's not a third-party trusted authority. When more than one CA is running on the BC we can achieve a decentralized CA along with the decentralized storage for the BCA platform. Currently only one CA is running on the network, but more than one is possible to cooperate in real life scenarios, where a trusted chain can be formed between the various CAs, for example at an institution level (one per HEI). The actual implementation differs from the above general one and is displayed in Fig. 6, where the trusted CA runs on one of the BC nodes, participating on the network.

To achieve the automation of assigning DCs to trainees and students but also enforce the rules that frames the allowed and prohibited interactions between BCA's participating actors, BC smart contracts (SC) are running on the background. A SC is a software program that runs on the BC network and contains conditions that must be met in order for specific actions to happen, such as storing a diploma in a student's EDP or controlling the process when the student consents upon to an institutions request to present their diploma to apply for a master course. SCs are functions that run on the background as services and they can be called directly or indirectly. For example, if a HEI other than the one that issued the diploma attempts to access a student's diploma without their consent, a SC is activated without a human interaction to prevent this from happening. In another situation, where the student wants to present their diploma to apply for a master course or for a job application, a human interaction is needed to trigger the appropriate SC. This happens through a software application, which is called DAPP. The DAPP makes possible for the BCA users to interact with the SCs and consequently the BC network through a user-friendly environment, hiding the complexity of the background technologies. The system is powered by the Hyperledger Indy BC solution (Linux Foundation), which is an open source framework, providing tools, libraries and reusable components for providing decentralized digital identities with focus on interoperability among the various connected applications and platforms [19]. It contains mechanisms for storing students' and alumni's verifiable DCs in the EDP and associate them with specific digital decentralized IDs called DIDs [20], keeping them secured, immutable and private.

2.4 Expected Impact

The piloting of the platform is expected to have strategic positive effect in the following areas:

- The trust between the training institutions as well as the companies (future employers of the trainees) regarding the knowledge and skills of the trainees. The security and availability of the data that is registered and exchanged will improve.
- The competitiveness of the educational institution based on the innovations that the project brings to the educational practice. The transparency of the whole process supported by BC technology will enhance the credibility of the institutions and the degrees they award. It aspires also to provide the framework for personalized education in HEIs and LLLIs, and enhance student mobility between educational institutions, as well as business confidence in

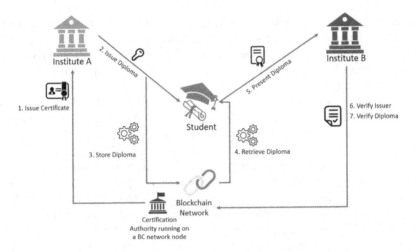

Fig. 6. Auto-verification of a diploma in BlockAdemiC project

the quality of education and the breadth of their knowledge and skills. Students and alumni will be better evaluated by the companies as prospective employees.

– In the educational part, to help in better quality learning and teaching practices, improved curricula, better guidance services for students in terms of their courses and studies either in the same institution or in other educational institutions. Teachers will have a more complete picture of their students' progress, so they will be able to intervene more sportively and effectively, but also in a more general context. Teachers and therefore the educational institution, will be in position to guide and support students during and after completion of courses, and faster (in terms of selecting subsequent courses or study programs at other institutions). Due to the transparency of the information, that information and the monitoring will be more effective for the student

– In the financial part to exploit economic benefits deriving from the BCA application adoption and usage by the stakeholders.

– In the social part to provide better educational services and opportunities in society and enhance the credibility of institutions to students and businesses. The project will also help build institutional trust, using the functions of BC immutability and timestamping. The service will provide employers and recruiters with reliable information about their degree and other skills they may have acquired during their careers.

3   Conclusions and Further Research

The academic and employment processes are being transformed due to the technological evolution. Along with this, students and job seekers need to prove their skills and special training in the digital world and educational institutions are called to provide these proofs, while maintain the integrity of this information. BC open source technologies support the integrity of information stored and exchanged, ensures the validity of the stored microcredentials and diplomas and protects against forgery of these digital credentials. That creates a secure, trusted and transparent environment where all of the education stakeholders take advantage form the automation of verification of diplomas, certificates, skills and microcredentials in general. With BCA, the decentralization of the information provides an additional layer of data redundancy eliminating the problems of service outages, single point of failure and data loss. Furthermore, by utilizing an EDP, a secure and organization independent personal storage for tokens is created which can accompany the holder throughout their education and working careers.

We acknowledge the limited testing of the system in a lab environment, and we anticipate to use it in production to confirm the expected outcomes presented in this paper. In order to explore the system's usefulness for the education ecosystem's end users a research questionnaire has been formed and distributed to selective HEIs at principals and students. Its results are encouraging and substantiates our expected results and they will be presented after piloting the system. Additionally, in the future, we intend to connect the identity management system we developed based on Hyperledger Indy with well established and legacy systems of HEIs and LLLIs in order to examine the compatibility and interoperability with these systems on site. In our next paper we will also present metrics regarding the efficiency of the system. Another research area that concerns us for BCA is the scalability of the solution and we will examine it in relation to the actual piloted environment. As mentioned before, the questionnaire provided us preliminary information for the necessity of a platform like BCA which provides an holistic solution from the educational process to the actual employment market and we are sure that our system will be well established among the different stakeholders.

Acknowledgements. This work was funded from the PA (Partnership Agreement for the Development Framework) 2014–2020 under project No. T2EDK-04180, project BlockAdemiC.

References

1. America's small businesses still can't find workers, but that's not their biggest problem. (n.d.). https://www.cnbc.com/2021/08/10/the-labor-shortage-isnt-main-streets-biggest-problem.html. Accessed 10 Feb 2022
2. Gresch, J., Rodrigues, B., Scheid, E., Kanhere, S.S., Stiller, B.: The proposal of a blockchain-based architecture for transparent certificate handling. In: Abramowicz, W., Paschke, A. (eds.) BIS 2018. LNBIP, vol. 339, pp. 185–196. Springer, Cham (2019). https://doi.org/10.1007/978-3-030-04849-5_16
3. Brown, M., Mhichil, M.N., Beirne, E., Mac Lochlainn, C.: The global microcredential landscape: charting a new credential ecology for lifelong learning. Journal Articles; Reports - Descriptive 27 (2021)

4. Pickard, L., Shah, D., De Simone, J.J.: Mapping microcredentials across MOOC platforms. Learn. MOOCS (LWMOOCS) **2018**, 17–21 (2018). https://doi.org/10. 1109/LWMOOCS.2018.8534617

5. Nofer, M., Gomber, P., Hinz, O., Schiereck, D.: Blockchain. Bus. Inf. Syst. Eng. **59**(3), 183–187 (2017). https://doi.org/10.1007/s12599-017-0467-3

6. Gleeson, J., Lynch, R., McCormack, O.: The European Credit Transfer System (ECTS) from the perspective of Irish teacher educators. Eur. Educ. Res. J. **20**(3), 365–389 (2021). https://doi.org/10.1177/1474904120987101

7. Greek National Qualifications Framework. https://nqf.gov.gr/en/index.php/ta-8-epipeda. Accessed 10 Feb 2022

8. Jirgensons, M., Kapenieks, J.: Blockchain and the future of digital learning credential assessment and management. J. Teach. Educ. Sustain. **20**(1), 145–156 (2018). https://doi.org/10.2478/jtes-2018-0009

9. Bahrami, M., Movahedian, A., Deldari, A.: A comprehensive blockchain-based solution for academic certificates management using smart contracts. In: 2020 10th International Conference on Computer and Knowledge Engineering (ICCKE), pp. 573–578 (2020). https://doi.org/10.1109/ICCKE50421.2020.9303656

10. Turkanović, M., Hölbl, M., Košič, K., Heričko, M., Kamišalić, A.: EduCTX: a block-chain-based higher education credit platform. IEEE Access **6**, 5112–5127 (2018). https://doi.org/10.1109/ACCESS.2018.2789929

11. BouSaba, C., Anderson, E.: Degree validation application using solidity and ethereum blockchain. SoutheastCon **2019**, 1–5 (2019). https://doi.org/10.1109/ SoutheastCon42311.2019.9020503

12. Gräther, W., Kolvenbach, S., Ruland, R., Julian, S., Ferreira Torres, C., Wendland, F.: Blockchain for education: lifelong learning passport. In: Prinz, W. (ed.) European Society for Socially Embedded Technologies. EUSSET, Bonn. 1st ERCIM Blockchain Workshop 2018. Proceedings of the Blockchain Engineering: Challenges and Opportunities for Computer Science Research, 8–9 May 2018, Amsterdam, Netherlands (2018)

13. Lesjak, M., Juvan, E., Ineson, E.M., Yap, M.H.T., Axelsson, E.P.: Erasmus student motivation: why and where to go? High. Educ. **70**(5), 845–865 (2015). https://doi. org/10.1007/s10734-015-9871-0

14. Kutty, R.J., Javed, N.: Secure blockchain for admission processing in educational institutions. In: 2021 International Conference on Computer Communication and Informatics (ICCCI), pp. 1–4 (2021). https://doi.org/10.1109/ICCCI50826.2021. 9402654

15. Lechner, C.M., Tomasik, M.J., Silbereisen, R.K.: Preparing for uncertain careers: how youth deal with growing occupational uncertainties before the education-to-work transition. J. Vocat. Behav. **95–96**, 90–101 (2016). https://doi.org/10.1016/ j.jvb.2016.08.002. ISSN 0001-8791

16. Pastore, F.: Why so slow? The school-to-work transition in Italy. Stud. High. Educ. **44**(8), 1358–1371 (2019). https://doi.org/10.1080/03075079.2018.1437722

17. Singh, M., Kaur, H., Kakkar, A.: Digital signature verification scheme for image authentication. In: 2015 2nd International Conference on Recent Advances in Engineering & Computational Sciences (RAECS), pp. 1–5 (2015). https://doi.org/10. 1109/RAECS.2015.7453277

18. Munivel, E., Ajit, G.M.: Efficient public key infrastructure implementation in wireless sensor networks. In: 2010 International Conference on Wireless Communication and Sensor Computing (ICWCSC), pp. 1–6 (2010). https://doi.org/10.1109/ ICWCSC.2010.5415904

19. Priya, N., Ponnavaikko, M., Aantonny, R.: An efficient system framework for managing identity in educational system based on blockchain technology. In: 2020 International Conference on Emerging Trends in Information Technology and Engineering (ic-ETITE), pp. 1–5 (2020). https://doi.org/10.1109/ic-ETITE47903.2020.469
20. Terzi, S., Savvaidis, C., Sersemis, A., Votis, K., Tzovaras, D.: Decentralizing identity management and vehicle rights delegation through self-sovereign identities and blockchain. In: 2021 IEEE 45th Annual Computers, Software, and Applications Conference (COMPSAC), pp. 1217–1223 (2021). https://doi.org/10.1109/COMPSAC51774.2021.00168

Open Source Discovery, Adoption, and Use: An Informal Perspective

Anthony I. Wasserman[✉] [iD]

Carnegie Mellon University – Silicon Valley, Moffett Field, CA 94035, USA
tonyw@acm.org

Abstract. While the number of open source software projects and related businesses has grown rapidly, it remains difficult for those unfamiliar with open source to take advantage of what it has to offer. Newcomers often have a difficult time finding suitable projects to use, understanding the rules imposed by their licenses, and making contributions of their own to an open source project. Large high-tech companies have led this rapid growth and have cooperated to produce "best practices" guidelines to help others improve their internal governance of open source practices and their participation in the larger community. However, many individuals and organizations that could benefit from open source software, as well as contribute to open source projects, face barriers to doing so. This paper takes their perspective and informally presents these key issues and roadblocks, as well as addressing some of the questions that they often raise in their efforts to find, adopt, and use open source software.

Keyword: Open source

1 Introduction

Open source is everywhere. Well-known projects such as Linux, Firefox, WordPress, VLC, MySQL, and cryptocurrency projects Bitcoin and Ethereum, are just a handful of open source projects used by millions of people. The company Synopsys [1] does inventories of software every year, and has discovered that more than 90% of all software products contain open source code, even if they themselves are not open source.

Software developers rely heavily on open source projects, whether that's Eclipse as a Java programming environment, or Docker or Kubernetes for containers and deployment. If you took away all the open source development tools, you wouldn't have much left. It's not hard to discover that almost every project has open source in it, even the entertainment systems on commercial airplanes, which run Linux.

Most of the people who come to conferences related to open source have a lot of experience with it. They know about the history of open source, when the Open Source Initiative (OSI) [2] was founded, and details about various open source licenses. They also know how people use GitHub to store their development

A. Cerone et al. (Eds.): SEFM 2021 Workshops, LNCS 13230, pp. 264–276, 2022.
https://doi.org/10.1007/978-3-031-12429-7_19

work, how to make a pull request, the names of the open source foundations and their major projects, and maybe how to look up the history of particular projects on OpenHub. They've been involved in using and contributing to open source projects, and know how open source communities work.

However, for people without that kind of experience, it's not so easy. They look at open source, and it's a completely new thing to them. The home page of GitHub [3] provides no help. One can see what projects are popular in GitHub, and follow a link to them. GitHub also shows a list of people who are "trending". Somebody who's just starting out and has been told to go to GitHub will look at that home page, and be tempted to run in the opposite direction. To them, the meanings of "trending developer" and "pull request" are a complete mystery. That issue shows why people in organizations often have difficulty finding free and open source software that meets their needs, even as they use it all the time without knowing it.

2   Barriers to Open Source Adoption and Use

There's also a history of people in the open source community not being welcoming to people they call "noobs". Those newcomers want to understand how open source works, to see what's available, and how open source can address their problems so they can use (and eventually contribute to) open source projects in addition to whatever they're doing now. But they are often discouraged by the difficulty of getting involved.

There has also been a longstanding problem with gender discrimination. Until recently, women were significantly underrepresented in the open source community. However, this issue is being actively addressed, both by the diminished roles of two of the worst male offenders and by the growing community of women holding leadership roles in open source projects and organizations.

Yet another barrier to open source adoption and use is the dominance of the English language on most of the larger projects. While the projects themselves support end users working in many different languages, the *lingua franca* among developers is English. (There are, of course, thousands of projects in other languages, but these tend to be more limited in their geographical reach.)

These barriers can be overcome, but they can make it harder for companies, non-profit organizations, and government agencies to get started. These organizations range from having no familiarity with open source all the way to having their own internal Open Source Program Office. The most experienced of these will have a team of people who manage their open source, keep track of the inventory of projects and versions that people are using, and make sure that they understand how open source is being used, both internally and in projects in which they participate. Many of these organizations release software for use by others. They often put it on GitHub, and they have their own section where people can go and and see that software to both use it and contribute to it.

The least experienced companies and organizations that are just starting out are learning as they go. Over a period of years, they develop their own expertise. It's not uncommon for an organization to take a 10 year journey from where they started out with all the management resistance to open source, to the point where they're now recognized as a significant contributor to the open source community.

These organizations, particularly when they're starting out, have several problems that they have to address. The first is that they don't consider open source options at all. They're used to talking to vendors, and they have relationships with the salespeople and technical support people of those vendors. When they're looking for something new, that's where they naturally look first. If they knew about the website alternativeto.net, they could type in the name of their favorite proprietary program and see some open source alternatives that often closely match the capabilities of the proprietary software and in some of the newer aspects of software do as well or better. Another thing is that these organizations don't really know how to do is evaluate possible candidate software. What they have always done in the past is to call up their favorite industry analyst, or look at some trade publication, or (prior to the pandemic) go to a trade show . They'll read product reviews, and they'll get their impressions. But that doesn't work very well for open source software, which tend to get very little coverage from industry analysts or journalists.

3   Types of Open Source Projects

Beyond access to the source code, there are other important qualities in evaluating proprietary vs. open source software. However, there are more than two categories to be considered. For open source software, there are distinctions between community, foundation, commercial, and open core projects.

3.1   Commercial Projects

Many commercial projects are offered by vendors who follow an open core approach, with a basic version of their product available under an OSI-approved open source license and an enhanced set of features available for a fee under a traditional commercial End User License Agreement. Users typically have the option of discussion boards for addressing questions and issues with the basic community version, and paid vendor support for the commercially-licensed version. These vendors promote the commercial version of the product, and it can be difficult for someone unfamiliar with open source software to find the community edition. However, the commercial approach may be the best choice for newcomers to open source, since it most closely resembles the way that they have acquired other software. The developers responsible for the project typically work for the vendor, which also has paid sales, marketing, and professional services teams.

A second category of commercial open source business merely supports one or more existing open source projects, perhaps by cloud hosting of a project and/or by offering training and support. Given the rules of open source licenses, there is nothing to prevent anyone at all from offering these services independently of the open source project itself. That freedom has caused problems for some open source projects, which have lost much of their revenue to larger, well-known hosting services. As a result, they have changed their licenses to "source available" and included restrictions that prevent third parties from hosting the software. The new licenses are not approved by the Open Source Initiative and are, strictly speaking, no longer open source.

3.2   Foundation-Based Projects

Many of the most widely used non-commercial open source projects are supported by foundations, which raise money from donors to support their work. Several of these foundations, particularly the Linux Foundation [4] and the Apache Foundation [5], support multiple projects and oversee their governance and development practices. Many foundation-sponsored projects are well-established and mature, with many thousands of users, extensive documentation, a responsive team of maintainers, and a dependable process for releasing enhancements and fixing bugs. The quality of the software is high, comparable if not superior to vendor-supported commercial products. The developers working on these projects may be paid by the foundation, but may also be paid by separate employers to work on the project as part (or all) of their job.

3.3   Community-Based Projects

Finally, community-based open source projects are independently developed and maintained, often without a dependable source of revenue. Contributors may work on these projects in their free time, which may vary depending on their personal situations. There is often a group of core maintainers (committers), supported by other volunteer contributors. Some of these projects have gone through many releases over many years and are of the highest quality, while others are no longer actively maintained. It is unusual for community-based projects to invest in sales and marketing efforts, so it is more difficult to find them and to evaluate their suitability. These projects are supported on associated discussion boards and, occasionally, on independent developer-focused sites, such as Stack Overflow [6].

In proprietary software development, the product direction is typically overseen by one or more product managers, whose job it is to translate market requirements into a set of features to be developed and to define the roadmap for successive releases of the product. Non-commercial open source projects, including foundation and community-based projects, frequently lack the development leadership provided by a product manager. Instead, these projects have one or more committers who determine product direction. In addition to writing pieces that they view as important, they may also receive contributions from volunteers, who have independently developed code and contributed it to the project. Such code is typically reviewed by the committer team for readability, overall quality, and relevance to the project. High-quality contributions are accepted into the project's main code base, while others may be rejected outright or returned to the contributor for rework and resubmission.

There's also an important difference between proprietary and open source projects in terms of scheduling. If a project relies heavily on volunteers, these volunteers sometimes find other things in their lives that interrupt them. It's not as easy to produce a schedule as with a commercial product. That said, there are many popular open source projects that come out of communities rather than from companies. For example, Scikit-learn [7] is very popular with Python developers. Shotcut [8] and OBS Studio [9] are high-quality open source video production tools that are competitive with the commercial products in the market.

These distinctions are important for organizations seeking to adopt open source software, both for their own internal use and for inclusion in a product that they will make available to others. Organizations seeking to redistribute code in this way must also become familiar with the various popular open source licenses, some of which impose restrictions on including open source software in a commercial product. Some open source projects, such as MySQL, [10] use dual licensing, offering both an approved open source license and a traditional commercial license, where the intended use of the software determines the appropriate license.

Needless to say, these categories introduce a new dimension of complexity to an organization's effort to use open source software. Because of these distinctions, many of the vendors of open core products downplay the open source aspects of their project since it is in their financial interest to sell their commercial project rather than to direct potential customers to the free open source version.

4   Intended Uses of Open Source Software

A key factor in any software evaluation project is the risk that it presents to the adopting organization. Software intended for business-critical and life-critical applications needs an assured and timely response, especially for bug fixes, security breaches, and other issues that can harm an organization and/or its customers. That's an important reason why companies, governments, and other organizations have historically relied on proprietary products and services from well-established commercial vendors. These business considerations also present a challenge for startup companies trying to compete with the larger vendors, whether or not their code is open source. The well-known vendors have professional sales teams that can address both the technical and business concerns of prospective customers. The evaluation and the associated sales process might involve comparative evaluation among several similar products, as well as a supported evaluation period and price negotiations.

Apart from these high-visibility, high-risk projects are less critical applications, often for internal use within the organization. In these situations, open source projects are often a superior option, particularly considering the total cost of ownership. If the open source software project is led by a vendor, then the vendor may have a community version of it. That doesn't come with paid support. But you can go to the website and exchange ideas with people and get help, as well as download and install new versions of that software. The vendor may try to convince its users to upgrade from the community version to a paid version, since that's how the vendor is going to monetize the work that they've put in on building that software.

Another place where open source can easily be used is as part of a hosted application, such as those found in e-commerce and informational sites. Not only is the business risk relatively low, but many of these projects are widely used. The best example is WordPress [11], which is used for about 1 of every 3 websites, more than 400 million sites in all [12]. Studies of these sites have shown that almost all of the vulnerabilities are associated with plug-ins, not with the WordPress content management system (CMS) itself. Also, JavaScript frameworks such as jQuery [13] and React.js [14] are extensively used in millions of web sites [15], making them a safe choice for organizations without much previous experience with open source. When these are used in hosted applications, they can be easily updated as needed.

5   Commercial Support and Open Source Projects

Many organizations have policies that require a source of support for any piece of software that is included in a product that is offered to its customers. Such a policy would seem to rule out the use of foundation-based and community-based open source projects. However, companies have been created to provide paid support for many of the most mature projects in those categories.

For example, Drupal [16] is a widely used open source CMS that was first released more than 20 years ago. It is now on its 9th major release, with hundreds of minor releases along the way. Beyond that, Drupal has attracted many developers who build add-ons and customized themes. They have also built a community of companies that do customized creation of applications built on Drupal; among these companies are Chapter Three, Pantheon, and Lullabot. An organization can contract their site development work to one of these (or similar) companies, and they will use Drupal and a well-tested set of add-ons. In addition, the founders of the Drupal project created Acquia [17], which offers hosted service and monitoring of these Drupal applications as paid services, along with marketing tools. In short, an organization wanting to use Drupal has the flexibility to handle all aspects of implementation and support itself, or to pay others to handle some or all of those details, all with open source software.

The Drupal/Acquia combination is not the only example of commercial support for open source projects. For example, Datastax provides support for the open source database Cassandra, while Aiven and Instacluster provide commercial hosting for Cassandra. Going one step further, 47deg [18] and OpenLogic [19] (owned by Perforce) offer commercial support for numerous popular open source projects, helping organizations with this important aspect of open source adoption. There is a Wikipedia article listing many pairings of open source projects with their source(s) of commercial support.

6   The Challenge of Evaluating Open Source Software

There are different types of open source software. Some open source software comes from companies that simply chose to make their source code available, while earning money from providing services or from having proprietary add-ons to their open source software. Organizations are familiar with calling a vendor and getting a trial of the software along with technical support to see how the software works, before they make a buying decision.

By contrast, if they try to do a search on something like "open source content management systems", they'll receive a huge list of results. Beyond that, many of the search results are ads from companies that have paid for certain keywords. Even worse, the first page(s) of results may be filled with lists of "the 10 best content management systems" as collected by third party companies in return for a payment from the developers of those content management systems. Many of the listed options will be from vendors of proprietary software, not open source systems. The search process, as it currently exists on the Web, may make it very difficult to identify the open source options because the creators of community-based open source projects rarely spend money to appear on one of the first pages of results provided by such searches. In fact, these lists of "best" products are a form of advertising.

The organization is then left to keep searching and to then find a way to evaluate the options. Evaluation of open source software is somewhat different than the evaluation of proprietary software. In a personal survey on software evaluation techniques some years ago, the following answers were the most common:

1. previous experience: knowing how to use it, knowing what works and what doesn't, knowing how to get product support.
2. published reviews, similar to those for hotels, restaurants, or movies. Some of the reviews are written by people who do that professionally.
3. contact a salesperson for a vendor and ask some questions.
4. conform to an organization's standard for that task, eliminating the need for an independent evaluation.
5. participate in a forum discussion, such as those on Stack Overflow, which is a good way to learn about particular kinds of software; many vendors have their own forums as well. It's not just open source projects that have forums.
6. do a personal trial, taking advantage of a free trial evaluation period offered by many vendors of proprietary software products. That makes it possible to get a sense of how it works and whether it's going to meet current needs.
7. ask a knowledgeable and experienced friend or colleague and accept their recommendation.

8. do a detailed internal evaluation of several products, performing competitive analysis among several candidates and scoring them against a set of criteria covering both product functionality and business risk.

Note that some of these techniques will work for both proprietary and open source software, while others are better suited for proprietary software. Oddly enough, almost no one responded that they reviewed the source code of an open source project to gain a good understanding of its quality, an aspect that is rarely available when assessing proprietary software products.

The evaluation might also include a check on the performance or security of the project, perhaps compliance with certain industry standards, including interfaces. In medical computing software, for example, in the United States, there are security regulations (HIPAA) that must be met. In summary, evaluation is a complicated process, and it's important to rule out projects that don't meet the requirements. There are also millions of open source projects that have been started and abandoned, and it's essential to rule out the "dead" projects in an evaluation.

All of these approaches can be used to find a worthy open source project to meet a specific need. The next question is "what does it mean to be worthy?" The minimum qualities of a worthy project is that it addresses the user needs, that problems are going to get fixed by project maintainers, and that there's a team of people actively releasing bug fixes and new versions. Beyond that are issues related to the size and culture of the community surrounding the project, especially its openness to new participants and contributors.

People who are familiar with open source can find information on the details of a project. They can easily see how well the project is managed. From the main project page (likely in GitHub) or the summary data in OpenHub [21], one can see who's working on it, when they have contributed and how many lines are written in which programming languages. One can also see who the committers are, how many versions have been released, how many downloads have been made, and how many stars have been awarded to the project by others. It's also possible to look at the Bugzilla [22] listing, for example, to see how many open bugs there are, how many have been assigned to people, all of which provides a sense of how active the project is and how well the project is being managed. But someone who is new to open source may not know how to find these details.

7   Quantitative Evaluation of Open Source

There have long been efforts across many different categories to do quantitative evaluations of products, giving a relative weight to the major features of a product and then associating a score for that feature in a product. That approach yields numeric scores, allowing the evaluator to compare multiple similar products with one another or to see if an individual product scores above or below a threshold level. In 2005, the Business Readiness Rating (BRR) [23] project was created to apply that evaluation methodology to open source software, using a set of relevant categories for evaluation. The overall goal was to help organizations and people with evaluation of open source software. A high score indicated its suitability for use in a critical project, while a low score meant that it should be used with caution or not at all.

While the BRR provided a well-defined quantitative evaluation framework, the project itself was unsuccessful, both for personal and technical reasons. Performing an evaluation was too labor-intensive, with very little automated support for the process. Evaluating several different candidate projects was simply too time-consuming. As noted above, most evaluations are done more informally. In 2012, a follow-on project, OSSpal [24], was started, with the aim of overcoming the major shortcomings of the earlier BRR project.

Instead of leaving the options in a software category completely wide open, OSSpal is based on curation of projects in many different categories, following the taxonomy of software developed by the International Data Corporation (IDC). The underlying thought was that evaluators want to know the 3–5 most worthy projects in a category, in terms of the quality of the software, the longevity of the project, available support, the quality of the code, documentation, and security. That information was stored on a website (https://osspal.org), which not only provided essential information and links related to a project, but also opportunity for registered users to post their own evaluations. The selection of projects in a category was based on the questions that a typical evaluator might ask: What does it do? What are the key functions and features? How well does it run? Does it scale? Is it reliable? Is it usable? Is it secure? Is it supported? Is it well documented? Evaluators could then look at these results based on the specific needs of their organization, then follow a link to the home page for the product being evaluated and download it.

The idea was to get a community-based evaluation score for each project, individual comments, in much the same way you can find reviews for other products, including proprietary software and automobiles.

Unfortunately, the OSSpal project has also been unsuccessful in practice. The effort to generate awareness of the project was insufficient, so the project never achieved the needed critical mass of user participation. Also, the IDC taxonomy is better suited for corporate applications rather than for tools used for software development and system infrastructure, which have many more open source development projects.

8   The Impact of Open Source Software

Looking back a decade or more, open source had a relatively minor position. Since then, however, changes in technology and in the business environment have made open source much more important. The leading tools for large scale data management, machine learning, and artificial intelligence are all open source. Much the same is true for tools used for integration and continuous deployment of applications, as well as for monitoring of cloud-based applications. Organizations seeking to develop and use modern applications will find it necessary to understand and use open source software.

On the business side, long-term industry leaders such as Microsoft and IBM have made major investments in open source, with Microsoft acquiring GitHub and IBM acquiring RedHat, the leading commercial vendor of Linux. In short, the world has changed.

That has also affected people's attitudes about open source. For a long time, people would be in either the open source camp or the proprietary camp. This author proposed the Wasserman Index, a measure of how much of your activities could be done entirely with open source software, i.e., without using proprietary applications from commercial vendors. For most people, the Index was very low, since there are very few open source mobile apps or open source games for game consoles. While there are many people who make a serious effort to avoid using proprietary software products, it isn't easy to do so, particularly when working in specialized domains.

These days, many individuals and organizations are looking for the best software that's available for the task at hand. Sometimes it's proprietary software that they license from a vendor, running it in-house on their own server. Maybe it's hosted software, such as salesforce.com, which is proprietary software, but which uses Open Source extensively. Similarly, they can opt for open source software. As mentioned above, WordPress is a great example of widespread open source adoption. As a result, many organizations, companies, and agencies deploy a mix of proprietary and open source software.

9   Why Participate in Open Source Projects

Newcomers to open source often ask why people volunteer their time to work on an open source project, and why companies pay people to work on open source projects. In the first case, people frequently volunteer their time to support causes. When there was a need for a service to help track victims of natural disasters and to help people find treatment for serious diseases, including COVID-19, many thousands of software developers started or joined projects to address those problems. What is now the Sahana EDEN project managed by the Sahana Foundation [25] was originally developed in response to the 2004 tsunami in the Indian Ocean. The United Nations Development Program has created a COVID-19 Open Source Digital Toolkit [26], and there are a myriad of other open source projects created to address humanitarian needs. But it's not just this type of application that attracts volunteers; many people want to make fixes or enhancements to existing projects, or to start their own project. Just as people volunteer for a wide range of activities and social causes, many software developers contribute their knowledge of software development to projects of personal interest.

As for corporate contributions to open source projects, a company might decide that a project is of value to them, so they're going to pay their own employees to work on it. That effort isn't always completely altruistic, as companies may have preferences about which features should be added or introduced to a project to make it more useful for their needs. Excellent examples of such corporate contributions can be found in projects such as the Linux kernel, the Eclipse development environment [27], and the Kubernetes project [28] for management of containers. Many companies also contribute to open source foundations as a way to be more involved in the open source community, both to support specific types of projects and as a way to make themselves more attractive to developers that they may want to hire.

10   Where to Learn More

Becoming knowledgeable about open source software requires learning about topics from law, e.g., licenses, to development methods and tools, just to name a couple. There are several different categories of education to learn more.

There are numerous courses on specific applications. People wanting to know how to use WordPress, for example, or Drupal, can easily find websites, books, individual courses and videos, and lots more. Classes are available on the project websites, on YouTube, or on some of the commercial course sites such as Coursera and Skillshare. Vendors of commercial open source often offer both in-person and video training on their products, sometimes charging a fee. In short, learning about a specific project is pretty straightforward.

It's more difficult to find broader coverage. There are relatively few degree-awarding academic programs, along with several university-based certificate programs that offer a set of courses related to open source. One issue in developing such programs is the breadth of topic material. This author started the FLOSSbok project [29], which is aimed at providing comprehensive information about open source software that can allow its users to find a wide variety of materials related to various aspects of open source development and use. For this project to be successful, it needs contributions from people whose expertise covers the wide range of relevant material.

For companies and other organizations, the work of the TODO Group [30] is particularly valuable. They present the open source practices for companies, and have produced a series of guides (currently 17) linked from their Groups page. They cover licensing, how to contribute to projects, and how to start projects, among other things. In addition, they have created a more complete course on how to start an open source project office, which is found on GitHub.

11   A Final Word

Open source software is similar to a participatory sport, in that sitting on the sidelines and watching is insufficient. One effective way to get started is to select a piece of open source software and use it. Firefox [32], WordPress, and LibreOffice [33], just to name a few, are mature and high-quality applications that not only perform well, but also give users confidence in using open source.

Initially, most people and organizations are simply users of open source projects, taking advantage of the work of others. But it's valuable to go beyond being an end user and to step up to becoming part of the community around any project of interest. Anyone can contribute requests for enhancements and participate in the discussion forums of a project. The next step is to be a contributor to the project, not necessarily making a code contribution, but answering questions on a forum or translating the natural language in the user interface. For people who become particularly knowledgeable about a particular project, there is often an opportunity to join a company that is providing commercial support for the project.

Open source is not going to replace proprietary software, but rather will co-exist with it. People routinely use some of each, and open source software is frequently included in commercial products, particularly applications hosted in the cloud. With that in mind, it's valuable to include open source projects in every effort to discover the best options for both personal and business use, and to enhance the ability of organizations to adopt and use open source when appropriate.

References

1. Synopsys, Open Source Security and Risk Analysis Report (2021). https://www.synopsys.com/software-integrity/resources/analyst-reports/open-source-security-risk-analysis.html
2. Open Source Initiative: https://opensource.org. Accessed 5 Apr 2022
3. GitHub: https://www.GitHub.com. Accessed 5 Apr 2022
4. Linux Foundation: https://linuxfoundation.org. Accessed 5 Apr 2022
5. The Apache Software Foundation: https://apache.org/. Accessed 5 Apr 2022
6. Stack Overflow: https://stackoverflow.com/. Accessed 5 Apr 2022
7. https://scikit-learn.org/stable/
8. Shotcut: https://www.shotcut.org/. Accessed 5 Apr 2022
9. OBS Open Broadcaster Software: https://obsproject.com/. Accessed 5 Apr 2022
10. MySQL: https://www.mysql.com/about/legal/licensing/oem/. Accessed 5 Apr 2022
11. WordPress: https://wordpress.org/. Accessed 5 Apr 2022
12. Website Builder: https://websitebuilder.org/blog/wordpress-statistics/. Accessed 5 Apr 2022
13. Jquery: https://jquery.com/. Accessed 5 Apr 2022
14. ReactJS: https://reactjs.org/. Accessed 5 Apr 2022
15. Sitepoint: https://www.sitepoint.com/most-popular-frontend-frameworks-compared/. Accessed 5 Apr 2022
16. Drupal: https://www.drupal.org/. Accessed 5 Apr 2022
17. Acquia: https://www.acquia.com/. Accessed 5 Apr 2022
18. Degrees: https://www.47deg.com/services/open-source-support/. Accessed 5 Apr 2022
19. OpenLogic: https://www.openlogic.com. Accessed 5 Apr 2022
20. Wikipedia: https://en.wikipedia.org/wiki/Listofcommercialopen-sourceapplicationsandservices. Accessed 5 Apr 2022
21. Synopsys Black Duck OpenHub: https://www.openhub.net. Accessed 5 Apr 2022
22. Bugzilla.org: https://www.bugzilla.org. Accessed 5 Apr 2022
23. Wasserman, A., Pal, M., Chan, C.: "Business Readiness Rating Project" BRR Whitepaper 2005 RFC 1. www.immagic.com/eLibrary/ARCHIVES/GENERAL/CMU,US/C050728W.pdf. Accessed 5 Apr 2022

24. Wasserman, A.I., Guo, X., McMillian, B., Qian, K., Wei, M.-Y., Xu, Q.: OSSpal: finding and evaluating open source software. In: Balaguer, F., Di Cosmo, R., Garrido, A., Kon, F., Robles, G., Zacchiroli, S. (eds.) OSS 2017. IAICT, vol. 496, pp. 193–203. Springer, Cham (2017). https://doi.org/10.1007/978-3-319-57735-7_18
25. Sahana Foundation: https://sahanafoundation.org/. Accessed 5 Apr 2022
26. COVID-19 Open Source Digital Toolkit: https://sgtechcentre.undp.org/content/sgtechcentre/en/home/featured-work/digital-tools-for-covid-19.html. Accessed 5 Apr 2022
27. Eclipse Foundation: https://www.eclipse.org/. Accessed 5 Apr 2022
28. Kubernetes: https://kubernetes.io/. Accessed 5 Apr 2022
29. FLOSSbok: https://flossbok.org. Accessed 5 Apr 2022
30. ToDo Group Guides: https://todogroup.org/guides/. Accessed 5 Apr 2022
31. OSPO 101: https://github.com/todogroup/ospo101. Accessed 5 Apr 2022
32. Firefox Browser: https://www.mozilla.org/en-US/firefox/new. Accessed 5 Apr 2022
33. Libre Office: https://www.libreoffice.org. Accessed 5 Apr 2022

DrPython–WEB: A Tool to Help Teaching Well-Written Python Programs

Tommaso Battistini[1], Nicolò Isaia[1], Andrea Sterbini[1(✉)] (iD),
and Marco Temperini[2(✉)] (iD)

[1] Computer Science Department, Sapienza University of Rome, Rome, Italy
sterbini@di.uniroma1.it
[2] Department of Computer, Control and Management Engineering,
Sapienza University of Rome, Rome, Italy
marte@diag.uniroma1.it

Abstract. A good percentage of students, while learning how to program for the first time in a higher education course, often write inelegant code, i.e., code which is difficult to read, badly organized, not commented. Writing inelegant code reduces the student's professional opportunities, and is an indication of a non-systematic programming style which makes it very difficult to maintain (or even understand) the code later, even by its own author. In this paper we present DrPython–WEB, a web application capable to automatically extract linguistic, structural and style-related features, from students' programs and to grade them with respect to a teacher-defined assessment rubric. The aim of DrPython–WEB is to make the students accustomed to good coding practices, and stylistic features, and make their code better. There are other systems able to perform code analysis through quality measures: the novelty of DrPython–WEB, with respect to such systems, is in that it analyzes also linguistic and stylistic features.

Keywords: Teaching programming · Feature extraction · Good coding practices · Python

1 Introduction

One of the main tasks of a computer programming course is to allow the students to reach an adequate level of skills, so to be able to produce well written programs, well organized, commented, readable, possibly efficient. The difficulty of accomplishing such a task is particularly felt in Higher Education in Computer Science, as students in that area will become, in a relatively close future, professionals with important responsibilities in private and public sectors [1–3].

Students' skills to produce programs showing good or even high "quality" are acquired through practice and are applied to various aspects of programming, ranging from the capability to define suitable algorithms to solve a given problem, through to the ability to design a program and the relevant data structures,

A. Cerone et al. (Eds.): SEFM 2021 Workshops, LNCS 13230, pp. 277–286, 2022.
https://doi.org/10.1007/978-3-031-12429-7_20

to practical coding abilities that allow a student to produce a readable program, i.e. a program whose instructions are 1) textually formatted in a readable fashion, 2) easy to interpret, as far as their purposes are concerned, and 3) as clearly commented as possible.

Both learning and teaching of Computer Programming are challenging tasks, when the traditional approach to education is used [4]. Hence, the availability of web-based automated support can be of great value, especially in Higher Education, where often direct interactions between a student who is solving a programming task, and a teacher who could help, are not easily achievable [5], especially in the case of Italian university courses where the student/tutor ratio is very high. E.g., in our courses (at Sapienza University of Rome) the usual number of students per teacher in the initial courses is 150, which is possibly due to 150 being the maximum allowed number of students per teacher by law.

Many researchers have recognized the impact of the students' programming style on their grades and have developed systems to recognize stylistic features of programs [6–8], which are usually mainly focused on programming constructs. We follow this line of research by considering also linguistic features to better recognize if the program, considered as a written text, is readable and easy to understand and to maintain.

Moreover, we want to use the results of our style/readability assessment as a didactic tool to accustom students to write readable code. To this aim we chose to use rubrics to automatically assess students' submissions, like it has been done for Appinventor [9,10]. The teacher defines the rubric to focus the students' attention on the most important programming, stylistic and linguistic features.

In this paper, we present a web-based system, DrPython–WEB, whose use could help a student improve her/his coding skills, by pointing out and recognizing the "elegance" of the student's code in an automated and real-time fashion.

By "elegance" we mean a subset of the several qualities of a program, mentioned earlier, related to structure, readability and maintainability. On these aspects DrPython–WEB focuses its program analysis, and evaluation. In particular, given a program, the analysis is performed on a set of features, extracted from the program (see later), as well as on the good naming quality of the identifiers (i.e., the names given by the programmer to certain structures of the program, such as types, variables, and functions).

DrPython–WEB is clearly inspired to the DrScratch system [11] and other rubric-based systems [9], which statically analyze the submitted Appinventor programs to extract program features and highlight higher levels of competency on different topics (e.g., usage of more complex data structures, complex conditionals, parallelism etc.).

Similarly, DrPython–WEB measures the usage of many Python constructs with the aim of recognizing more expert programmers, and produces a teacher-defined rubric-based assessment to invite students to learn and to use the more advanced Python constructs.

Beside the focus on programming competency, we want also to accustom students to write good-style code (modularized, readable, well documented). During our lessons we use always a "describe first - implement later" development methodology and with DrPython–WEB we want to assess both the expertise and the style of the submitted homeworks.

The novelty of DrPython–WEB, w.r.t. other feature extraction systems [8], is in its analysis of linguistic features, type of keywords used in the documentation, their semantic distance from the exercise topics, use of self-explanatory identifiers.

We developed DrPython–WEB with a twofold aim: on the one hand we would like to encourage students to practice and improve their coding style; on the other hand we wanted to support both student's awareness and teacher's assessment procedures, by providing them with visual summaries of data, reporting the elements on which the overall evaluation of the code was based.

We haven't yet used the system in class but we are going to experiment with it in the next courses. Therefore we cannot yet present a comprehensive analysis of the actual effects of its use for the students and teachers.

So, in this paper we present the system, and its features, showing how we used it on a relatively large dataset of programs (produced by students during a recent edition of a course on *Basics in Programming* held at our University). Such dataset is made of programs produced to solve tasks related to the several mandatory homework requested during the course, and the solutions submitted for final exams. The programs available spans 4 years and 2 parallel courses, with 4 mandatory homeworks and 4 optional, for an average of 350 students a year, and an approximate total of 10000 homework submissions and 1000 exam submissions. These homework and exams, until now, have been graded w.r.t. their correctness through unit tests. A small bonus was awarded to students that implement faster and/or less intricate (with smaller cyclomatic complexity) code. With DrPython–WEB we want to start awarding a bonus also to more elegant code.

The main goals in this paper are then the following:

Goal 1: *To show that DrPython–WEB can automatically extract the stylistic features of a program, and assess their usage to push students towards a better programming style.*
We will see that DrPython–WEB is able to 1) perform an automatic check of the hundreds of programs in our sample, 2) analyze, in such programs, the coding qualities we associated above to "elegance", and 3) express a quality grade for each program.

Goal 2: *To show how DrPython-WEB supports personalization of the assessment depending on the teacher's preferences, each stage in the course, or just the specific assignment's characteristics.*
In this respect, we will see that the analysis performed by DrPython–WEB can be configured by the teacher, who is able to finely-tune the assessment by specifying in a rubric her/his preferences about the features to be taken into consideration, and their weight in the computation of the overall quality

grade. The metric for the overall grade is the result of the weights chosen by the teacher for the rubric's features. Notice that, as features are normally difficult to normalize, so we usually award bonuses by first ranking the grades obtained and then by selecting the highest performing 50% of them.

In particular, the possibility to configure many assessment rubrics allows the teacher to adapt the analysis of a given batch of programs, depending on the relevant characteristics of a given task, and/or the aspects to be taken care of at a given point-in-time of the course.

We plan to add the DrPython–WEB rubric-based stylistic self assessment to our Q2A-I system [12]. In Q2A-I students self-assess their python exercises, which are tested with unit tests, and receive bonuses for faster solutions and/or for less intricate programs (with lower cyclomatic complexity). In Q2A-I students participate in a formative peer-assessment phase where they suggest each other how to improve their algorithms by reading and commenting each other's algorithm descriptions.

We are still working on the analysis to understand if the already available data, collected in the previous years, shows that the linguistics features are linked to the exercise and exam grades (notice that in the previous years we have not asked the students to write nice code because of the difficulty of automatically checking for stylistic properties). No easy linear or monotonic relation seems to arise from our initial analysis yet. This is expected, as the student's population is made of several groups of students with different skills and behaving differently.

In the following sections we will:

1. Present the software library *DrPython*, which we developed to provide core functionalities for the analysis of a program: using these new functionalities DrPython–WEB was developed.
2. Present the use of DrPython–WEB on a set of sample programs, in order to show the characteristics of the system and see its potential application on the field.
3. Present some conclusions, submitting that DrPython–WEB, although subject to further improvements, can be an effective means to help the students to improve their coding style.

2   DrPython: Feature Extraction Module

DrPython–WEB is based on the feature extraction library we developed for this task (named DrPython). We chose to base the system on feature extraction because features allow for an expandable and easy to understand definition and description of assessment rubrics. Others have used feature extraction on Python programs (e.g. [8]) to extract textual and structural features from programs. To these type of features we add linguistic features (see below).

We developed the DrPython library to analyze the student's program and algorithm description to recognize/extract three types of features:

- **Code syntax features:** the number of specific language constructs in the program (functions, classes, super-classes of each class, methods, try-except, list-comprehensions, if-then-else, generators, lambda, recursive functions, variables, arguments),
- **Code quality measures:**
  - McCabe's cyclomatic complexity [13], that captures how much a function control flow is intricate,
  - Halstead's measures [14], that captures a function's conceptual complexity from its vocabulary size and number of operators used,
  - *Code smells*[1], i.e., code structures that often imply bad coding practices.
- **Linguistic features:**
  - *Good identifiers*, i.e., self-explanatory names that convey the meaning of their function. This relieves the programmer from having to recall what type of data is in a variable and what its place is in the algorithm, as well as the action performed by a function/method,
  - *Good documentation practices* i.e., using comments and doc-strings to describe the reason for particular programming choices. This helps the reader to better understand the meaning of the algorithm implemented.
  - The usage of *pertinent keywords* related to the exercise description both in comments/doc-strings or in the algorithm description. This allows DrPython to automatically check (roughly) if the documentation is adequate to the task.

The code syntax features are extracted/counted by means of the **redbaron**[2] source code analysis library that allows to easily query the code structure for specific constructs. Redbaron queries use a syntax similar to CSS selectors (as it's done in jQuery w.r.t. the DOM of HTML pages). This in turn will allow us to easily expand in future the set of code syntax features extracted.

The code quality measures are computed by means of the **radon**[3] library which computes the code metrics: Mc'Cabe cyclomatic complexity of each function, Halstead measures, SLOC, comment count and other simple code metrics.

Finally, to extract the linguistic features DrPython uses the automatic term extraction module **pyATE** [15] to select the 25 highest ranked keywords returned by its **Combo Basic** algorithm [16], and the text analysis library **spacy**[4] to analyze the documentation/comments and the algorithm description. To decide if a particular identifier used by a student is of good/medium/bad quality, DrPython performs the following steps:

- It extracts the pertinent keywords with pyATE from the teacher's exercise task description
- It decomposes the identifier into its component words
- It compares the words (by means of spacy semantic similarity and the Word-Net semantic network) to grade their similarity to the pertinent keywords

---

[1] https://wiki.c2.com/?CodeSmell, accessed 1/11/21.
[2] https://redbaron.readthedocs.io, accessed 1/11/21.
[3] https://radon.readthedocs.io, accessed 1/11/21.
[4] https://spacy.io, accessed 1/11/21.

- It classifies the identifier in the top/medium/bad group depending on having its max similarity to a keyword above 90%, between 40% and 90% or lower than 40%, respectively. We have initially chosen the thresholds to split identifiers in the top/medium/bad classes as the 90% and 40% values, i.e. we consider a very good identifier to be very similar to the exercise keywords, a good identifier sufficiently similar, and a bad identifier rather dissimilar from the keywords. A more detailed study of the data to find the best threshold will follow.

DrPython can be used both as a stand-alone program, to be run from the command line, or integrated in the DrPython–WEB web-based application described below.

For example, with DrPython one could analyze many student files and collect all extracted features as a CSV file and study, for example:

- How the extracted features correlate with each other or with other data (exam grades or readability judgements manually collected)
- How different assessment rubrics will produce different grade distributions

To make the assessment rubrics easier to use, and to automate the submission and assessment of the programs, we have developed the web-based application (DrPython–WEB).

3   Dr.Python-WEB: The System

The DrPython–WEB system allows the teacher to define one or more **assessment rubrics** to grade the submitted programs/algorithms depending on the features extracted, in order to encourage students to use more readable Python constructs, a better linguistic style, and to better modularize their code.

DrPython–WEB is a classic LAMP[5] based web-application written in Python where:

- The teacher defines assessment rubrics depending on the exercise and/or the course phase.
  We want to allow the teacher to define different rubrics in different phases of the course (or even for specific exercises). This way the teacher would be free to, for example, assign more weight to course topics/python constructs that have been recently explained, or to python constructs that are particularly effective to solve efficiently the specific exercise.
- The students submit their code to get the style assessment grade and compare their results with each others'.
- The teacher has an overall view of the students' leaderboard and a detailed view of all submitted programs and assessment results, and thus has the ability to finely tune the rubric in case the assessment is distorted, and update the assessment.

---

[5] LAMP=Linux, Apache, MySQL, PHP/Perl/Python.

Homework 1

Edit template   Show me all student results   Show me all student evaluations

(Not mapped)

Feature	Weight value	Minimum range	Maximum range
Effort	1.5	5.0	100.0
Effort	1.0	100.0	500.0
Effort	0.5	500.0	5000.0
Cyclomatic complexity	0.8	2.0	15.0
Cyclomatic complexity	0.5	15.0	30.0
% Good Identificators	1.2	70.0	100.0

Fig. 1. Assessment rubric that awards more points for lower cyclomatic complexity, lower Halstead's Effort and high percentage of good identifiers: see further explanations in the following text.

Assessment rubrics are defined by the teacher by specifying what are the features assessed, and what are their weights associated to given ranges of their values.

In Fig. 1 we show an assessment rubric that awards more points to a lower *Halstead's effort* ('Effort' in the figure), to a lower *cyclomatic complexity*, and to a higher *percentage of good identifiers* depending on the ranges observed for their values. Notice that, in general, each feature ranges over non-normalized interval of values. We can imagine that a teacher would built a rubric like the one in figure to spur students to: modularize their program into smaller less complex functions (with lower cyclomatic complexity), to write more readable code (using mainly self-explanatory identifiers), and with a less complex algorithm (with lower Halstead's effort).

Notice that an assessment rubric can assign different points to different ranges of feature values extracted, as shown in the figure, where we show three different ranges for the Halstead's Effort measured. This way, the teacher could associate to each feature a weight function with complex shape.

The teacher can update the rubric by either changing the ranges of application or the points given for each feature/range rule or by adding/removing new feature/range/points rules to the rubric.

As the features measured are normally heterogeneous, we plan to study the distribution of the extracted features over our dataset to propose standardized ranges to the teachers and help them during rubric definition.

All the programs are made available to the other students after the submission deadline. After assessment the students' results and programs are shown and linked in the DrPython–WEB leaderboard, so that each student can compare their program style with others, as shown in Fig. 2

		Exercise 1						
		% Good Identificators		Cyclomatic complexity		Effort		
Student	Name	Feature value	Feature evaluation	Feature value	Feature evaluation	Feature value	Feature evaluation	Total
student0	program01.py	85.71	1.2	3.0	0.8	49.76	1.5	3.5
student2	program01.py	20.0	0.0	2.0	0.8	6.96	1.5	2.29
student3	program01.py	42.85	0.0	3.0	0.8	241.76	1.0	1.8
student4	program01.py	11.11	0.0	5.0	0.8	143.25	1.0	1.8
student6	program01.py	25.0	0.0	4.0	0.8	30.31	1.5	2.29
student7	program01.py	11.11	0.0	4.0	0.8	68.33	1.5	2.29
student8	program01.py	80.0	1.2	4.0	0.8	167.59	1.0	3.0
student12	program01.py	60.0	0.0	3.0	0.8	182.64	1.0	1.8
student13	program01.py	60.0	0.0	4.0	0.8	61.02	1.5	2.29
student14	program01.py	23.07	0.0	6.0	0.8	52.0	1.5	2.29
student15	program01.py	20.0	0.0	4.0	0.8	48.0	1.5	2.29
student16	program01.py	20.0	0.0	4.0	0.8	15.5	1.5	2.29
student17	program01.py	50.0	0.0	4.0	0.8	30.31	1.5	2.29
student18	program01.py	50.0	0.0	3.0	0.8	474.54	1.0	1.8
student19	program01.py	60.0	0.0	4.0	0.8	69.3	1.5	2.29

Fig. 2. Leaderboard example, showing the features checked for this exercise and the points assigned according to the previous assessment rubric. Notice that the only features shown are those included in the assessment rubric.

4   Conclusions and Future Work

In Computer Science Education, and in particular in teaching/learning of computer programming, the development of severals students' skills deserves and needs support by automated systems; among them *written communication, testing, project management, teamwork, sheer programming skills* appear to be prominent [3, 17].

In previous work we have dealt with some of these aspects, such as collaboration on common goals, self- and automated evaluation, peer learning [12, 18]. In this paper we have focused on students code analysis, aimed at evaluating programming style, and presented a novel library (DrPython) which extracts structural, quality and linguistic features from the programs and documentation submitted by students. DrPython is used within the novel DrPython–WEB application, that allows the teacher to build assessment rubrics specific to both the point in time, during the course, and/or the specific exercise.

We did not present an experimental in-depth evaluation, as we plan to use the DrPython–WEB system in our next courses to collect data on the student's submissions and check that its usage can improve the student's program quality. Actually we have collected data from uses of incremental versions of the system during previous courses, yet not in a sufficiently structured manner to allow for formal analysis. The further experimentation will allow us to see if the correlation between features and grades will improve (and in what shape) when the system is in-place, with respect to the data collected in earlier courses.

In further work we will try to detect cheating patterns (i.e., when students try to gain points with simple strategies without actually improving their programming style) and to make the feature extraction more robust/precise with respect to cheating. On a side note, we are actually not dealing with cheating on too harsh terms, as experience has shown that in several cases the effort to "fool the teacher" could increase a student's technical programming skills.

We also plan to collect readability assessments from the students during the course, in order to study how the exercise readability improves in time and how the code readability perception of the students changes while they are learning. From the collected data we intend to study if we can define a program readability measure that takes into consideration the linguistic features too.

Finally, we intend to study how the readability of a program is related to its grade, and/or to the grade received in the final lab-based exam. One aim of this work is to integrate the use of DrPython, and of the the DrPython–WEB rubric-based stylistic evaluation, in our previously mentioned systems, in order to support the student activity in collaborative projects [18], and the individual self-assessment of programming homeworks [12].

References

1. Breuker, D.M., Derriks, J., Brunekreef, J.: Measuring static quality of student code. In: Proceedings of the 16th Annual Joint Conference on Innovation and Technology in Computer Science Education, pp. 13–17 (2011)
2. Lu, Y., Mao, X., Wang, T., Yin, G., Li, Z.: Improving students' programming quality with the continuous inspection process: a social coding perspective. Front. Comput. Sci. **14**(5), 1–18 (2019). https://doi.org/10.1007/s11704-019-9023-2
3. Radermacher, A, Walia, G., Knudson D.: Investigating the skill gap between graduating students and industry expectations. In: Proceedings of the 36th International Conference on Software Engineering Companion, pp. 291–300 (2014)
4. Feldman, Y.A.: Teaching quality object-oriented programming. Technol. Educ. Resour. Comput. **5**(1), 1 (2005)
5. Chen, W.K., Tu, P.Y.: Grading code quality of programming assignments based on bad smells. In: Proceedings of the 24th IEEE-CS Conference on Software Engineering Education and Training, p. 559 (2011)
6. Chakraverty, S., Chakraborty, P.: Tools and techniques for teaching computer programming: a review. J. Educ. Technol. Syst. **49**(2), 170–198 (2020)

7. Rogers, S., Garcia, D., Canny, J.F., Tang, S., Kang, D.: ACES: automatic evaluation of coding style. Doctoral dissertation, Master's thesis, EECS Department, University of California, Berkeley (2014)

8. Wang, A., Alsam, A., Morrison, D., Strand, K. A.: Toward automatic feedback of coding style for programming courses. In: 2021 International Conference on Advanced Learning Technologies (ICALT), pp. 33–35 (2021)

9. Alves, N.D.C., von Wangenheim, C.G., Hauck, J.C.R., Borgatto, A.F.: a large-scale evaluation of a rubric for the automatic assessment of algorithms and programming concepts. In: Proceedings of the 51st ACM Technical Symposium on Computer Science Education. Association for Computing Machinery, New York, NY, USA, pp. 556–562 (2020)

10. Caiza, J.C., Del Alamo, J.M.: Programming assignments automatic grading: review of tools and implementations. In: 7th International Technology, Education and Development Conference (2013)

11. Moreno-León, J., Robles, G.: Dr. Scratch: a web tool to automatically evaluate Scratch projects. In: Proceedings of the Workshop in Primary and Secondary Computing Education, pp. 132–133 (2015)

12. Papandrea, S., Sterbini, A., Temperini, M., Popescu, E.: Q2A-I: a support platform for computer programming education, based on automated assessment and peer learning. In: Hancke, G., Spaniol, M., Osathanunkul, K., Unankard, S., Klamma, R. (eds.) ICWL 2018. LNCS, vol. 11007, pp. 3–13. Springer, Cham (2018). https://doi.org/10.1007/978-3-319-96565-9_1

13. McCabe, T.J.: A complexity measure. IEEE Trans. Softw. Eng. (4), 308–320 (1976) https://doi.org/10.1109/tse.1976.233837

14. Halstead, M.H.: Elements of Software Science. Elsevier North-Holland, Inc., Amsterdam (1977). ISBN 0-444-00205-7

15. Lu, K: Python automated term extraction (version v0.5.3). Zenodo (2021). https://doi.org/10.5281/zenodo.5039289

16. Astrakhantsev, N.: Methods and software for terminology extraction from domain-specific text collection. Ph.D. thesis, Institute for System Programming of Russian Academy of Sciences (2015)

17. Radermacher, A.: Evaluating the gap between the skills and abilities of graduating computer science students and the expectation of industry, Master's thesis, North Dakota State University (2012)

18. Sterbini, A., Temperini, M.: Collaborative projects and self evaluation within a social reputation-based exercise-sharing system. In: Proceedings-2009 IEEE/WIC/ACM International Conference on Web Intelligence and Intelligent Agent Technology-Workshops, WI-IAT Workshops, pp. 243–246 (2009)

Formal Methods Communities of Practice: A Survey of Personal Experience

Jonathan P. Bowen[1]([✉]) and Peter T. Breuer[2]

[1] School of Engineering, London South Bank University, Borough Road,
London SE1 1AA, UK
jonathan.bowen@lsbu.ac.uk
[2] Hecusys LLC, Atlanta, GA, USA
http://www.jpbowen.com

Abstract. This paper surveys certain Communities of Practice (CoP) in the field of formal methods for software engineering, especially with respect to state-based notations, using personal knowledge and experience. The multiple communities involved with formal methods are examined here as related CoPs. In this context, the CoPs are open communities encouraging participation by all those interested both in research and application. The authors have been involved with formal methods over several decades and for most of their careers, and it is hoped that the observations in this paper may help future community building to further the development of formal methods, and software engineering in general. The paper also relates the concepts of Networks of Practice (NoP) and Landscapes of Practice (LoP) to formal methods research and practice, and gives a brief introduction to the possibility of visualizing formal methods CoPs. A substantial bibliography is included at the end of the paper.

1 Introduction

The motivation of this paper is to record experience of developing communities of formal methods researchers with the aim of aiding current and future researchers in successfully growing their own communities around their research interests. Formal methods researchers can be very involved with the specifics and theoretical aspects of their research without considering how it may be used in practice at some point in the future. This problem continues to this day. In particular, technology transfer issues require different skills than a researcher may possess. For ultimate success in deployment, a community with a range of skills and backgrounds is needed, with the ability and motivation to communicate between each other.

The research methodology in this paper has been to draw on a career lifetime's experience of community building in the field of formal methods to consider how research communities may be built over decades. The paper uses the

A. Cerone et al. (Eds.): SEFM 2021 Workshops, LNCS 13230, pp. 287–301, 2022.
https://doi.org/10.1007/978-3-031-12429-7_21

social science framework of a *Community of Practice* (CoP), an approach that aids in understanding the evolution of communities that are based around an area of developing knowledge [78,95,96]. The examples provided are intended to help formal methods researchers in considering the state and development of their own communities of fellow researchers and practitioners.

Section 2 gives an overview of experience leading to this paper, with some specific community-building examples. Section 3 presents the main results of the paper around the framework of a CoP with examples drawn from experience in developing formal methods communities. Section 4 extends the CoP concept to multiple CoPs in a *Network of Practice* (NoP) and more widely to a *Landscape of Practice* (LoP), with some examples in the formal methods domain. Section 5 briefly provides an introduction to possibily visualizing formal methods CoPs in order to gain an idea of their nature and structure. Some conclusions on the potential impact of the investigation and ideas for future work are provided in Sect. 6. The paper includes a substantial bibliography that could itself be used for investigating community structures and interdependence, based on co-author relationships for example.

2   Background

> ... *a job is about a lot more than a paycheck. It's about your dignity. It's about respect. It's about your place in your community.*
>
> – Joe Biden

By way of background, the first author of this paper has been involved in building and investigating communities [30], both in the area of formal methods [43], especially the Z notation [49], and also in museum-related [6,59] and arts-related [36,50] contexts. This has been facilitated by the increasing possibility of worldwide virtual communities without geographic bounds [12]. Bowen originally became active in formal methods community building at the Oxford University Computing Laboratory's Programming Research Group (PRG) in the late 1980s. He was a Research Officer working on formal methods [69] and specifically the Z notation [25,64] at the time. He also became involved with the European ESPRIT **ProCoS I** and **II** projects on *Provably Correct Systems*, led by Tony Hoare at Oxford, Dines Bjørner at DTH in Denmark, and others in the early 1990s [7,22,46].

The subsequent **ProCoS-WG** Working Group of 25 partners around Europe existed to organize meetings and workshops in the late 1990s [23]. The **ProCoS-WG** final report in 1998 [47] presented comments by members of the group, including those who joined after the start of its formation. For example, Prof. Egon Börger [38] of the University of Pisa in Italy participated at many **ProCoS-WG** meetings. He was an invited speaker at the ZUM'97 conference [45] and, with Jean-Raymond Abrial and Prof. Hans Langmaack of the University of Kiel, he organized an important set of case studies formalizing a Steam Boiler problem in a variety of formal notations [4], including a number of contributions by **ProCoS-WG** members. He used **ProCoS-WG** to present his work on the correctness theorem for a general compilation scheme for compiling Occam programs to Transputer code [17]. The influence of the **ProCoS** initiative has continued for decades after the original projects and Working Group [33,68].

The BCS-FACS (Formal Aspects of Computing Science) Specialist Group forms another example of a formal methods community in the United Kingdom. Such a community depends on a core committee to keep it active. Although it has been in existence since 1978, there was a period when activities declined in the 1990s. However, from the early 2000s, a new committee was formed, leading to a renewal of activities. For example, in December 2003, the BCS-FACS Workshop *Teaching Formal Methods: Practice and Experience* was held at Oxford Brookes University [9]. The group also holds regular evening seminars mainly at the BCS London office and selected talks have appeared as chapters in an edited book [11].

In 2008, the newly formed *Abstract State Machines, B and Z: First International Conference, ABZ 2008* started in London, UK, edited by Egon Börger (ASM), Michael Butler (B-Method), myself (Z notation), and Paul Boca as a local organizer [15,16]. This was an extension of the earlier ZB conferences, formed from a combination of previously separate B and Z conferences. In 2011, a special issue of selected and extended papers from the ABZ 2008 conference was produced for the *Formal Aspects of Computing* journal [14]. More recently, a 2018 book on *Modeling Companion for Software Practitioners* using the ASM approach has appeared [18,34].

3 Communities of Practice

> *Everything we do is practice for something greater than where we currently are. Practice only makes for improvement.*   – Les Brown

A *Community of Practice* (CoP) [95] is a social science concept useful for modelling the collaborative activities of professional communities [12] with a common goal over time [96,99]. It can be relevant in a variety of contexts, for example, agile methods [75,101], student teaching in higher education [80–82,84], and developing large organizations [88]. Information on the successful creation of a CoP is available [94]. CoPs are typically open communities and it is in this context that they are discussed in this paper.

A CoP modelling approach can be used in various scenarios, for example, in the context of this paper, formal methods communities [31,33,49]. A CoP consists of:

1. A **domain** of knowledge and interest. In the case of formal methods, this is the application of mathematical approaches to computer-based specification, modelling and development.
2. A **community** based around this domain. For formal methods, like other academically-based disciplines, this includes conference organizers and programme committee members that are interested in formal methods as core

facilitators, conference presenters and delegates as participants, as well as other researchers and practitioners involved with developing and using formal methods.

3. The **practice** undertaken by the community in this domain, developing its knowledge, sometimes formalized as a *Body of Knowledge* (BoK) [49] – see also Sect. 4. The formal methods community encourages the transfer of research ideas into practical use [65,66]. Some formal methods approaches have been used in industrial-scale software-based projects, although information on these can be difficult to promulgate due to commercial sensitivities and Non-Disclosure Agreements.

There are various stages in the development of a CoP [95]:

1. **Potential:** There needs to be an existing network of people to initiate a CoP. In the case of formal methods as a whole, researchers interested in theoretical computer science, especially discrete mathematics and logic, were the starting point. For example, the initial Z meetings were held with informal proceedings [19,20] in Oxford, due to the location of the Programming Research Group there.

2. **Coalescing:** The community needs to establish a rhythm to ensure its continuation. In the case of many successful formal methods, a regular specialist workshop is typically established initially. For the Z notation, a more formal Z User Meeting (ZUM) was established, together with a Z User Group (ZUG) established in 1992 [48]. Initially, meetings were in the United Kingdom, but it then became an international conference in 1995 [42]. Online information was maintained, initially as a FTP service with an associated Z FORUM electronic mailing list [26].

3. **Maturing:** The community must become more enduring. An initial workshop series may become a more formal conference series and establish itself internationally. With maturity, there may be merging with other formal methods. For example, the International Conference of Z Users became the ZB conference in 2000 [41], combined with the B-Method, and then the ABZ conference, combining ASM, the B-Method, and the Z notation in a single conference in 2008 [15]. This conference has continued through to the present [86]. Another sign of maturity is the production of a standard, e.g., the international ISO/IEC standard for the Z notation [70]. The FTP service became a website and the Z FORUM mailing list was linked with the `comp.specification.z` newsgroup [26] (now part of Google Groups).

4. **Stewardship:** The community needs to respond to its environment and develop appropriately. A particular formal methods community should interact with related organizations, e.g., those associated with similar formal methods. Overall, Formal Methods Europe (FME, https://www.fmeurope.org) has acted as a stewarding organization internationally, developing beyond the bounds of Europe, and organizing the regular FME conference from 1993 [103], becoming the FM conference more recently, and following on from the original VDM conferences.

5. **Legacy:** All CoPs eventually end; if successful they morph into further communities. State-based formal methods communities such as those around ASM, B, VDM, Z, etc., have coalesced around the ABZ conference, which continues to this day, as noted above [86]. The various CoPs around these approaches are at different levels of development with respect to their CoP evolution. Currently, as of 2021, there is much activity around the B-Method and the related Event-B. Research around ASM is also still active. However, research on VDM and Z is now somewhat dormant. Exactly how all these related communities will continue is something that is worth considering and planning for at the appropriate time.

It remains to be seen precisely what legacy the various state-based formal methods, especially those associated with the ABZ conference, leave in the future. For the moment, the various communities continue to come together through the ABZ conference, as well as other more informal and individual interactions.

It is interesting to reflect on the occurrence of various formal methods and tools in the titles of papers in the two most recent ABZ conference proceedings for 2020 [87] and 2021 [86], as reported by Bowen [37]. Event-B is the most popular formal method, with 14 papers. 11 papers mention the Rodin tool, which provides Event-B support. There are nine papers with ASM in the title (including two mentioning the associated ASMETA toolset). Alloy, a Z-like language with tool support, is mentioned in three paper titles, as is the ProB tool providing tool support for B. The Atelier B, UML-B, and UPPAAL tools are each mentioned in one title. TLA, VDM, and Z are not mentioned in any paper titles. So, the "A" (ASM and Alloy) and "B" (mainly Event-B with the associated Rodin and ProB tools) in conference title "ABZ" are still active with respect to research, especially strongly in the cases Event-B/Rodin and ASM. However, the "Z" part of the conference has essentially disappeared. That said, Z is still an inspiration for some formal methods research and is still used in industrial projects, even if not widely publicized. Tools are increasingly important for industrial use of formal methods at scale.

4   Networks and Landscapes of Practice

The idea of a *Community of Practice* (CoP) was originally introduced in 1991 [78]. Since then, the concept has been extended, firstly to the notion of a *Network of Practice* (NoP) in 2000 [56] and most recently to a *Landscape of Practice* (LoP) in 2014 [97,98].

In a CoP, researchers and practitioners are connected and mutually engaged [85]. An NoP connects several CoPs where there are connections but not necessarily deep mutual engagement. For example, state-based formal methods have connections between them, but they each form separate CoPs.

Some key state-based formal methods CoPs are summarized in Fig. 1, together forming an NoP. There have been a number of comparative studies covering various formal methods approaches. For example, a 1996 Steam Boiler

Control case study competition book demonstrated different formal methods [4,5]. A 2001 book [58] (second edition in 2006 [61]) presented the questions that should be answered in developing an example invoicing case study using a variety of formal methods. These indicate the connections between these CoPs, but the contributions tend to be submitted independently by members of the relevant CoP. The ABZ conference [86,87] has also been important in cementing ties between the state-based formal methods CoPs (Fig. 1), helping to form an NoP. Similarly, FME and its associated FM (formerly FME [103]) conference have also helped significantly in forming a formal methods NoP internationally. On a smaller scale, the BCS-FACS Specialist Group mentioned earlier has helped to play a similar role in the United Kingdom [9–11].

Alloy: Daniel Jackson [73].

ASM: (Abstract State Machines) Egon Börger & Yuri Gurevich [13,18].

B-Method & Event-B: Jean-Raymond Abrial [2,3].

TLA: (Temporal Logic of Actions) Leslie Lamport [77].

VDM: (Vienna Development Method) Dines Bjørner & Cliff Jones [74].

Z: Jean-Raymond Abrial & Mike Spivey [1,70,89].

Fig. 1. Some key state-based formal methods CoPs with their major progenitors/promulgators.

Formal methods also overlap with CoPs from other areas that require a rigorous basis to their approaches. For example, formal methods are beneficial in compilers [51,62,63], Hardware Description Languages (HDL) [55,104], Human-Cyber-Physical Systems (HCPS) [79], logic programming [21,24,27], safety-critical systems [28,93], security [53,54], software maintenance [39,40,91], software testing [76,92], etc. That said, there can be some resistance to accepting formal methods in some communities within software engineering as a whole [8,44,67]. Education and training are important aspects with respect to the acceptance and promulgation of formal methods [29,32,57].

An LoP is even wider than an NoP, crossing boundaries and peripheries between CoPs with more widely related interests. Typically an LoP identifies communities following the same *Body of Knowledge* (BoK), making up the collection of activities, concepts, and terms associated with a professional domain, normally defined by the relevant professional association or learned society. In the case of formal methods, there have been some efforts to define specific BoKs [49], for example, specifically for model checking in software development [90] and in the railway domain [60].

Formal methods CoPs by and large form part of the field of software engineering, although the approach can be applied to computer hardware design as well. Software engineering has a BoK originally developed by the IEEE Computer Society and issued as an ISO/IEC international standard, first in 2005

[71] (187 pages), with an updated version issued in 2015 [72] (336 pages), known as "SWEBOK". The SWEBOK standard covers software requirements, design, testing, etc., all areas where formal methods can be an appropriate approach, especially in safety and security-critical systems.

5   Visualization of Communities

Visualization of CoPs can aid in the identification of knowledge innovation and transfer between CoPs [83]. It is possible to visualize and formalize communities, especially those that are online [30,50]. In addition, patterns in citations can be investigated formally [52]. Co-author and citation relationships can help in understanding the structure of a CoP and its interconnections within an NoP or LoP. In the mathematical LoP, the Erdös number (the distance through co-authorship from the highly collaborative 20th-century mathematician Paul Erdös) is often used as a measure for a particular mathematician's involvement in the mathematical research community [50]. This measure applies to formal methods researchers as well, being one of the more mathematical areas of computer science [30].

There are various visualization tools online that allow graphical views of co-authors and cited authors. As an example, consider the case of a 2009 formal methods survey in the *ACM Computing Surveys* journal [102], using the online *Connected Papers* tool (https://www.connectedpapers.com), as illustrated in Fig. 2. It is possible to identify clusters of citations by the co-author (top and right of Fig. 2). In the righthand cluster, two of the co-authors have collaborated significantly. These two clusters represent two different formal methods CoPs. The size of the circle for a citation indicates the number of citations that it has received (i.e., broadly, its importance) and the darker the colour, the later the date of the citation.

There is much scope for further research on the visualization of formal methods and other CoPs, which could easily fill a whole paper.

6   Conclusion

> *Education is for improving the lives of others and for leaving your community and world better than you found it.*
>
> – Marian Wright Edelman

As discussed in this paper, there are a number of competing state-based formal methods for modelling computer-based systems. Communities associated with formal methods have developed since the 1980s around the Z notation, the B-Method, Event-B, ASM, and other paradigms. Each approach has its own advantages and disadvantages, which are beyond the scope of this paper to provide in detail. Each also has its own community of adherents, that have now somewhat merged with the establishment of the ABZ conference in 2008 [15].

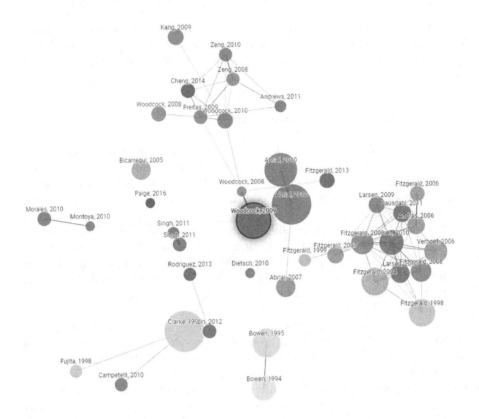

Fig. 2. A visualization of the citations in a 2009 formal methods survey paper [102].

By their nature, formal methods communities tend to be open communities, with participation by both researchers and practitioners actively encouraged.

The various interrelated formal methods communities may be seen as examples of Communities in Practice (CoP) in action. CoPs can potentially merge and create new CoPs. For example, the B-Method and then Event-B were developed after the Z notation largely by the same progenitor, Jean-Raymond Abrial, with some in the Z community subsequently becoming part of the B community. These interrelated formal methods CoPs form a Network of Practice (NoP).

From experience, a successful Community of Practice depends on people with different skills for success, be it for ideas, vision, organization, etc. Typically, a small number of key personnel are needed for the successful launch of any new formal methods community. Figure 1 provided some key initial personnel for a selection of state-based formal methods. A successful CoP then needs to reach a critical mass in size, following some of the developments covered in this paper.

The earliest formal methods communities were formed around VDM and Z, which grew up in parallel, although Z concentrated on formal specification at a high level with little tool support, whereas VDM also considered refinement towards program code more explicitly. These are now in the late stages of a CoP. ASM developed separately somewhat later and its flexibility has proved useful on modelling systems at a high level. Due to the lack of refinement and tool support in Z, the B-Method and then Event-B were developed to more explicitly handle these aspects with some compromise on the high-level nature of the language. Alloy also provides tool support using a Z-like language that is useful as a prototyping tool, perhaps for more intricate or critical parts of a large Z specification that could benefit from closer investigation for example. Judging by the activities reported in recent ABZ conferences, Event-B is the most active CoP at the moment. Future formal methods CoPs need to ensure good industrial-strength (and ideally open) tool support for success.

This paper has considered formal methods CoPs, and briefly discussed associated Networks of Practice (NoPs), an overall Landscape of Practice (LoP), and visualization of these communities. Further research could be usefully undertaken in these additional areas. Bibliographies such as the one at the end of this paper and other more substantial databases such as Google Scholar (https:// scholar.google.com) could be used for investigating formal methods community structures and interdependences, based on co-author and citation relationships.

Predicting the future is always difficult, but formal methods communities have been successful enough to leave their mark on the computer science community as a whole. Certainly, the most active formal methods CoPs have shifted from consideration of fundamental ideas to tool support, enabling better potential for industrial usage. The experience of the authors is mainly with state-based formal methods, but the examples in this paper may also be applicable to model-checking communities, for example. In any case, the authors hope that the perspective presented here will help future researchers in developing a Community of Practice based on their own research.

Acknowledgements. The authors are grateful to many formal methods colleagues for inspiration and collaboration over the years. Jonathan Bowen thanks Museophile Limited for financial support. The reviewers provided some helpful comments, especially with respect to landscapes of practice.

We dedicate this paper to our former colleague, Sergiy Vilkomir (1956–2020) [92, 93], of Kharkiv, Ukraine [35, 100].

References

1. Abrial, J.R.: Data semantics. In: Klimbie, J.W., Koffeman, K.L. (eds.) IFIP TC2 Working Conference on Data Base Management, pp. 1–59. Elsevier, North-Holland (1974)
2. Abrial, J.R.: The B-Book: Assigning Programs to Meanings. Cambridge University Press (1996)
3. Abrial, J.R.: Modeling in Event-B: System and Software Engineering. Cambridge University Press (2010)
4. Abrial, J.-R., Börger, E., Langmaack, H. (eds.): Formal Methods for Industrial Applications. LNCS, vol. 1165. Springer, Heidelberg (1996). https://doi.org/10.1007/BFb0027227

5. Abrial, J.-R., Börger, E., Langmaack, H.: The steam boiler case study: competition of formal program specification and development methods. In: Abrial, J.-R., Börger, E., Langmaack, H. (eds.) Formal Methods for Industrial Applications. LNCS, vol. 1165, pp. 1–12. Springer, Heidelberg (1996). https://doi.org/10.1007/BFb0027228

6. Beler, A., Borda, A., Bowen, J.P., Filippini-Fantoni, S.: The building of online communities: an approach for learning organizations, with a particular focus on the museum sector. In: Hemsley, J., Cappellini, V., Stanke, G. (eds.) EVA 2004 London Conference Proceedings, pp. 2.1–2.15. EVA Conferences International, University College London, UK (2004). https://arxiv.org/abs/cs/0409055

7. Bjørner, D., et al.: A ProCoS project description: ESPRIT BRA 3104. Bull. Eur. Assoc. Theor. Comput. Sci. **39**, 60–73 (1989). http://researchgate.net/publication/256643262

8. Black, S.E., Boca, P.P., Bowen, J.P., Gorman, J., Hinchey, M.G.: Formal versus agile: survival of the fittest. Computer **42**(9), 37–45 (2009). https://doi.org/10.1109/MC.2009.284

9. Boca, P.P., Bowen, J.P., Duce, D.A. (eds.): Teaching formal methods: practice and experience. In: Electronic Workshops in Computing (eWiC), FACS, BCS (2006). http://www.bcs.org/ewic/tfm2006

10. Boca, P.P., Bowen, J.P., Larsen, P.G. (eds.): FACS 2007 Christmas Workshop: Formal Methods in Industry. In: Electronic Workshops in Computing (eWiC), FACS, BCS (2007). http://www.bcs.org/ewic/fmi2007

11. Boca, P.P., Bowen, J.P., Siddiqi, J.I. (eds.): Formal methods: state of the art and new directions. Springer, London (2010). https://doi.org/10.1007/978-1-84882-736-3

12. Borda, A., Bowen, J.P.: Virtual collaboration and community. In: Information Resources Management Association (ed.) Virtual Communities: Concepts, Methodologies, Tools and Applications, chap. 8.9, pp. 2600–2611. IGI Global (2011)

13. Börger, E.: The origins and development of the ASM method for high-level system design and analysis. J. Univ. Comput. Sci. **8**(1), 2–74 (2002). https://doi.org/10.3217/jucs-008-01-0002

14. Börger, E., Bowen, J.P., Butler, M.J., Poppleton, M.: Editorial. Formal Aspects Comput. **23**(1), 1–2 (2011). https://doi.org/10.1007/s00165-010-0168-x

15. Börger, E., Butler, M., Bowen, J.P., Boca, P. (eds.): ABZ 2008. LNCS, vol. 5238. Springer, Heidelberg (2008). https://doi.org/10.1007/978-3-540-87603-8

16. Börger, E., Butler, M.J., Bowen, J.P., Boca, P.P. (eds.): ABZ 2008 Conference: Short Papers. BCS, London, UK (2008)

17. Börger, E., Durdanovic, I.: Correctness of compiling Occam to Transputer code. Comput. J. **39**(1), 52–92 (1996)

18. Börger, E., Raschke, R.: Modeling companion for software practitioners. Springer, Heidelberger (2018). https://doi.org/10.1007/978-3-662-56641-1

19. Bowen, J.P. (ed.): Proceedings of Z users meeting. Oxford University Computing Laboratory, Rewley House, Wellington Square, Oxford, UK (1987). https://doi.org/10.13140/RG.2.2.20103.34724

20. Bowen, J.P. (ed.): Proceedings of the third annual Z users meeting. Oxford University Computing Laboratory, Rewley House, Wellington Square, Oxford, UK (1988). http://researchgate.net/publication/2526997

21. Bowen, J.P.: From programs to object code and back again using logic programming: compilation and decompilation. J. Softw. Maint. Res. Pract. **5**(4), 205–234 (1993). https://doi.org/10.1002/smr.4360050403

22. Bowen, J.P.: A ProCoS II project description: ESPRIT Basic Research project 7071. Bull. Eur. Assoc. Theor. Comput. Sci. **50**, 128–137 (1993). http://researchgate.net/publication/2521581

23. Bowen, J.P.: A ProCoS-WG working group description: ESPRIT basic research 8694. Bull. Eur. Assoc. Theor. Comput. Sci. **53**, 136–145 (1994)

24. Bowen, J.P.: Rapid compiler implementation. In: He [62], chap. 10, pp. 141–169

25. Bowen, J.P.: Formal specification and documentation using Z: a case study approach. International Thomson Computer Press (1996). http://researchgate.net/publication/2480325

26. Bowen, J.P.: Comp.specification.z and Z FORUM frequently asked questions. In: Bowen, J.P., Fett, A., Hinchey, M.G. (eds.) ZUM 1998. LNCS, vol. 1493, pp. 407–416. Springer, Heidelberg (1998). https://doi.org/10.1007/978-3-540-49676-2_25

27. Bowen, J.P.: Combining operational semantics, logic programming and literate programming in the specification and animation of the Verilog hardware description language. In: Grieskamp, W., Santen, T., Stoddart, B. (eds.) IFM 2000. LNCS, vol. 1945, pp. 277–296. Springer, Heidelberg (2000). https://doi.org/10.1007/3-540-40911-4_16

28. Bowen, J.P.: The ethics of safety-critical systems. Commun. ACM **43**(4), 91–97 (2000). https://doi.org/10.1145/332051.332078

29. Bowen, J.P.: Experience teaching Z with tool and web support. ACM SIGSOFT Softw. Eng. Notes **26**(2), 69–75 (2001). https://doi.org/10.1145/505776.505794

30. Bowen, J.P.: Online communities: visualization and formalization. In: Blackwell, C. (ed.) Cyberpatterns 2013: Second International Workshop on Cyberpatterns - Unifying Design Patterns with Security, Attack and Forensic Patterns, pp. 53–61. Oxford Brookes University, Abingdon, UK (2013). http://arxiv.org/abs/1307.6145

31. Bowen, J.P.: A relational approach to an algebraic community: from Paul Erdős to He Jifeng. In: Liu, Z., Woodcock, J., Zhu, H. (eds.) Theories of Programming and Formal Methods. LNCS, vol. 8051, pp. 54–66. Springer, Heidelberg (2013). https://doi.org/10.1007/978-3-642-39698-4_4

32. Bowen, J.P.: The Z notation: whence the cause and whither the course? In: Liu, Z., Zhang, Z. (eds.) SETSS 2014. LNCS, vol. 9506, pp. 103–151. Springer, Cham (2016). https://doi.org/10.1007/978-3-319-29628-9_3

33. Bowen, J.P.: Provably correct systems: community, connections, and citations. In: Hinchey, M.G., Bowen, J.P., Olderog, E.-R. (eds.) Provably Correct Systems. NMSSE, pp. 313–328. Springer, Cham (2017). https://doi.org/10.1007/978-3-319-48628-4_13

34. Bowen, J.P.: Egon Börger and Alexander Raschke: modeling companion for software practitioners. Formal Aspects Comput. **30**(6), 761–762 (2018). https://doi.org/10.1007/s00165-018-0472-4

35. Bowen, J.P.: In memoriam: a tribute to five formal methods colleagues. FACS FACTS **2020**(1), 13–29 (2020). https://doi.org/10.13140/RG.2.2.13481.62560

36. Bowen, J.P.: A personal view of EVA London: past, present, future. In: Weinel, J., Bowen, J.P., Diprose, G., Lambert, N. (eds.) EVA London 2020: Electronic Visualisation and the Arts. pp. 8–15. Electronic Workshops in Computing (eWiC), BCS, London, UK (2020). https://doi.org/10.14236/ewic/EVA2020.2

37. Bowen, J.P.: ABZ 2021 conference report. FACS FACTS **2021**(2), 65–70 (2021). https://www.bcs.org/media/7577/facs-jul21.pdf

38. Bowen, J.P.: Communities and ancestors associated with Egon Börger and ASM. In: Raschke, A., Riccobene, E., Schewe, K.-D. (eds.) Logic, Computation and Rigorous Methods. LNCS, vol. 12750, pp. 96–120. Springer, Cham (2021). https://doi.org/10.1007/978-3-030-76020-5_6

39. Bowen, J.P., Breuer, P.T., Lano, K.C.: A compendium of formal techniques for software maintenance. Softw. Eng. J. **8**(5), 253–262 (1993). https://doi.org/10.1049/sej.1993.0031

40. Bowen, J.P., Breuer, P.T., Lano, K.C.: Formal specifications in software maintenance: from code to Z++ and back again. Inf. Softw. Technol. **35**(11–12), 679–690 (1993). https://doi.org/10.1016/0950-5849(93)90083-F

41. Bowen, J.P., Dunne, S., Galloway, A., King, S. (eds.): ZB 2000. LNCS, vol. 1878. Springer, Heidelberg (2000). https://doi.org/10.1007/3-540-44525-0

42. Bowen, J.P., Hinchey, M.G. (eds.): ZUM 1995. LNCS, vol. 967. Springer, Heidelberg (1995). https://doi.org/10.1007/3-540-60271-2

43. Bowen, J.P., Hinchey, M.G.: Formal methods. In: Gonzalez, T., Díaz-Herrera, J., Tucker, A.B. (eds.) Computing Handbook, Computer Science and Software Engineering, pp. 1–25. Chapman and Hall/CRC Press, 3rd edn (2014). https://doi.org/10.1201/b16812-80

44. Bowen, J.P., Hinchey, M.G., Janicke, H., Ward, M., Zedan, H.S.M.: Formality, agility, security, and evolution in software engineering. In: Hinchey, M.G. (ed.) Software Technology: 10 Years of Innovation in IEEE Computer, chap. 16, p. 384. Wiley-IEEE Press (2018). https://doi.org/10.1002/9781119174240.ch16

45. Bowen, J.P., Hinchey, M.G., Till, D. (eds.): ZUM 1997. LNCS, vol. 1212. Springer, Heidelberg (1997). https://doi.org/10.1007/BFb0027279

46. Bowen, J.P., Hoare, C.A.R., Langmaack, H., Olderog, E.R., Ravn, A.P.: A ProCoS II project final report: ESPRIT basic research project 7071. Bull. Eur. Assoc. Theor. Comput. Sci. **59**, 76–99 (1996). http://researchgate.net/publication/2255515

47. Bowen, J.P., Hoare, C.A.R., Langmaack, H., Olderog, E.R., Ravn, A.P.: A ProCoS-WG working group final report: ESPRIT working group 8694. Bull. Eur. Assoc. Theor. Comput. Sci. **64**, 63–72 (1998). http://researchgate.net/publication/2527052

48. Bowen, J.P., Nicholls, J.E. (eds.): Z user workshop, London 1992. In: Proceedings of the Seventh Annual Z User Meeting, London, 14–15 December 1992. Springer, London (1993). https://doi.org/10.1007/978-1-4471-3556-2

49. Bowen, J.P., Reeves, S.: From a community of practice to a body of knowledge: a case study of the formal methods community. In: Butler, M., Schulte, W. (eds.) FM 2011. LNCS, vol. 6664, pp. 308–322. Springer, Heidelberg (2011). https://doi.org/10.1007/978-3-642-21437-0_24

50. Bowen, J.P., Wilson, R.J.: Visualising virtual communities: from Erdős to the arts. In: Dunn, S., Bowen, J.P., Ng, K.C. (eds.) EVA London 2012: Electronic Visualisation and the Arts, pp. 238–244. Electronic Workshops in Computing (eWiC), BCS (2012). http://arxiv.org/abs/1207.3420

51. Breuer, P.T., Bowen, J.P.: A PREttier compiler-compiler: generating higher-order parsers in C. Softw. Pract. Experience **25**(11), 1263–1297 (1995). https://doi.org/10.1002/spe.4380251106

52. Breuer, P.T., Bowen, J.P.: Empirical patterns in Google Scholar citation counts. In: SOSE 2014: IEEE 8th International Symposium on Service Oriented System Engineering, pp. 398–403. IEEE (2014). https://doi.org/10.1109/SOSE.2014.55

53. Breuer, P.T., Bowen, J.P.: Fully encrypted high-speed microprocessor architecture: the secret computer in simulation. Int. J. Crit. Comput.-Based Syst. **9**(1–2), 26–55 (2019). https://doi.org/10.1504/IJCCBS.2019.098797

54. Breuer, P.T., Bowen, J.P., Palomar, E., Liu, Z.: On security in encrypted computing. In: Naccache, D., et al. (eds.) ICICS 2018. LNCS, vol. 11149, pp. 192–211. Springer, Cham (2018). https://doi.org/10.1007/978-3-030-01950-1_12

55. Breuer, P.T., Madrid, N.M., Bowen, J.P., France, R., Petrie, M.L., Kloos, C.D.: Reasoning about VHDL and VHDL-AMS using denotational semantics. In: Borrione, D. (ed.) Conference on Design, Automation and Test in Europe, pp. 346–352. ACM (1999). https://doi.org/10.1109/DATE.1999.761144

56. Brown, J.S., Duguid, P.: The Social Life of Information. Harvard Business Review Press (2000)

57. Cerone, A., Roggenbach, M., Schlingloff, B.H., Schneider, G., Shaikh, S.A.: Teaching formal methods for software engineering - ten principles. Informatica Didactica **9** (2015). https://www.informaticadidactica.de/uploads/Artikel/Schlinghoff2015/Schlinghoff2015.pdf

58. Frappier, M., Habrias, H. (eds.): Software specification methods: an overview using a case study. FACIT, Springer, London (2001). https://doi.org/10.1007/978-1-4471-0701-9

59. Giannini, T., Bowen, J.P. (eds.): Museums and Digital Culture. SSCC, Springer, Cham (2019). https://doi.org/10.1007/978-3-319-97457-6

60. Gruner, S., Kumar, A., Maibaum, T.: Towards a body of knowledge in formal methods for the railway domain: identification of settled knowledge. In: Artho, C., Ölveczky, P.C. (eds.) FTSCS 2015. CCIS, vol. 596, pp. 87–102. Springer, Cham (2016). https://doi.org/10.1007/978-3-319-29510-7_5

61. Habrias, H., Frappier, M. (eds.): Software Specification Methods. ISTE (2006). https://doi.org/10.1002/9780470612514

62. He, J. (ed.): Provably correct systems: modelling of communication languages and design of optimized compilers. International Series in Software Engineering, McGraw-Hill (1994)

63. Jifeng, H., Bowen, J.: Specification, verification and prototyping of an optimized compiler. Formal Aspects Comput. **6**(6), 643–658 (1994). https://doi.org/10.1007/BF03259390

64. Henson, M.C., Reeves, S., Bowen, J.P.: Z logic and its consequences. Comput. Inform. **22**(3–4), 381–415 (2003)

65. Hinchey, M.G., Bowen, J.P. (eds.): Applications of Formal Methods. Series in Computer Science, Prentice Hall International (1995)

66. Hinchey, M.G., Bowen, J.P. (eds.): Industrial-strength formal methods in practice. In: FACIT, Springer, London (1999). https://doi.org/10.1007/978-1-4471-0523-7

67. Hinchey, M.G., Bowen, J.P., Glass, R.L.: Formal methods: point-counterpoint. Computer **29**(4), 18–19 (1996). https://doi.org/10.1109/MC.1996.10044

68. Hinchey, M.G., Bowen, J.P., Olderog, E.-R. (eds.): Provably Correct Systems. NMSSE, Springer, Cham (2017). https://doi.org/10.1007/978-3-319-48628-4

69. Hinchey, M., Bowen, J.P., Rouff, C.A.: Introduction to formal methods. In: Rouff, C.A., Hinchey, M., Rash, J., Truszkowski, W., Gordon-Spears, D. (eds) Agent Technology from a Formal Perspective. NASA Monographs in Systems and Software Engineering. Springer, London (2006). https://doi.org/10.1007/1-84628-271-3_2

70. ISO: Information technology - Z formal specification notation - syntax, type system and semantics. International Standard 13568, ISO/IEC (2002)

71. ISO/IEC JTC 1/SC 7: Software engineering - guide to the software engineering body of knowledge (SWEBOK). Tech. Rep. ISO/IEC TR 19759:2005, ISO/IEC (2005). https://www.iso.org/standard/33897.html

72. ISO/IEC JTC 1/SC 7: Software engineering - guide to the software engineering body of knowledge (SWEBOK). Tech. Rep. ISO/IEC TR 19759:2015, ISO/IEC (2015). https://www.iso.org/standard/67604.html

73. Jackson, D.: Software Abstractions: Logic, Language, and Analysis. The MIT Press (2011)

74. Jones, C.B.: Systematic Software Development Using VDM. Series in Computer Science, Prentice Hall International (1986)

75. Kahkonen, T.: Agile methods for large organizations-building communities of practice. In: Agile Development Conference, pp. 2–10. IEEE (2004). https://doi.org/10.1109/ADEVC.2004.4

76. Kapoor, K., Bowen, J.P.: Test conditions for fault classes in Boolean specifications. ACM Trans. Softw. Eng. Methodol. **16**(3), 1–12 (2007). https://doi.org/10.1145/1243987.1243988

77. Lamport, L.: The temporal logic of actions. ACM Trans. Program. Lang. Syst. **16**(3), 872–923 (1994). https://doi.org/10.1145/177492.177726

78. Lave, J., Wenger, E.: Cambridge University Press (1991)

79. Liu, Z., Bowen, J.P., Liu, B., Tyszberowicz, S., Zhang, T.: Software abstractions and human-cyber-physical systems architecture modelling. In: Bowen, J.P., Liu, Z., Zhang, Z. (eds.) SETSS 2019. LNCS, vol. 12154, pp. 159–219. Springer, Cham (2020). https://doi.org/10.1007/978-3-030-55089-9_5

80. Mavri, A., Ioannou, A., Loizides, F.: Cross-organisational communities of practice: enhancing creativity and epistemic cognition in higher education. Internet High. Educ. **49**, 100792 (2021). https://doi.org/10.1016/j.iheduc.2021.100792

81. McDonald, J., Cater-Steel, A. (eds.): Communities of Practice. Springer, Singapore (2017). https://doi.org/10.1007/978-981-10-2879-3

82. McDonald, J., Cater-Steel, A. (eds.): Implementing Communities of Practice in Higher Education. Springer, Singapore (2017). https://doi.org/10.1007/978-981-10-2866-3

83. Novak, J., Wurst, M.: Collaborative knowledge visualization for cross-community learning. In: Tergan, S.-O., Keller, T. (eds.) Knowledge and Information Visualization. LNCS, vol. 3426, pp. 95–116. Springer, Heidelberg (2005). https://doi.org/10.1007/11510154_6

84. Nunes, R.R., et al.: Enhancing students' motivation to learn software engineering programming techniques: a collaborative and social interaction approach. In: Antona, Margherita, Stephanidis, Constantine (eds.) UAHCI 2015. LNCS, vol. 9177, pp. 189–201. Springer, Cham (2015). https://doi.org/10.1007/978-3-319-20684-4_19

85. Pyrko, I., Dörfler, V., Eden, C.: Communities of practice in landscapes of practice. Manag. Learn. **50**(4), 482–499 (2019). https://doi.org/10.1177/1350507619860854

86. Raschke, A., Méry, D. (eds.): ABZ 2021. LNCS, vol. 12709. Springer, Cham (2021). https://doi.org/10.1007/978-3-030-77543-8

87. Raschke, A., Méry, D., Houdek, F. (eds.): ABZ 2020. LNCS, vol. 12071. Springer, Cham (2020). https://doi.org/10.1007/978-3-030-48077-6

88. Schulte, B.: The Organizational Embeddedness of Communities of Practice. F, Springer, Wiesbaden (2021). https://doi.org/10.1007/978-3-658-31954-0

89. Spivey, J.M.: The Z Notation: A Reference Manual, 2nd edn. Series in Computer Science, Prentice Hall International (1992)

90. Taguchi, K., Nishihara, H., Aoki, T., Kumeno, F., Hayamizu, K., Shinozaki, K.: Building a body of knowledge on model checking for software development. In: IEEE 37th Annual Computer Software and Applications Conference, pp. 784–789 (2013). https://doi.org/10.1109/COMPSAC.2013.129

91. van Zuylen, H.J. (ed.): The REDO Compendium: Reverse Engineering for Software Maintenance. Wiley (1993)

92. Vilkomir, S.A., Bowen, J.P.: From MC/DC to RC/DC: formalization and analysis of control-flow testing criteria. Formal Aspects Comput. **18**(1), 42–62 (2006). https://doi.org/10.1007/s00165-005-0084-7

93. Vilkomir, S.A., Bowen, J.P., Ghose, A.K.: Formalization and assessment of regulatory requirements for safety-critical software. Innov. Syst. Softw. Eng. **2**, 165–178 (2006). https://doi.org/10.1007/s11334-006-0006-8

94. Webber, E.: Building Successful Communities of Practice. Blurb (2016)

95. Wenger, E.: Communities of Practice: Learning, Meaning, and Identity. Cambridge University Press (1998)

96. Wenger, E., McDermott, R.A., Snyder, W.: Cultivating Communities of Practice: A Guide to Managing Knowledge. Harvard Business School Press (2002)

97. Wenger-Trayner, E., Fenton-O'Creevy, M., Hutchinson, S., Kubiak, C., Wenger-Trayner, B. (eds.): Learning in Landscapes of Practice: Boundaries, Identity, and Knowledgeability in Practice-Based Learning. Routledge (2014)

98. Wenger-Trayner, E., Wenger-Trayner, B.: Learning in Landscapes of Practice. Routledge (2014)

99. Wenger-Trayner, E., Wenger-Trayner, B.: Introduction to communities of practice: a brief overview of the concept and its uses (2015). https://wenger-trayner.com/introduction-to-communities-of-practice/

100. Wikipedia: Sergiy Vilkomir. Wikipedia: The Free Encyclopedia. https://en.wikipedia.org/wiki/Sergiy_Vilkomir

101. Wohllebe, A., Götz, M.: Communities of practice for functional learning in agile contexts: definition approach and call for research. Int. J. Adv. Corp. Learn. **41**, 62–69 (2021). https://doi.org/10.3991/ijac.v14i1.21939

102. Woodcock, J., Larsen, P.G., Bicarregui, J., Fitzgerald, J.: Formal methods: practice and experience. ACM Comput. Surv. **41**(4), 1–36 (2009). https://doi.org/10.1145/1592434.1592436

103. Woodcock, J.C.P., Larsen, P.G. (eds.): FME 1993. LNCS, vol. 670. Springer, Heidelberg (1993). https://doi.org/10.1007/BFb0024633

104. Huibiao, Z., Bowen, J.P., Jifeng, H.: From operational semantics to denotational semantics for Verilog. In: Margaria, T., Melham, T. (eds.) CHARME 2001. LNCS, vol. 2144, pp. 449–464. Springer, Heidelberg (2001). https://doi.org/10.1007/3-540-44798-9_34

Learning from Mistakes in an Open Source Software Course

Olzhas Zhangeldinov$^{(\boxtimes)}$

Nazarbayev University, Astana, Kazakhstan
olzhas.zhangeldinov@nu.edu.kz

Abstract. Open Source Software development includes many peculiarities, which may not be apparent at first sight. Committing to Open Source projects may be a difficult task without a prior knowledge. This paper describes the experience of a student that took a course on Open Source Software. Much attention was paid to mistakes that were made during the course, and a reflection on decisions was conducted.

Keywords: Open source software · Learning experience · Oss challenges

1 Introduction

Open source projects are widely used in almost every software product today since even proprietary products depend on open source compilers, libraries, or tools [1]. A programmers' involvement in the open source community is an important part of their learning. This paper describes my experience gained attending the Open Source Software course taken in the Spring semester of the 2020–2021 academic year at Nazarbayev University.

1.1 Course Description

The learning outcomes of the course belong to one of these two parts: theoretical and practical. Theoretical is understanding the concept of open source software and its aspects, while practical is learning how to participate in the development of open source software. The theoretical part included lectures about the history and current state of open source software. It was assessed by quizzes, a midterm exam, and a short paper about a specific topic related to the Open Source Software course. The contribution part required that students would choose their role and project, commit to the project and submit deliverables about the work done. The deliverables included short weekly reports about commits and pull requests. Besides the weekly reports, the students had to deliver more detailed reports with their reflections on the work done and decisions.

This paper evolved from the final report written for the course. The paper includes more details about the course and its contribution part. This part

A. Cerone et al. (Eds.): SEFM 2021 Workshops, LNCS 13230, pp. 302–311, 2022.
https://doi.org/10.1007/978-3-031-12429-7_22

involved serious struggles for me and other students, and the reasons for this will be discussed in this paper. A thorough self-assessment was made at the end of the course, and the main mistakes were identified. The paper will demonstrate the mistakes and describe lessons learned from these mistakes.

1.2   Background and Expectations

When the course started, my background in software engineering included the development of HTTP servers and web applications. The projects I had built heavily depended on open source packages. Therefore, I was already familiar with the packages' installation procedures and bug reporting. Additionally, I had worked on projects that involved several other people. Hence, I also had the experience of software development in a team before the course started.

Contributing to the open source community attracted me since I started using the open source packages. These packages make a great impact on software engineering, and being a part of such world-changing technologies was a very attractive perspective.

I had the skills needed for the project development, but I was completely new to contributing to the open source community. Therefore, I expected the course to help me understand and apply conventions specific to the development of Open Source projects. Learning these conventions would help me to commit to the projects later, which was a good motivation to take the course.

2   Project Selection

The selection of a project to commit to, despite my expectations, was a very challenging task. I did not complete this task as successfully as I wanted to. The reasons for the bad project choice will be discussed in later parts of the paper.

At the beginning of the university course, the lectures included generic rules about choosing a good project to contribute to. They intended to select a project with an easy way to have the contribution accepted. Such projects would allow passing the practical part of the course, where students needed to report on their commits to their selected projects. Lectures informed students about several metrics that would help them to identify a project that needs intense development. The metrics included the number of commits, frequency of closing pull requests, number of contributors, and, finally, maturity of the project.

The project selection process consisted of two stages. For the first selection, we needed to identify five possible options, and for the second, after a week to weigh our decisions, we had to pick one as a final choice. The first decision took into account criteria presented during lectures before the selection. After receiving additional information the following week, we could change our final decision. This two-staged process had a positive impact on the project selection experience. During the period between the stages, I could assess the choices more mindfully and resolve possible hesitations.

2.1 The First Selection

In the first stage of our selection, the students selected five projects as the first choice and four other possible options. I decided that I should include projects with different motivations, sizes, and familiarity. The reason for this was the initial uncertainty about the course and how the choice would affect my course experience. I sorted them by the increasing level of contribution difficulty. Nevertheless, I tried to pick projects with the best numbers for the metrics given above.

Laravel. As the first choice, I picked the project I wanted to commit to the most because I had already been its passive user and planned to contribute there. Thus, I chose the Laravel [2] project, which is a back-end framework for web servers. I was the most familiar with this project among other options. Moreover, I was interested in this project the most because of my wide usage of the package in several projects.

NextCloud. NextCloud [3] is an application to store and manage files on the cloud. The project was written in already familiar for me PHP [4] and JavaScript [5]. This project would be a good option if I decided to conduct more research with comfortable programming languages.

Graphana. As the next level of difficulty, I included Graphana [6]. This project focuses on visualization of large volumes of data. The main focus of this project is a web representation using TypeScript [7]. The server-side computations are made with Golang [8].

Zola. I was eager to use Rust [9] language because I considered it a promising language. However, writing projects in Rust is a challenging task because the language is hard to master. Therefore, I chose a small project named Zola [10], which is a static website generator. Despite the project's field being familiar to me, I would still need to learn Rust.

Nushell. Another project written in Rust is Nushell [11]. It is a Linux shell with an improved user interface. This project was the most difficult to learn. Not only was I not proficient enough for the Rust language, but also I had never built a command line application. I included this in case there would not be enough things to learn in other projects.

2.2 Final Decision

The course lectures, besides other hints, advised choosing relatively new projects with enough contributors and active daily work. The projects were compared to each other, focusing on these details. The comparative table (Table 1) illustrates the differences between the projects. I tried to pick projects with as many activities as possible. However, projects with high activity are also very mature

Table 1. Comparison of open source projects from the first choice.

Project name	# of commits	Closed PR frequency	Maturity	# of contributors
Laravel	32,000	7 per day	8 years	2600
NextCloud	60,000	9 per day	8 years	785
Graphana	32,000	13 per day	7 years	1590
Zola	1600	1 per week	4 years	265
Nushell	3500	3 per week	2 years	261

projects. Lectures warned about low activity in projects with a long lifetime, and firstly I considered Zola and Nushell as good options. Nevertheless, their frequency of closed pull requests was drastically lower than in old projects. For the practical part of the course, students had to write weekly reports about the work done, preferably commit every week, and interact with the community. I was afraid that the regularity of several PRs per week would not guarantee that my requests would be closed before the weekly report submission. Neither did it assure enough communication for the report. Therefore, I ended up excluding relatively new projects from my options.

The most mature projects left were Laravel, NextCoud, and Graphana. All the metrics were acceptable for each project. Therefore, the difference in these metrics should not affect a student's experience as a contributor. Therefore, I decided to choose Laravel, because I was the most familiar with it. It would be easier for me to find tasks to complete since I was using the framework in real projects.

3   Project Description

3.1   Governance

The leadership model of Laravel is "Benevolent Dictator" with Taylor Otwell [12] playing the most influential role in the project development. He decides on what features to include and the overall development path of the project.

There is also a group named The Laravel Framework [13], which includes core developers of the project. They have designated roles and more weight on their claims. Their work is not limited to the framework, and only five out of 30 of the most active Laravel contributors belong to the group [14].

3.2   Community

The community is structured hierarchically so that the ideas flow from the bottom layers (passive, active users) to higher (core developers and then the maintainer).

Ideas are usually presented as pull requests because the community promotes adding some proof of concept along with the suggestions. The feed of a pull request tracks discussions of an idea. Therefore, every step of the contribution is performed at the project's GitHub repository.

3.3 Licence

The project is licensed under the MIT license [15]. It is a common choice for vendored packages. The commercial use of the package is allowed, as well as the modification for any purpose, and sharing of the package (distribution).

4 Technical Aspects

4.1 Architecture

The architecture of the framework is split based on classes, interfaces, and traits. The higher-level division is modular. The modules contain specific classes consumed by a kernel of the framework (service providers, contracts, console commands). The modules are exposed to a user via Facades [16] for neat usage.

4.2 Related Modules

I have done my work within specific modules. The module choice was based on the complexity of the module. I tried to avoid working with modules that implied using drivers for other programs, e.g. Eloquent module for database interaction because I did not have enough knowledge about them.

The work I have done is related to several modules, namely Routing, Foundation\Console, and Foundation\Auth.

The Routing module maps HTTP routes to specific controllers. The Foundation namespace includes many submodules that provide boilerplate functions for most back-end servers. The Foundation\Console module contains console commands for automatic configuration and code generation for the server.

The Foundation\Auth module provides some shortcuts for user authentication and access groups (Fig. 1).

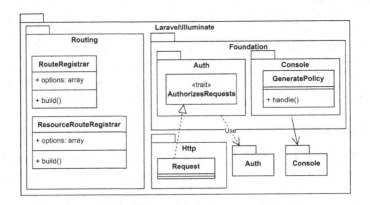

Fig. 1. Package diagram of related modules

4.3   Project Status

At the time of my project selection, the project had entered a soft freeze mode. New features were rarely accepted during this period to maintain the code base properly. This circumstance imposed additional challenges for a successful contribution to the project. I missed the project status from my attention during the project selection. Therefore, I was limited in the number of ways in which I could contribute to the project. There were not many unsolved bugs either, because developers focused their attention on them instead of features.

5   Role and Work Done

Initially, I chose a Developer role. After the first pull request, I realized that the project paused accepting new features, and entered a soft-freeze mode. The maintainer declined most of the proposals replying with the same notification about the soft-freeze. Due to this, I started to look for bugs and missing features that intuitively should be present. Therefore, at the end of the semester, I was playing both roles of Developer and Tester.

5.1   The First Commit: Minor Feature to Group Routes by a Common Controller Class

For my first commit, I have implemented a minor feature. The feature was an additional option for grouping routes in the Routing module, which was inspired by Ruby on Rails. I forked the project, made a new branch, committed my changes to the branch, and posted a pull request [17]. Despite the work being minor, the maintainer declined the pull request because it was not important enough for the soft-freeze status of the project.

5.2   Testing Modules and Bug Fix

Then I decided to switch my priorities in the development from generating new ideas to testing the existing features. However, since the project has a large community and the last version was in the soft freeze for a while, major bugs were already fixed. I decided to test the Policy feature because it was one of my favorite features. I found a bug in the code that generates a stub for a new policy. The code is in the Foundation\Console module. Duplicated lines in the stub did not collapse because the line matcing regular expression did not account for Windows' carriage return \r symbol in the line-feed. I attempted to make the matching of the line-feed platform-independent [18]. However, another contributor pointed out that my solution would break the behavior in some projects. I tested his proposed solution, copied the code, and made a pull request that was merged in a few hours [19].

5.3   Adding Compatibility Between Old and New Features

I realized that there were not many known bugs left in the project. I scanned the recently merged pull request to find possible bugs there. I found a recently added Routing method *missing*. The function allows adding a custom callback, which is called when a model bound to the route is not found in the database (e.g. users/1 - binds User with id = 1). When I was testing it, I decided that it should be compatible with the *resource* method added a long time ago. The *resource* method automatically defines a set of CRUD routes for a model. I implemented a wrapper for the *missing* method when building *resource* routes. My decision was not a mistake, and the pull request with this modification was merged in the latest version branch and released in the next minor release [20].

5.4   Interaction with Community

There was little interaction with the community during my contribution process. My first PR was declined by the maintainer a few hours after I posted it. In this timeframe, only one contributor asked me a question about route caching. I checked the functionality when they pointed it out and reported that the caching worked as expected.

The new feature was merged very fast. There was no following discussion likewise.

Most interactions with the community were during the submitting the bug fix. However, it was not a full-fledged discussion since the contributor only provided better code than mine.

Therefore, a fast activity in the project possibly implies that there might not be a space for communication. Probably, picking slower-paced development will result in a higher quality experience received from contributing to the open source.

6   Lessons Learned

There are some lessons about Open Source Software that I took from contributing to the project. The contribution to the Open Source has many aspects. They must be taken into account when trying to help to develop a project. The lessons fall into several categories that embrace specific details of interacting with the Open Source community. Most of the new information was from the mistakes that I made because of a lack of experience. These lessons might guide new students of the course to proper decisions. They might help them pass the practical part of the course.

6.1   Identifying a Good Project to Contribute

One of the main mistakes was choosing a project without a need for active help. I expected that I would commit changes that I needed personally for my projects.

My thought was that they would be helpful for others, and adding new features for users sounded interesting to me, especially when considering such a large project as Laravel, which thousands of projects depend on. Laravel is developed very fast and releases new major versions every several months. However, I was wrong when I supposed that it would not be a problem to propose new features to the project during the code freeze.

I analyzed the number of contributors and the frequency of pull requests. However, these factors were not enough for selecting the project in my case. Despite the high frequency of merges and many contributors, committing to the project was a big challenge. I could observe the reasons even before I selected the project. However, I have learned it too late and decided not to switch the project in the middle of the semester. My main mistake was the lack of attention to specific details about the project.

The first and the most significant detail is the status of the project. Since I started committing to the project in a soft-freeze mode, I had few chances of merging and being helpful as a developer. The status of the project plays a big role in choosing a project for contribution. The main reason is that if the project does not accept any major changes, you will not be helpful as a developer.

The status of the project can be identified by recently closed pull requests. Although I looked at the list of the pull request, I did not pay attention to its content every time. However, those PRs contained important information that can inform a reader about the project's status. Many of the pull requests that were proposed in the last month, were typically closed with the reply from Taylor Otwell that proposals of new features are not accepted because they want to preserve code maintainability [21].

With this experience of choosing the wrong project to contribute to, I learned a lesson to investigate deeper whether the project needs my help in the specific role or not.

6.2   Writing Pull Requests

After writing and reading some pull requests, I understood that it would be better if pull requests obeyed some basic rules that would help other people in the community. One of the most important rules is the proportion of the number of words in PR and the amount of changed code.

My first pull request contained a lot of words and redundant examples, although the commit changed a very small portion of the source code. When I read other pull requests and compared them to mine, I understood that a PR should give hints about how many changes the pull request brought. It might be too troublesome for developers to open the *Changes* section and scan all the code to get an estimate of the work done. Thus, it is better to follow a good proportion between PR length and the number of changes.

The proportion is specific for every community. Some projects encourage developers to write long PRs with all descriptions and comments, while Laravel usually follows the format of a short message, without many details and examples.

6.3   Finding Tasks

Since the project did not accept many changes, I mainly searched for any task that I could do. I was not an experienced developer in the project, thus I did not understand complex parts of the code. Therefore, I needed to find something simple and useful.

After some weeks of trying, I understood the concept that should assist in finding the tasks by yourself. Since no major changes were accepted, only minor fixes and additions should be considered. However, if there is no repository of pending tasks that should be completed, it might be even harder to understand what a project needs.

However, I managed to bring a commit that was merged to the main branch. The commit included the extension of a relatively new feature to integrate it with an old method. I learned from this that projects may often lack coherence between features. Commits that allow several features to work together are a good contribution that does not conflict with a soft freeze.

Therefore, I learned to adapt to the workflow that takes place in the project. This made my commits more valuable and gave them more chances to be merged.

7   Conclusion

Although I used to develop the project with the help of the Open Source Software course, I had not contributed to any of them before taking the course. Nevertheless, taking this course became a good motivation for taking the first steps in developing an Open Source community. I chose a project that I felt attracted to and started committing to it. Some mistakes were made, which led to several difficulties with decisions on further actions. However, the mistakes allowed me to learn essential lessons about interaction with the Open Source community:

- A contributor should analyze a project before committing to it. The project's peculiarities might not be suitable for a desired role in the development.
- The current status of the project's development and possible code freezes might hinder the contributor's engagement.
- Even during a freeze mode, the contributor should adapt to the environment and identify the work that the project needs. Soft freezes still allow bug fixes and minor contributions like the integration of recently added features with old functionality.
- A project's common conventions should be followed when interacting with the community.

Taking these lessons into account might allow other students to acquire a better learning experience and more community interactions.

Acknowledgements. Professor Antonio Cerone was the lecturer of the university course introducing Open Source. John Noll, the OpenCERT PC member, helped me with the English language in the paper.

References

1. Wang, H., Chen, W.: Open source software adoption: a status report. IEEE Softw. **18**(2), 90–95 (2001)
2. Laravel Repository. https://github.com/laravel/framework. Accessed 5 Oct 2021
3. Nextcloud Repository. https://github.com/nextcloud/server. Accessed 5 Oct 2021
4. Bakken, S.S., Suraski, Z., Schmid, E.: PHP Manual, vol. 1. iUniverse Incorporated, Bloomington (2000)
5. Flanagan, D.: JavaScript: The Definitive Guide. O'Reilly Media Inc., California (2006)
6. Graphana Repository. https://github.com/grafana/grafana. Accessed 5 Oct 2021
7. Bierman, G., Abadi, M., Torgersen, M.: Understanding typescript. In: European Conference on Object-Oriented Programming, pp. 257–281. Springer, Berlin, Heidelberg (2014). https://doi.org/10.1007/978-3-662-44202-9_11
8. Meyerson, J.: The go programming language. IEEE Softw. **31**(5), 104–104 (2014). https://doi.org/10.1109/MS.2014.127
9. Matsakis, N.D., Klock, F.S., II.: The rust language. ACM SIGAda Ada Lett. **4**(3), 103–104 (2014)
10. Zola Repository. https://github.com/getzola/zola. Accessed 5 Oct 2021
11. Nushell Repository. https://github.com/nushell/nushell. Accessed 5 Oct 2021
12. Taylor Otwell GitHub page. https://github.com/taylorotwell. Accessed 5 Oct 2021
13. Taylor Otwell GitHub page. https://github.com/laravel. Accessed 5 Oct 2021
14. Laravel Contributors page. https://github.com/laravel/framework/graphs/contributors. Accessed 15 Apr 2021
15. Laravel MIT License File. https://github.com/laravel/framework/blob/9.x/LICENSE.md. Accessed 15 Apr 2021
16. Gamma, E., Helm, R., Johnson, R., Vlissides, J.: Design Patterns: Elements of Reusable Object-Oriented Software. Addison-Wesley Longman Publishing Co., Inc, USA (1995)
17. [8.x] Route group for same controller. https://github.com/laravel/framework/pull/36213. Accessed 5 Oct 2021
18. [8.x] Make user policy command fix (Windows). https://github.com/laravel/framework/pull/36445. Accessed 5 Oct 2021
19. [8.x] Make user policy command fix (Windows) - fixed version. https://github.com/laravel/framework/pull/36464. Accessed 5 Oct 2021
20. [8.x] Add resource missing option. https://github.com/laravel/framework/pull/36562. Accessed 5 Oct 2021
21. Response about soft-freeze. https://github.com/laravel/framework/pull/36213#issuecomment-776738560. Accessed 5 Oct 2021

ASYDE 2021 - 3rd International Workshop on Automated and verifiable Software sYstem DEvelopment

ASYDE 2021 Organizers' Message

ASYDE 2021 provided a forum for researchers and practitioners to propose and discuss automated software development methods and techniques, compositional verification theories, integration architectures, flexible and dynamic composition, and automated planning mechanisms.

During the last three decades, automation in software development has gone mainstream. Software development teams strive to automate as much of the software development activities as possible. Automation helps, in fact, to reduce development time and cost, as well as to concentrate knowledge by bringing quality into every step of the development process.

Realizing high-quality software systems requires producing software that is efficient, error-free, cost-effective, and that satisfies customer requirements. Thus, one of the most crucial factors impacting software quality concerns not only the automation of the development process but also the ability to verify the outcomes of each process activity and the goodness of the resulting software product. Realizing high-quality software systems requires producing software that is efficient, error-free, cost-effective, and that satisfies evolving requirements.

For this year's edition, ASYDE solicited two types of contributions. Besides traditional technical papers, it included short papers, which covered tool demonstration papers, position papers, and well-pondered and sufficiently documented visionary papers. ASYDE received a total of three submissions. From these submissions, ASYDE accepted two papers. All papers were peer-reviewed by at least three Program Committee members.

We were very pleased to welcome Patrizio Pelliccione, Professor in Computer Science and Software Engineering at Gran Sasso Science Institute (GSSI), Italy, as the ASYDE 2021 keynote speaker. The title of the keynote talk was "Software Certification: Lessons Learned from the Development of a Mechanical Ventilator for COVID-19".

We would like to thank the ASYDE Steering Committee for the support provided during the ASYDE 2021 organization. A special thanks goes to the SEFM 2021 organizers who have been supportive and, in particular, to the Workshop Chair, Antonio Cerone (Nazarbayev University, Kazakhstan), who was always willing to consider or accept new suggestions and ideas, reacting quickly and positively to the ASYDE 2021 requests.

March 2022

Marco Autili
Alessio Bucaioni
Claudio Pompilio
Marjan Sirjani

Organization

Program Committee Chairs

Marco Autili   University of L'Aquila, Italy
Alessio Bucaioni   Mälardalen University, Västerås, Sweden
Claudio Pompilio   University of L'Aquila, Italy
Marjan Sirjani   Mälardalen University, Västerås, Sweden

Steering Committee

Farhad Arbab   Centrum Wiskunde & Informatica,
     The Netherlands
Marco Autili   University of L'Aquila, Italy
Federico Ciccozzi   Mälardalen University, Sweden
Dimitra Giannakopoulou   NASA, USA
Pascal Poizat   Paris Nanterre University, France
Massimo Tivoli   University of L'Aquila, Italy

Program Committee

Luciano Baresi   Polytechnic University of Milan, Italy
Steffen Becker   University of Stuttgart, Germany
Antonio Brogi   University of Pisa, Italy
Giovanni Denaro   University of Milano-Bicocca, Italy
Antinisca Di Marco   University of L'Aquila, Italy
Amleto Di Salle   University of L'Aquila, Italy
Predrag Filipovikj   Scania Group, Sweden
Francesco Flammini   Linnaeus University, Sweden
Ludovico Iovino   Gran Sasso Science Institute, Italy
Ehsan Khamespanah   University of Tehran, Iran
Elena Lisova   Mälardalen University, Sweden
Marina Mongiello   Polytechnic University of Bari, Italy
Cristina Seceleanu   Mälardalen University, Sweden
Catia Trubiani   Gran Sasso Science Institute, Italy
Apostolos Zarras   University of Ioannina, Greece

A Probabilistic Model Checking Approach to Self-adapting Machine Learning Systems

Maria Casimiro[1,2]([✉]), David Garlan[1], Javier Cámara[3], Luís Rodrigues[2], and Paolo Romano[2]

[1] Institute for Software Research, Carnegie Mellon University, Pittsburgh, PA, USA
[2] INESC-ID, Instituto Superior Técnico, Universidade de Lisboa, Lisboa, Portugal
maria.casimiro@tecnico.ulisboa.pt
[3] ITIS Software, Universidad de Málaga, Málaga, Spain

Abstract. Machine Learning (ML) is increasingly used in domains such as cyber-physical systems and enterprise systems. These systems typically operate in non-static environments, prone to unpredictable changes that can adversely impact the accuracy of the ML models, which are usually in the critical path of the systems. Mispredictions of ML components can thus affect other components in the system, and ultimately impact overall system utility in non-trivial ways. From this perspective, self-adaptation techniques appear as a natural solution to reason about how to react to environment changes via adaptation tactics that can potentially improve the quality of ML models (e.g., model retrain), and ultimately maximize system utility. However, adapting ML components is non-trivial, since adaptation tactics have costs and it may not be clear in a given context whether the benefits of ML adaptation outweigh its costs. In this paper, we present a formal probabilistic framework, based on model checking, that incorporates the essential governing factors for reasoning at an architectural level about adapting ML classifiers in a system context. The proposed framework can be used in a self-adaptive system to create adaptation strategies that maximize rewards of a multi-dimensional utility space. Resorting to a running example from the enterprise systems domain, we show how the proposed framework can be employed to determine the gains achievable via ML adaptation and to find the boundary that renders adaptation worthwhile.

Keywords: Machine-learning based systems · Self-adaptation · Probabilistic model checking · Architectural framework

1 Introduction

Machine learning (ML) is present in most systems we deal with nowadays and is not a trend that will vanish in the years to come. Like all other components in a

Support for this research was provided by Fundação para a Ciência e a Tecnologia (Portuguese Foundation for Science and Technology) through the Carnegie Mellon Portugal Program under Grant SFRH/BD/150643/2020 and via projects with references POCI-01–0247-FEDER-045915, POCI-01–0247-FEDER-045907, and UIDB/50021/2020.

system, ML components can fail or simply produce erroneous outputs for specific inputs [8,14,15]. This problem is exacerbated by the fact that the environments in which the ML components operate may be different from those that the component may have been trained on [31]. When such a situation occurs, the system is likely to suffer from a problem known as data-set shift [29]. Since ML components rely on input data to learn representations of the environment, data-set shift can cause accuracy degradations which ultimately affect system utility. Hence an important requirement for systems with ML components is to be able to engineer those systems in such a way as to be able to adapt the ML components when it is both possible and beneficial to do so.

Current approaches to architecture-based self-adaptive systems provide a useful starting point. Following the well-known MAPE-K pattern [17], a system is monitored to produce updated architectural models at run time, which can then be used to determine whether and how a system might be improved through the application of one or more tactics. In support of this approach, there are a variety of tactics that might be brought to bear on ML-based systems including model retraining [32], various incremental model adjustments [20], data unlearning [5], and transfer learning [16,27] techniques.

However, deciding both whether and how to adapt an ML-based system is non-trivial. In particular, typically there are costs as well as benefits for applying tactics. Determining whether the benefits of improving an ML component outweigh its costs involves considerations of timing, resources, expected impact on overall system utility, the anticipated environment of the system, and the horizon over which the benefits will be accrued. Moreover, in practice there is often considerable uncertainty involved in all of these factors.

This paper proposes a probabilistic framework based on model checking to reason, in a principled way, about the cost/benefits trade-offs associated with adapting ML components of ML-based systems. The key idea at the basis of the proposed approach is to decouple the problems of i) modelling the impact of adaptation on the ML model's quality (e.g., expected accuracy improvement after retrain) and ii) estimating the impact of ML predictions on system utility. The former is tackled by incorporating in the framework the key elements that capture relevant dynamics of ML models (e.g., the expected dependency between improvement of model's quality and availability of new training data). The latter is solved by expressing inter-component dependencies via an architectural model, enabling automatic-reasoning via model checking techniques.

We resort to a running example from the enterprise systems domain to showcase how to instantiate the proposed framework via the PRISM model checker. Finally, we present preliminary results that show how system utility can be improved through the adaptation of ML components.

The remainder of this document is organized as follows: Sect. 2 motivates the need for the proposed framework and highlights existing challenges; Sect. 3 presents the proposed framework and Sect. 4 shows how it can be applied. Section 5 evaluates the framework, Sect. 6 overviews related work, Sect. 7 discusses existing limitations and future work and Sect. 8 concludes the paper.

2   Motivation

In this work we focus on self-adaptation of ML-based systems which are composed of both ML and non-ML components. For example, fraud detection systems rely on ML models to output the likelihood of a transaction being fraudulent and on rule-based models (non-ML component) to decide whether to accept/block/review a transaction based on the ML's output [2]. Similarly, cloud configuration recommenders rely on ML models to select the platform (non-ML component such as a virtual machine) on which users should deploy their jobs. These recommenders are typically guided by user-defined objective functions such as minimizing execution time [1,6].

There are two key requirements associated with reasoning about self-adaptation of ML components. First, it is necessary to understand if and how ML predictions affect overall system utility. Second, it is necessary to estimate the costs and benefits of the available adaptation tactics. Let us now discuss the key challenges associated with each requirement.

i) **Impact of Machine Learning Predictions.** A key problem that needs to be addressed to enable automatic reasoning on the dynamics of ML-based systems is determining to what extent incorrect predictions will impact overall system utility. In fact, this is not only application but also context dependent. For example, in cloud configuration recommenders, when the relative difference in job execution speed between the available cloud configurations is low, ML mispredictions have little impact on system utility [6]. Similarly, in a fraud detection system, the impact of mispredictions is different in periods with higher volumes of transactions, in which it is critical to maximize accepted transactions, while accurately detecting fraud [2].

ii) **Estimating Costs and Benefits of Adaptation Tactics.** Predicting the time/cost and benefits of ML adaptation tactics is far from trivial. This prediction is strongly influenced both by the type of models and their settings (hyperparameters and execution infrastructure), and by the input data employed in the adaptation process. For instance, in the case of a tactic that triggers the retrain of an ML model, the benefits of tactic execution are dependent on the data available for the process – data more representative of the current environment contributes to higher benefits. Differently, if the adaptation tactic consists of querying a human (human-in-the-loop tactic), the benefits are now dependent on human expertise. Their execution latency and economic cost are also likely to be different and affected by factors that are inherently tactic dependent, e.g., the retraining time is affected by the amount of available training data, whereas the latency of a human-in-the-loop tactic may depend on the complexity of the problem the human is required to solve.

In this work, we argue that by leveraging formal methods we can instantiate the problem of reasoning about the need for adaptation at a general architectural level. The framework we propose allows us to abstract away from system-specific issues and instead instantiate the decision of whether to adapt ML components

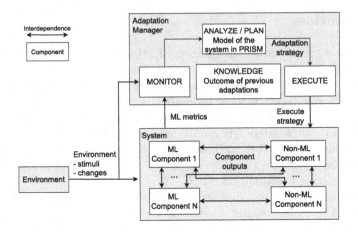

Fig. 1. Framework modules and inter-dependencies.

as a general decision that relies on the key factors of ML-based systems. The next section introduces our formal framework.

3   Framework for Self-adaptive ML-Based Systems

This section describes a generic framework that can be used to derive formal models of self-adaptive ML systems. The resulting models can then be utilized to enable automatic reasoning via probabilistic model checking tools such as PRISM [19]. In fact, the use of such tools in the self-adaptive systems (SAS) domain is not new [4,25,26]. Conceptually, the proposed framework can be regarded as a specialization of the frameworks already proposed in this field, which targets a specific class of "managed" systems: systems containing ML-based components. As such, in the following sections our focus will be on how to capture the most relevant dynamics of ML-components via abstract and generic models that can be easily extended and customized to specific use cases.

3.1   Architectural Overview

As in typical frameworks for SAS, our framework requires specifying the behavior of the following modules: environment, system, and the adaptation manager (Fig. 1). Next, we discuss how each of these modules is modelled in the proposed framework. For the ML component, we further describe its internal state, how it evolves, and the methods exposed by its interface to allow for inter-component interactions.

Environment. In the environment component, it is necessary to consider the types of stimuli to which the system responds/reacts. Additionally, since our goal is to reason about the impact of environment changes, these must also be modelled. Examples of environment stimuli are for instance the transactions that a fraud detection system has to classify or the jobs received by a scheduler.

Adaptation Manager. As in typical SAS, the adaptation manager contains a repository of adaptation tactics. However, and differently from previous work in this domain, we consider adaptation tactics that directly actuate over the ML component [7], such as retraining the ML component. Each adaptation tactic is specified by: (i) a precondition that triggers its execution and which generally depends on the state of the system and the environment; (ii) the effects on the targeted components. We model adaptation tactics as a tuple composed of tactic cost and tactic latency. This division allows us to study the impact of the different dimensions of tactics on overall system utility. For example, consider a retrain tactic. It has a latency associated, since retraining a model takes a non-negligible amount of time. If that tactic is executed in cloud environments it also has a monetary cost, which depends both on its latency, and on the underlying cloud platform selected for the execution. By leveraging model checking tools such as PRISM [19] we can thus explore alternative adaptation policies with the objective of identifying the one that, for example, maximizes system utility.

System. The key novelty of our framework is that it enables reasoning about the adaptation of ML-based systems, which we abstractly define as systems that comprise two types of components: ML-based and non-ML based components (Fig. 1). An ML-based component is used to encapsulate an ML-model that can be queried and updated (e.g., retrained). Non-ML based components are used to encapsulate the remaining functional components of the system being managed/adapted. Our framework is agnostic to the modelling of the application-dependent dynamics of non-ML based components, which can be achieved by resorting to conventional techniques already proposed in the SAS literature [4,25,26]. However, we require non-ML components to interact with ML components in two ways: i) pre-processing data to act as input to the ML component or using the ML component's outputs to perform some function ii) affecting system utility by adding negative/positive rewards upon completion of a task. For example, in an ML-based scheduling system, once a job completes its execution on the selected cloud platform (a non-ML based component), it triggers the accrual of a reward (e.g., dependent on the execution time of the job) on system utility.

As for modelling the ML components, our design aims to ensure the following key properties: (i) **generic** – designed to be applicable to offline and online learning, supervised, unsupervised and semi-supervised models, different types of ML models (e.g., neural networks, random forests); (ii) **tractable** – designed to be usable by a probabilistic model checker like PRISM, having a high level of abstraction to aid systematic analysis via model checking; (iii) **expressive** – designed to capture key dynamics of ML models that are general across ML models; (iv) **extensible** – designed to be easily extended to incorporate additional adaptation tactics and customized to capture application specific dynamics. The following section introduces the proposed modelling approach for ML components.

3.2   Machine Learning Components

We consider ML components that solve classification problems, i.e., whose possible outputs are defined over a discrete domain. We argue that this assumption is not particularly restrictive given that any regression problem (in which the ML model's output domain is continuous) can be approximated via a classification problem by discretizing the output domain. We abstractly formalize the behavior of an ML based component by specifying (i) its state, (ii) the set of events that change its state, (iii) the logic governing how the internal state evolves due to each possible event.

Machine Learning Component State. The state of an ML component is characterized by two elements: a confusion matrix and the set of new data (knowledge) which encodes information regarding the data accumulated so far by the ML component. The confusion matrix is a tabular way to represent the quality of a classifier. We opt for using the confusion matrix to abstract over the internal dynamics of the specific model being used while still capturing the quality of its predictions. More in detail, it is a matrix $n \times n$ where n represents the number of output classes. In cell (i, j) the confusion matrix maintains the probability for an input of class i to be classified by the model as belonging to class j. In fact, due to how it is constructed, it provides access to metrics such as false positives/negatives, which in turn constitute the basis to compute alternative metrics, like f-score or recall, that can be of interest for specific applications/domains (e.g., relevant for fraud detection systems [2]). Our framework supports the specification of multiple confusion matrices, which can be of interest when there are several types of input to an ML model (e.g., if a scheduler receives different job types, some easier to classify than others, this could be modelled by associating a different confusion matrix to each job type).

The second element of the ML component's state represents the knowledge maintained by the ML component. This is characterized by the data which the model has already used for training purposes, and the new data that is continually gathered during operation and that represents the current state of the environment the system is operating in. However, the representation of this knowledge is application dependent. While in simpler use-cases it may be enough to simply maintain a counter for the inputs that arrive in the system, in more complex scenarios the data gathered during execution may encode important information to characterize the environment (e.g., measures of dataset shift [29]). This knowledge can be used by the adaptation tactics when these are executed over the ML component.

Machine Learning Component Interface. In order for the other components in the system to interact with the ML component, we propose a general interface that enables this interaction. Generally, any ML component that supports adaptation requires three key methods: *query*, *update_knowledge*, and *retrain*. Clearly, new methods can be added to this interface to tailor the framework to specific application requirements. For example, to capture manipulations of the dataset (e.g., sub-sampling certain types of inputs) or to support further adaptation tactics (e.g., unlearning).

As the name suggests, *query* is used to solicit a prediction from the ML component. The internal behavior of the ML component when issuing predictions is abstracted away by reasoning only over the likelihoods specified in the confusion matrix. More precisely, the output event produced by executing a query is a probabilistic event, which can assume any of the possible output classes of the classifier, with the probability given by the classifier's confusion matrix.

The method *update_knowledge* should be called when the system has reacted to an event and thus there is new data to be accounted for. The framework can keep track either of the pair $\langle ML\ input,\ ML\ prediction \rangle$ or of the triple $\langle ML\ input,\ ML\ prediction,\ real\ output \rangle$. The selection of either option is domain dependent. For example, in the fraud detection domain, knowing the actual value of a transaction (legitimate or fraudulent) is not always possible [28]. The pair is thus required when this is the case. Finally, *retrain* corresponds to retraining the ML model resorting to the data stored in the ML component's state.

Any method has an associated cost and latency that directly impact system utility. These are captured by the framework as follows. The latency is used to determine after how many units of time the effects of each method are applied to the component (and to the system). The cost of executing each method (when the method has a cost) is discounted from the system's utility.

Machine Learning Component State Evolution. When any of the previously described interface methods is triggered, the state of the ML component can be altered. Since *query* consists only of asking the ML model for a prediction, it does not alter the component's state. *update_knowledge* changes the knowledge of the ML component by adding instances to that set. Finally, *retrain* changes both elements of the state: the confusion matrix and the knowledge are updated to reflect the execution of the adaptation tactic. In the case of a retrain adaptation tactic the data used for executing the tactic is updated in the knowledge such that it is no longer considered new data. At the same time, the confusion matrix is updated to reflect the current performance of the model after having been retrained. In fact, we propose a simple model that aims to capture the improvements to the confusion matrix given by the execution of a retrain adaptation tactic. The rationale behind the proposed model is that the larger the number of new samples seen since the last training (i.e., new data), the larger should be the expected reduction in the misclassification rate. Specifically, the confusion matrix should be updated as follows. The diagonal is incremented by a factor $\delta = (100 - cell_{ii}) * new\_data * impact\_factor$ that is proportional to the model's loss and to the amount of new data (e.g., number of new samples). The *impact_factor* allows for flexibility in different types of retrain (e.g., when the hyper-parameters of the model are also updated, the benefits may be higher). The remaining cells in the same row should then be updated as $cell_{ij} = cell_{ij} - \delta cell_{ij}/(1 - cell_{ii})$. The non-diagonal cells are thus reduced proportionally to δ while ensuring that no cell gets a value lower than 0,

Table 1. Visualization of an update to a confusion matrix. We assume *new_data* = 6 and *impactFactor* = 0.1 which yields $\delta = (100 - 95) \times 6 \times 0.1 = 3$. P1 and P2 stand for Platform 1 and Platform 2, respectively.

Pred. \ Real	P1	P2
P1	95	5
P2	5	95

P \ R	P1	P2
P1	$95 + 3$	$5 - \frac{3 \times 5}{(100-95)}$
P2	$5 - \frac{3 \times 5}{(100-95)}$	$95 + 3$

P \ R	P1	P2
P1	98	2
P2	2	98

(a) Initial confusion matrix. (b) Confusion matrix update. (c) Final confusion matrix.

and that the total reduction on non-diagonal cells is equal to δ. Table 1 provides an example of a confusion matrix being updated.

Dealing with Uncertainty. As shown by recent work in SAS, capturing uncertainty and including it when reasoning about adaptation contributes to improved decision making [4,25,26]. Uncertainty can affect a range of components, including non-ML components (e.g., the execution time of a job on a specific cloud platform is unknown), ML components (e.g., in the fraud domain, as there is no real time access to real labels of transactions, we cannot measure the *current* model's performance, but at most estimate it [28]), and adaptation tactics (e.g., with 90% probability retrain is expected to reduce the misclassification rate, however in the remaining 10% of the cases the misclassification rate remains the same). In the proposed framework, uncertainty regarding a specific component or event can be naturally integrated by defining the affected state of the component/event via discrete distributions built leveraging historical data. Uncertainty can thus be conveniently captured by expressing the outcome of an uncertain action (or state) via a probabilistic event.

4   Model Checking the Need for Adaptation

This section exemplifies, based on a running example, how to leverage the proposed framework to reason about whether to adapt ML components. By introducing in the formal model non-deterministic choices between the tactics available for execution, a model-checking based approach can be used to determine, at any time, which adaptation tactic to enact in order to maximize system utility. We implement our framework using the PRISM [19] tool, which allows to model non-deterministic phenomena via probabilistic methods such as Markov Decision Processes (MDPs) and generate optimal strategies for reward-based properties. PRISM has been extensively used by the literature on Self-Adaptive Systems (SAS) to reason about adaptation trade-offs [9,21,25,26].

We start by introducing a simple use-case, which was selected to exemplify the framework. The following sections then describe the modules of the PRISM model in more detail. Due to lack of space, we do not provide details about the implementation of the framework and of the running example in PRISM. However, we are working on a technical report that includes these details and will make it available in the near future.

Running Example. Consider a system that receives jobs and has to select a platform for them to execute. As it is often the case in practice, we assume that the execution time of a job on a given platform depends on the job's characteristics, i.e., it may execute faster on a platform than on another [1,6]. Each time a job completes, the system receives a fixed reward. As such, in a given period, the system will strive to complete as many jobs as possible by selecting the platform that can execute each incoming job in the shortest amount of time, so as to accrue the maximum benefits possible. For example, the system could receive different data analytic jobs with diverse characteristics (e.g. neural network (NN) training, data stream processing) [1,6]. The system then relies on an ML component to decide the best platform for a specific job to execute in. For instance, the training of a neural network can be offloaded to GPUs or CPUs. While both platforms allow the system to complete its task (i.e., execute the job) one platform may be more efficient (lower latency) than the other, thus allowing the system to complete more jobs in a given horizon. We are interested in scenarios in which the type of job generated by the environment is altered, for example due to data-set shift [29], thus leading the ML model to lose accuracy [18].

Machine Learning Adaptation. To equip the system with adaptation capabilities, so that it can deal with environment changes and accuracy fluctuations of the ML-based predictor, for instance due to unknown jobs (e.g., unseen NN topology), we consider that each time a new job arrives, the adaptation manager can decide between simply querying the current ML model (i.e., no adaptation/tactic *nop*) or adapting it (tactic *retrain*), in order to increase its accuracy and maximize the likelihood of executing the job in the preferred platform. The retrain adaptation tactic consists of incorporating additional training data in a new version of the model [32]. However, the execution of this tactic requires a non-negligible time interval for its effects to manifest themselves in the system, and has a monetary cost (e.g., if retrain is performed in the cloud) [1,6]. As such, overall system utility is defined as the sum of the benefits of completing jobs minus the cost of executing a tactic (tactic *nop* has no cost).

4.1   Modelling the Components of the Framework

Environment. We model the environment as generating two types of job (J1 and J2) according to probability *pJob1* (or *pJob2=1-pJob1*). Although it could be trivially extended to generate more job types, having only two is enough for the purpose of reasoning about whether to adapt the ML component.

Adaptation Manager. The adaptation manager is responsible for triggering adaptations. In the running example, two tactics are available to be executed: *nop* (no operation) and *retrain*. Whenever the system receives a new job generated by the environment, the pre-condition for the tactics' execution becomes true. At

Table 2. Discretized distribution of job latencies for both job types and for each platform and corresponding likelihoods. Each cell in each matrix has the probability of the corresponding PRISM transition. That is, in Table 2a with probability 18% the predicted latency for a job is 3 on platform 1 (P1) and 8 on platform 2 (P2). If, for this situation, the ML component is very accurate, it will select platform 1 to deploy the job, since P1 has the lowest latency. The difference between job types lies in the accuracy of the ML model for each.

(a) Job latency depends on the platform in which it is executed. Thus, in this case, ML accuracy has an impact on system utility.

(b) The diagonal accounts for more than half of the probability, thus it is more likely that the latency of a job will be the same regardless of the platform, which means that ML accuracy should have no impact on system utility.

P2 lat. P1 lat.	prob.	6 20%	8 30%	10 50%
3	60%	12%	18%	30%
5	30%	6%	9%	15%
7	10%	2%	3%	5%

P2 lat. P1 lat.	prob.	3 15%	5 70%	7 15%
3	15%	2.25%	10.5%	2.25%
5	70%	10.50%	49.0%	10.50%
7	15%	2.25%	10.5%	2.25%

this point, the model checker, when asked to synthesize optimal policies, decides whether to adapt or do nothing and triggers the corresponding tactic in the ML component. The latency of the tactic is accounted for by the ML component during tactic execution. The tactic's cost is subtracted from the system's rewards when the job completes its execution.

Non-Machine-Learning Component. The non-ML component, which is responsible for simulating the execution of the jobs has two possible platforms at its disposal. When the environment generates a new job, and after the adaptation manager has selected the adaptation tactic to execute, the executor is ready to deploy the job on the selected platform. To simulate the execution, it needs to know the latency of the job. As there is intrinsic uncertainty in determining job execution latency, we assume the existence of historical data which can be used to construct distributions of possible execution latencies. These distributions can be discretized (as shown in Table 2) and fed to PRISM so that model checking is feasible and this uncertainty is explicitly modeled. Whenever a job completes, the utility of the system is updated by adding the job completion reward and subtracting the tactic execution cost.

Machine Learning Component. In this use-case, since there are two possible execution platforms and two input job types, we define two binary confusion matrices: one for each type of job. As for the knowledge element of the state of the ML component, in order to model this aspect, we count the inputs of each type that are received and use this count as a proxy for the amount of information encoded in these new inputs. This count is increased whenever an input arrives and reset whenever the tactic is executed. This information then contributes to the impact of the retrain adaptation tactic on the ML component. The increase in model's accuracy is computed as described in Sect. 3.2.

4.2 Collecting Rewards

Since the system receives a fixed reward whenever it completes a job, its goal is to maximize the number of jobs completed in a given time period, while simultaneously minimizing the costs spent on retraining. This requires seeking an adequate trade-off between investing time and resources to retrain the model and reasoning about the expected impact of the current accuracy on system utility. Since model checking tools require the specification of properties in order to compute optimal policies, we verify with PRISM a property that maximizes system utility when the state "end" is reached, i.e., when the time period expires.

5 Results

In this section, we evaluate whether the proposed framework can reason about the trade-offs of ML component adaptation. Specifically, to understand whether adding self-adaptation functionalities to ML-based systems translates into increased benefits to the system, we investigate two research questions:

RQ1 – What are the estimated utility gains achievable through ML adaptation?
RQ2 – Under what conditions does the framework determine that ML adaptation improves overall system utility?

Experimental settings. In our experiments we varied the following parameters: **(i)** the retrain cost; **(ii)** the retrain latency (1, 5, 10); and **(iii)** the probability that the environment generates each type of job (from 0 to 1 with 0.1 increments). Throughout all experiments we set to 100% the probability of the environment generating a new job. Since we are interested in modelling environment changes, we assume that the ML model has better knowledge for one type of job than for the other, which is assumed to be the environment change. Thus, the ML model has an accuracy of 95% for jobs of type 1 and an accuracy of 50% for jobs of type 2 (these correspond to symmetric confusion matrices), and the impact factor is set to 0.1 for both types of jobs. Job latencies and uncertainty in each platform are set according to Table 2.

Results. Figure 2 shows, for an execution context in which ML accuracy affects system utility, the utility gains achievable due to ML adaptation. The difference between plots corresponds to the latency of execution of the retrain tactic. We can see that, regardless of how this parameter is set, adapting the ML component improves system utility in specific areas of the space. Determining the boundary that divides the areas of the space in which it is worth/not worth adapting is thus a critical aspect. Our framework is capable of determining this boundary, which we show in the following paragraphs.

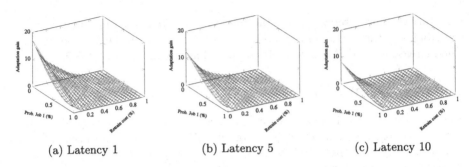

(a) Latency 1   (b) Latency 5   (c) Latency 10

Fig. 2. Utility gains achievable due to ML adaptation when the system operates in an execution context in which ML accuracy impacts system utility.

The plots in Fig. 3 represent, for different execution contexts and tactic latency, the conditions of the environment in which adapting the ML model improves overall system utility. We now focus on Fig. 3a and analyze the impact of the retrain cost and job probability variables. As expected, for low values of retrain cost, the framework tells us that adaptation is always worth it, regardless of the environment stimuli. However, as the cost starts to increase, adaptation is no longer the optimal action when the environment is more likely to generate jobs of type 1. This is due to the fact that, since the ML model has a good knowledge for jobs of type 1, the costs of adaptation outweigh its benefits. Differently, when the environment generates more jobs of type 2 (prob. job 1 < 50%), the tolerated cost of adaptation tactics increases. As tactic latency increases (Figs. 3b and 3c), we see that in order for adaptation to be worth it, its cost must also be lower than in scenarios with lower latency.

The difference between the figures in the top row (Figs. 3a, 3b,3c) and the figures in the bottom row (Figs. 3d, 3e,3f) is the execution context of the system, that is, the latencies of the jobs in each platform and their probabilities (which are set according to Table 2). For the bottom row it is more likely that a job has the same latency regardless of the platform (Table 2b). In such a situation, having an inaccurate ML model has little impact on system utility. The comparison between top and bottom rows demonstrates this effect: for the bottom row plots, adaptation pays off only in very few scenarios and only when tactic cost is low. In fact, we see that when latency is high (Fig. 3f), the cost of retrain has to be close to zero for adaptation to provide benefits.

Overall, our preliminary results confirm that adaptation of ML components in systems improves overall system utility (RQ1). Our framework also shows that, in this concrete example, if the costs of adaptation are too high, adaptation pays off if: **(i)** the environment starts generating more jobs that are less known (probability of job 1 lower than 0.5), or **(ii)** the tactic has a low latency. The framework further shows that in execution contexts in which ML accuracy is not expected to impact system utility, ML adaptation rarely pays off (RQ2).

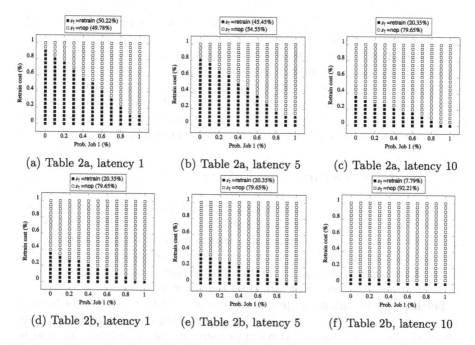

(a) Table 2a, latency 1   **(b)** Table 2a, latency 5   **(c)** Table 2a, latency 10

(d) Table 2b, latency 1   **(e)** Table 2b, latency 5   **(f)** Table 2b, latency 10

Fig. 3. Areas of the space in which ML adaptation improves overall system utility. Black squares correspond to configurations in which it is worth adapting. White squares correspond to configurations in which having the option to adapt does not improve utility. The top row corresponds to the execution context of Table 2a and the bottom row to the execution context of Table 2b.

6   Related Work

Existing frameworks for reasoning about self-adaptation are mostly focused on dealing with uncertainty [4,25,26] and on analyzing human involvement and the benefits of providing explanations [9,21–23]. Differently, the framework we propose is concerned with analyzing the trade-offs of adapting an ML component of a system when it has a negative impact on system utility.

In the literature on self-adaptive systems ML has been leveraged by recent works to improve different stages of the MAPE-K loop, such as the Plan and the Analysis stages [13]. Works focusing on collective SAS have also researched the most common learning strategies employed in these systems, finding that reinforcement learning is the most common [10,12]. Differently from both these lines of work, our framework is focused on single-agent systems that rely on ML in order to work as expected. Our framework, which leverages self-adaptation to improve AI in systems, is aligned with the vision of Bureš [3], that argues for a new self-adaptation paradigm in which self-adaptation and AI benefit and enable one-another. This is also aligned with the vision for continual AutoML [11], which advocates for architectures that can manage ML systems and adapt them in the face of adversities (such as uncertainty, data-set shift, outliers).

The literature on continual/lifelong learning addresses problems which self-adaptive systems also face [24, 30]. Specifically, this branch of literature focuses on open-world problems (i.e., unexpected changes can occur) and on how to learn new tasks from past tasks. In fact, as discussed in our prior work, the techniques developed in this field can be thought of as tactics of self-adaptive systems to adapt ML models [7]. Instead, in this work, we present an actual model-checking based framework to reason about whether the benefits of executing an adaptation tactic outweigh its costs and improve overall system utility.

7   Threats to Validity and Future Work

Our framework relies on the assumption that we have access to historical data, however this might not be the case for all systems. Specifically, for new deployments of systems (e.g., new client of a fraud detection company) there is an initial period during which data must be collected. Nonetheless, we believe this assumption is acceptable since this data collection period is also required to have an initial training set with which to train the ML component.

Further, in this work we adopt a pragmatic strategy for modelling the benefits of a retrain tactic, which corresponds to instantiating a parametric model that defines the benefits of retrain as being proportional to the model's loss and to the amount of new data gathered since the previous execution of the tactic. However, this corresponds to a simple model of the benefits of adaptation. In fact, not only is it possible that this relationship is not linear, but also there are likely other factors that influence the benefits of a tactic and that should be considered. Examples of such factors are the latency of a tactic and the context of the system upon the tactic's execution. Taking the example of the retrain tactic, intuitively one would expect the benefits provided by such a tactic to increase as its execution latency also increases, i.e., when the system is retrained for a longer period. However, it is also expected that the achievable accuracy will plateau at some point, thus not corresponding to a linear relationship. As future work, we plan to study how the benefits of different adaptation tactics vary based on parameters such as tactic execution latency, application and application context, to extend our current model of adaptation benefits to account for these factors.

To address possible scalability issues and generalize the proposed framework to more job types, one can change the PRISM model to reason about aggregated metrics instead of job specific metrics. This can be achieved for example by evaluating the system in terms of correctly scheduled jobs per time interval (user-defined). This would require the system to monitor the expected number of jobs generated per time interval and the confusion matrix to model the probabilities of a job being correctly or incorrectly scheduled (independently of their type).

Finally, since in the presented use case we considered a single adaptation tactic (model re-train), we plan to conduct the aforementioned study on several adaptation tactics. This corresponds to extending the repertoire of tactics considered by the framework and also entails the extension of the framework to account for dynamic environments.

8 Conclusion

In this paper we proposed a framework to reason about the need to adapt ML components of ML-based systems. Resorting to a running example, we showed how to instantiate the framework in a practical setting and how system utility can be improved through the adaptation of ML components. We further demonstrated how the adaptation decision boundary is affected by environment changes and execution context. As next steps, we plan to investigate more fine-grained approaches to modelling the effects of retraining ML models, as well as to extend our framework to consider additional adaptation tactics (e.g., unlearning and human-in-the-loop).

References

1. Alipourfard, O., et al.: Cherrypick: adaptively unearthing the best cloud configurations for big data analytics. In: Proceedings of NSDI (2017)
2. Aparício, D., et al.: Arms: Automated rules management system for fraud detection. arXiv preprint. arXiv:2002.06075 (2020)
3. Bureš, T.: Self-adaptation 2.0. In: Proceedings of SEAMS (2021)
4. Cámara, J., et al.: Reasoning about sensing uncertainty and its reduction in decision-making for self-adaptation. Sci. Comput. Program. **167**, 51–69 (2018)
5. Cao, Y., Yang, J.: Towards making systems forget with machine unlearning. In: Proceedings of IEEE S& P (2015)
6. Casimiro, M., et al.: Lynceus: cost-efficient tuning and provisioning of data analytic jobs. In: Proceedings of ICDCS (2020)
7. Casimiro, M., et al.: Self-adaptation for machine learning based systems. In: Proceedings of SAML. LNCS, Springer (2021)
8. Cito, J., Dillig, I., Kim, S., Murali, V., Chandra, S.: Explaining mispredictions of machine learning models using rule induction. In: Proceedings of ESEC/FSE (2021)
9. Cámara, J., Moreno, G., Garlan, D.: Reasoning about human participation in self-adaptive systems. In: Proceedings of SEAMS (2015)
10. D'Angelo, M., et al.: On learning in collective self-adaptive systems: state of practice and a 3d framework. In: Proceedings of SEAMS (2019)
11. Diethe, T., et al.: Continual learning in practice. Presented at the NeurIPS 2018 Workshop on Continual Learning (2019)
12. D'Angelo, M., et al.: Learning to learn in collective adaptive systems: mining design patterns for data-driven reasoning. In: Proceedings of ACSOS-C (2020)
13. Gheibi, O., Weyns, D., Quin, F.: Applying machine learning in self-adaptive systems: A systematic literature review. arXiv preprint. arXiv:2103.04112 (2021)
14. Gu, T., et al.: Badnets: evaluating backdooring attacks on deep neural networks. IEEE Access **7**, 47230–47244 (2019)
15. Huang, L., et al.: Adversarial machine learning. In: Proceedings of AISec (2011)
16. Jamshidi, P., et al.: Transfer learning for improving model predictions in highly configurable software. In: Proceedings of SEAMS (2017)
17. Kephart, J.O., Chess, D.M.: The vision of autonomic computing. Computer **36**(1), 41–50 (2003)
18. Koh, P.W., et al.: Wilds: A benchmark of in-the-wild distribution shifts. In: Proceedings of ICML. PMLR (2021)

19. Kwiatkowska, M., Norman, G., Parker, D.: PRISM 4.0: verification of probabilistic real-time systems. In: Gopalakrishnan, G., Qadeer, S. (eds.) CAV 2011. LNCS, vol. 6806, pp. 585–591. Springer, Heidelberg (2011). https://doi.org/10.1007/978-3-642-22110-1_47

20. Li, J., Hu, M.: Continuous model adaptation using online meta-learning for smart grid application. IEEE Trans. Neural Netw. Learn. Syst. **32**(8), 3633–3642 (2021)

21. Li, N., Adepu, S., Kang, E., Garlan, D.: Explanations for human-on-the-loop: a probabilistic model checking approach. In: Proceedings of SEAMS (2020)

22. Li, N., Cámara, J., Garlan, D., Schmerl, B.: Reasoning about when to provide explanation for human-in-the-loop self-adaptive systems. In: Proceedings of ACSOS (2020)

23. Li, N., Cámara, J., Garlan, D., Schmerl, B., Jin, Z.: Hey! preparing humans to do tasks in self-adaptive systems. In: Proceedings of SEAMS (2021)

24. Liu, B.: Learning on the job: online lifelong and continual learning. In: Proceedings of the AAAI Conference on Artificial Intelligence, vol. 34 (2020)

25. Moreno, G.A., et al.: Proactive self-adaptation under uncertainty: a probabilistic model checking approach. In: Proceedings of ESEC/FSE (2015)

26. Moreno, G.A., et al.: Uncertainty reduction in self-adaptive systems. In: Proceedings of SEAMS (2018)

27. Pan, S.J., Yang, Q.: A survey on transfer learning. IEEE TKDE **22**(10), 1345–1359 (2009)

28. Pinto, F., et al.: Automatic model monitoring for data streams. arXiv preprint. arXiv:1908.04240 (2019)

29. Quionero-Candela, J., et al.: Dataset Shift in Machine Learning. The MIT Press, Cambridge (2009)

30. Silver, D.L., Yang, Q., Li, L.: Lifelong machine learning systems: beyond learning algorithms. In: 2013 AAAI spring symposium series (2013)

31. Varshney, K.R., Alemzadeh, H.: On the safety of machine learning: cyber-physical systems, decision sciences, and data products. Big Data **5**(3), 246–255 (2017)

32. Wu, Y., Dobriban, E., Davidson, S.: DeltaGrad: rapid retraining of machine learning models. In: Proceedings of ICML (2020)

Integration of COTS Processing Architectures in Small Satellites for Onboard Computing Using Fault Injection Testing Methodology

Jose-Carlos Gamazo-Real$^{(\boxtimes)}$, Juan Rafael Zamorano-Flores ,
and Ángel Sanz-Andrés

University Institute of Microgravity "Ignacio Da Riva" (IDR), Universidad Politécnica de
Madrid, 28031 Madrid, Spain
{josecarlos.gamazo,juanrafael.zamorano,angel.sanz.andres}@upm.es

Abstract. Over the last decade, the interest of the space industry has increased towards smaller missions with reduced instruments. Advances in miniaturization technologies for electronics have increased the development of small spacecrafts, such as nanosatellites and microsatellites, from commercial-off-the-shelf (COTS) devices of low size, weight, and power. However, the effects of harsh space environment conditions on COTS electronic devices limit their use in high-performance multicore and heterogenous architectures, such as onboard edge computing based on Graphics Processing Units (GPU) for Artificial Intelligence (AI) applications. This article analyses the main considerations for adopting the Fault Injection Testing (FIT) methodology for software-intensive developments based on COTS, primarily intended to test embedded faults that emulate potential faults triggering to allow verification of fault detection, isolation, reconfiguration, and recovery capabilities in fault-tolerant and safety-critical systems. Two variants of techniques can be considered in the FIT methodology, based on hardware and software, but it is the software FIT variant (SFIT) that provides an inexpensive and time-efficient framework to replicate the fault triggering effects by injecting faults into code, data, and interfaces. A proposal is presented to apply the FIT methodology in the testing of future versions of the UPMSat-2 microsatellite payload at Spanish University Institute of Microgravity "Ignacio Da Riva" (IDR/UPM). The scope of the proposed methodology aims to facilitate the future validation of satellite real-time control systems, designed with model-driven engineering processes, and the future onboard integration of high-performance COTS GPUs to process sensor-fusion payload data and to implement auxiliary controllers with AI computing techniques.

Keywords: Fault Injection Testing (FIT) · Commercial-off-the-self (COTS) · Small satellite · Onboard computing · Graphics Processing Unit (GPU) · Fault-tolerant · Safety-critical · Real-time control system · Sensor-fusion payload data

A. Cerone et al. (Eds.): SEFM 2021 Workshops, LNCS 13230, pp. 333–347, 2022.
https://doi.org/10.1007/978-3-031-12429-7_24

1 Introduction

The satellite launch mass gradually dramatically increased in early 2000, such as Envisat Earth Observation satellite or Cassini planetary exploration missions, envisioned to reduce the cost per weight of payload launched and to increase synergistic measurements by incorporating multiple instruments in a single satellite. However, issues related to micro-vibrations, electromagnetic compatibility, and the different maturity levels of instruments could create significant engineering problems during the development. Furthermore, over the last decade the space industry has experienced an increased interest towards smaller missions with reduced instruments or single sensors, and to build these small spacecrafts from readily available, low cost, low power and compact commercial-off-the-shelf (COTS) components. The small satellites or nanosatellites, were initially envisioned as educational tools or low-cost technology demonstrators, such as the Cube-Sats [32, 37], with several studies to increase the onboard autonomy and data processing [7]. In the last ten years, more than eight hundred nanosatellites had been launched [33] and most of them based in the CubeSat standard [31].

In the development of CubeSats missions, the COTS technology has played a relevant role with respect to radiation-hardened (rad-hard) space qualified devices. Rad-hard devices often lag several generations behind their terrestrial counterparts with regards to computational resources. As the data volume and bottleneck are now increasing exponentially, more sophisticated processing algorithms will need to be implemented onboard, such as those based on Artificial Intelligence (AI), so the onboard processing resources are becoming a growing priority [13]. To cover those needs it is important to minimize the size, weight, and power ratios, so the devices for reconfigurable and heterogeneous computing has increased their relevancy, such as Field-programmable Gate Arrays (FPGA), general-purpose processors (GPP), and Graphics Processing Units (GPU) [5]. The increasing demands of onboard sensor and autonomous processing has focused the research on GPUs in space, which has added remarkable advances in AI processing. To develop more capable science instruments that capture larger volumes of data onboard [7], several worldwide actions have appeared such as European Space Agency (ESA) Φ-Lab for AI for Earth Observation [15, 16].

The short development life cycle and the low cost of small satellites have motivated the growing number of missions, but faster and cheaper space projects do not guarantee the success in orbit. The lack of good practices on design, assembly and tests has been considered as one of the major causes to nanosatellite mission failures, so the use of verification and validation techniques are required, such as the CubeSat standard [31]. A remarkable test method is the Fault Injection Testing (FIT) to perform verification and validation activities in a low or non-invasive manner. FIT was first employed in the 1970s to assess the dependability of fault-tolerant computers, but not until the mid-1980s academia began actively using fault injection to conduct experimental research [11]. This method considers the fault tolerant computing as the main need for space computers and checks the performance of fault tolerant strategies, such as the information redundancy given by error detection and correction coding (EDAC), or cyclic redundancy check. A real example is the fault emulator mechanism used in the NanoSatC-Br2 CubeSat project for robustness testing of interoperable software-intensive subsystems [4]. Some studies consider that FIT is an effective solution to the problem validating highly reliable computer systems [11].

With this approach, this article covers architectural and software aspects for the use of FIT in the integration of COTS devices in small satellites, with a special interest on GPUs for onboard AI processing. The presented work analyses the main effects of the space environments over COTS components as a limitation to their use, and drawn the availability of COTS GPUs for heterogenous computing in space projects despite their limitations. The description of FIT as testing methodology is presented and the proposal for its application in future versions of the UPMSat-2 microsatellite is described, where the control system of the current operational version was designed under a model-driven engineering perspective [29]. In addition, the article proposes a generic test system based on FIT for the development of space systems, where the manufacturers' proprietary tools that are specific to each device or system have typically been used.

The remainder of this article is structured as follows. In Sect. 2, the main effects of the space environment over electronics devices is explained, followed by the analysis of these effects over COTS devices and some examples of the specific behaviour of COTS GPUs in space. Then, in Sect. 3, a description of the typical software fault tolerant techniques in COTS is included, followed by the description of the FIT methodology with emphasis on the software variant. In Sect. 4, the UPMSat-2 project is described from an architectural and software perspective, and a proposal for application of the software FIT methodology in future versions of the UPMSat-2 is presented as a complement to the existing testing tools and as an attempt to reduce dependency on using manufacturers' proprietary tools in low-cost small space missions. Finally, the conclusions and future work are drawn in Sect. 5.

2   Motivation Scenario: COTS Processors for High-Performance Processing in the Space Environment

2.1   COTS Devices in Space

In traditional onboard data handling and processing systems, a qualified rad-hard onboard computer (OBC) is connected to redundant subsystems with a high integrity bus, such as MIL-STD-1553 or RS-422 [13], as in the UPMSat-2 studied in the Sect. 4. Rad-hard devices are designed to provide levels of immunity to the radiation effects, typically the total ionizing dose (TID) exceed 100 krad up to over 1 Mrad, single event upset (SEL) immunity above 75 MeVcm2/mg, and non-destructive single event effects (SEE) that typically appear once in 20 years [7]. Recent trends in the space industry require a more complex data handling, autonomous decisions, and massive signal processing, which opens the way to the COTS devices as competitors of conventional rad-hard ones, so currently ESA and United States are moving to accept COTS devices in space [30]. However, when non-rad-hard COTS devices are deployed in space can be severely harmed by harsh environmental conditions of extreme temperatures, high levels of ionizing radiation, etc. as they typically operate in a range of -30 to $+70$ °C and will suffer from TID and SEEs owing to cosmic particles. Space avionics are typically tolerant to 100–300 krad TID, but radiation tolerance is 10–50 krad in low Earth orbits (LEO) which can be tolerable by COTS for a short period of time [35] and COTS devices can be validated at lower cost with extensive radiation testing [19].

2.2 High-Performance Processing in Space with COTS GPUs

Trends in space missions show the generalization of computing requirements, so new generation flight computing systems must provide a heterogeneous architectural support by combining different technologies of processors, taking into account that a budget of 10 W is enough to deploy many-core DSPs, GPUs, or FPGAs to accelerate a conventional space-qualified CPU by one to three orders of magnitude [27, 30]. As hardware accelerators, the FPGAs normally provide an efficient data interface to payloads and GPUs implement a highly parallel Single-Instruction Multiple-Data architecture for onboard processing tasks [13, 14]. However, GPUs are not currently manufactured to provide radiation tolerance at the silicon gate level, which is an obstacle in their deployment for space systems. To overcome these problems, techniques of Radiation Hardening by Software Design could be implemented to ensure that radiation effects are mitigated [39, 41], such the NASA studies with Nvidia and AMD GPUs [7, 8].

3 Software-Based FIT Methodology

3.1 Description of the FIT Methodology

Scalable architectures and multicore processors have a range of capabilities for fault tolerance and recovery needed in radiation fields to support software-intensive onboard deployments. Typically, one of the most common methodologies for characterising the error resilience of a system is to perform beam testing, which uses a radiation source to radiate a physical system under test (SUT) and analyse the probability that a hardware gate-level error propagates to a software application. However, the radiation rates are difficult to control and experiments can be extremely costly. It limits the validation of small satellite missions in the so-called New Space industry [1], such as the projects of Airbus OneWeb [2], or SpaceX Starlink [40] to provide Internet access globally with a fleet of low-cost and high-performance satellites manufactured at high volumes.

Considering the increasing use of small satellite missions, such as nanosatellites or microsatellites, a faulty behaviour of New Space payload subsystems can be expected [4], as the satellite integration process includes a growing number of software functions. With this perspective, this article considers the FIT methodology as a valid approach for identification and comprehension of faulty events when the behaviour of a subsystem is characterised in the prototype, operation and integration phases [11]. FIT consists of testing the faults embedded in the system that emulate the triggering of potential faults to verify the system capabilities of fault detection, isolation, reconfiguration, and recovery [3], and it is mentioned by some safety standards, such as ISO 26262 and NASA-GB-8719.13. Although recent standards recommend fault injection, it is not yet clear how to use this approach effectively because those standards do not provide detailed information [12], even in safety standards such as DO-178B/C. This article proposes a realistic FIT implementation that offers some guidance.

A fault is a deviation in a hardware or software component from its intended function, and can arise during all stages in a system evolution. There are two sets of techniques in the FIT methodology that focused on hardware and interfaces (HFIT) or on software (SFIT). The HFIT variant related to the injection of faults in the interface are for robustness tests and represent a good option for space systems [12]. The SFIT variant provides an inexpensive and time-efficient framework for replicate the effects of physical hardware faults, such as bus and memory faults, by intentionally altering instructions in a controlled manner, or by changing the data at the application level (data corruption). The approaches normally considered for SFIT are code, data, and interface fault injection [4], as shown in Fig. 1. The SFIT becomes a feasible way to achieve better quality in small satellite missions, but it is limited to manipulating only visible states and software accessible blocks in the system architecture [14].

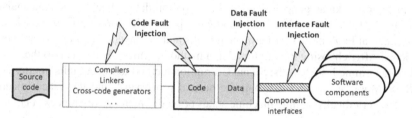

Fig. 1. Typical SFIT approaches.

3.2 Software Techniques for Fault Tolerance in COTS Devices

The COTS components could relatively replace rad-hard components by means of software techniques that mitigate the risks posed by the radiation environment, such as:

- **Generic Fault Tolerance.** They include EDAC and redundancy-based mechanisms for both memory and computational blocks [14]. From redundancy-based techniques, DMR or TMR can be used [38].
- **Algorithm-based Fault Tolerance.** These techniques consider the encoding of the data and changes in the algorithm to provide a greater error resilience, but they are not applicable for all algorithms, which promotes the use of the Generic ones [14].
- **Reboot.** A technique to mitigate the effects of SEE is to perform frequent reboots, but the frequency varies with the technology from weeks to days [7].
- **Middleware.** It is software layer between the application and the operating system to provide resource, fault, and power management [7]. It is an option used to overcome the reboot limitation, such as the Troxel Aerospace middleware [42].

3.3 Deployment of SFIT for Fault-Tolerant and Safety-Critical Systems

The SFIT techniques are attractive because they do not require expensive hardware and the realism of fault injection is an important condition for reproducing the faulty behaviours. Studies observed the same trend in the distribution of faults where algorithm faults dominate, assignment and checking faults have approximately the same weight, and interface and function faults are less frequent [12]. The objective of a FIT methodology is not only the estimation of coverage and latency parameters for analytical dependability models, but also the evaluation of other measures such as reliability, availability, and mean time between failures that are the basic characterisation of fault-tolerant and safety-critical systems.

Software faults are caused by the incorrect specification, design, or coding of a program. Although software does not physically break after being installed in a computer system, latent faults or bugs in the code can surface during operation especially under unusual workloads and extreme environments such as space. For this reason, SFIT is used primarily for testing software that implements fault-tolerance mechanisms. Figure 2 shows a generic FIT environment, which typically consists of the target system plus a fault injector, workload generator, controller, monitor, data collector, and data analyser [21]. The deployment of a FIT methodology requires selection of a fault model, such as the Stuck-at Fault model for permanent faults and the Inversion model for studies of transients. After choosing the fault model, a method to inject the faults into the system shall be determined, such as the Signal Corruption method to inject faults in interfaces, the State Mutation method that halt the normal processing, and the Trace Injection method to periodically sample machine states or record memory references [11].

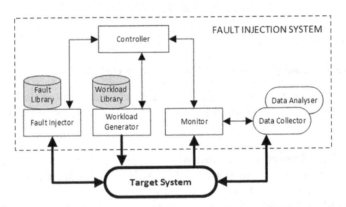

Fig. 2. Generic architecture of a FIT environment [21].

The development of fault-injection tools for SFIT accelerates the injection and measurement processes, enabling accurate results to be obtained in the shortest time using a formalised methodology in conducting automated experiments on a variety of systems. Table 1 shows examples of tools for SFIT automation developed in different institutions and companies, mainly universities due to the scientific and innovative nature of this methodology.

Table 1. Fault-injection tools developed for SFIT automation [11, 12, 21].

Tool	Description	Developer	Characteristics
FERRARI	Fault and Error Automatic Real-Time Injection	University of Texas (USA)	Software traps to inject CPU, memory, and bus faults
FTAPE	Fault Tolerance and Performance Evaluator	University of Illinois (USA)	Fault injection as bit-flips into registers in CPU modules and memory locations
DOCTOR	Integrated Software Fault Injection Environment	University of Michigan (USA)	Fault injection on CPU, memory, and network with timeouts, traps, and code
XCEPTION	Enhanced Automated Fault-Injection Environment	University of Coimbra (Portugal)	Inject faults using debugging and performance monitoring in processors
FIAT	Fault-Injection-Based Automated Testing Environment	Carnegie Mellon University (USA)	Set and clear bytes in the memory images of programs
DEPEND	Dependability and performability evaluation	University of Illinois (USA)	Fault-tolerant analysis with permanent, transient, and workload fault injection
REACT	Reliable Architecture Characterization Tool	Univ. Massachusetts-Texas (USA)	Abstracts multiprocessor architectural level. Life testing with fault injection
SAFE	SoftwAre Fault Emulator	University of Naples (Italy)	Injection of software faults in C/C++ programs for dependability assessment

4 Case Study: UPMSat-2 Microsatellite

4.1 System Overview

UPMSat-2 is a micro-satellite of an approximate mass of 50 kg and an external dimension of $0.5 \times 0.5 \times 0.6$ m. The system architecture of the spacecraft includes a magnetic attitude control system, voltage, intensity and temperature sensors for monitoring platform, and experiments and other devices to properly operate the satellite. The UPMSat-2 is conceived as a technology demonstrator with a payload of experiments from research groups and industry. There is an OBC that carries out the data handling and control functions of the satellite and the experiments, which is based on a Gaisler LEON3 processor and includes peripherals such as EEPROM and analog/digital ports.

The main functions to be carried out by the on-board computer are summarised in its software control architecture of Fig. 3. It is remarkable the telemetry and telecommand (TMTC) to maintains the communications with the ground station by means of the telecommunications hardware (TMC) and the attitude determination and control system (ADCS) to keep the proper orientation of the satellite with respect to the Earth.

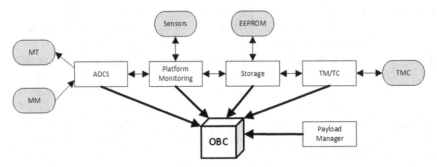

Fig. 3. UPMSat-2 software architecture.

4.2 Model-Based Development Process

The onboard software has been developed using a model-based systems engineering (MBSE) process consisting of a series of models that are progressively refined until the implementation code can be generated, as shown in Fig. 4. To support the development process, the TASTE toolset [36] was used. The design includes two main models, such as the Platform Independent Model (PIM) and the Platform Specific Model (PSM). PIM represents the intended behaviour of the system without considering the platform-specific details and includes several model views, such as data view based on ASN.1 [24], functional view based on SDL [25], and interface view that use the language AADL [17]. From PIM, the PSM considers the characteristics of the OBC with the deployment and concurrency views. The code was generated in Ada 2005 with Ravenscar profile for tasking [9] and a runtime based on GNAT/ORK+ kernel for LEON3 [28] was used.

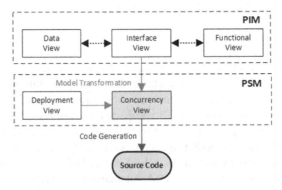

Fig. 4. UPMSat-2 model-based software development process

4.3 Validation Approach

The validation of the software and hardware was carried out using a V-model. Because a model-based development process was used, it allowed for model validation through simulation. Regarding the embedded systems, as the real space environment was not available, a Software Validation Facility (SVF) based on Hardware-In-the-Loop (HIL) allowed application software execution with a simulated environment. The ADCS software is executed on the embedded computer, while the system sensors, actuators, and the spacecraft dynamics are simulated on a development computer. The final SVF included systems and experiments from different companies, such as a SSBV reaction wheel, a Bartington magnetometer, or an IDR/UPM thermal controller. The Fig. 5 shows the general view of the platform and the SVF setup with the satellite electronics box (EBOX) connected to different systems by using a breakout board. Several tools were used to analyse response times, such as RapiTime [6] and MAST [18].

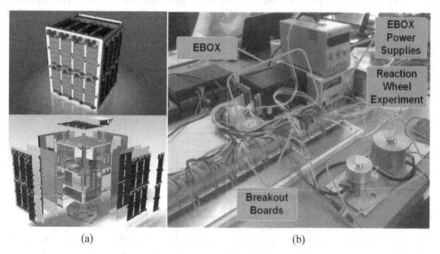

(a)   (b)

Fig. 5. a) UPMSat-2 platform. b) SVF setup with EBOX connected to other systems.

4.4 Application of SFIT: Integration of COTS Processing Devices

Proposal of Test System. The proposed SFIT methodology for future versions of the UPMSat-2 is under consideration to be implemented, which could serve as a complement to the existing testing tools and as an attempt to reduce dependency on using manufacturers' proprietary tools for low-cost small space missions in the New Space era. The interest of SFIT is based on an inexpensive, minimally invasive and time-efficient framework for replicate system faults by only influencing over application or interface levels. The application of the methodology in the integration of COTS devices for AI processing has been analysed by some authors, such as Iorgulescu [23] and Harrison [20], to comply with standards such as DO-178B/C, ECSS-E-ST-40C, or from the International Council of Systems Engineering (INCOSE) [26].

The proposed test system shown in Fig. 6 represents a SFIT environment consisting of a fault injector, a data collector, and a controller, which is based on the generic architecture explained above in the Fig. 2 and that could also be include some low-invasive features of other existing fault-injection tools listed in Table 1. A middleware layer (hypervisor) provides independency from the hardware and enable the use of a specific OS for (quasi) real-time and deterministic operations. The fault injector includes fault models that emulates system failures, a library with the components associated with each fault model, and script-based sequencer to dynamically program the behaviour of the faults injected depending on the functionality to be tested. According to the test procedure implemented in a script, the test system will inject or not faults on the messages exchanged with the SUT through its communication interface ports that can be a software-intense subsystem such as a satellite processing unit or a payload controller. These faults emulate controlled failures and errors in the SUT, avoiding application program modifications to inject faults in compile-time of the embedded code for real-time validation, thus improving the development and validation cycle. This approach can be compatible with MBSE and embedded testing standards such as ISO/IEC 9646.

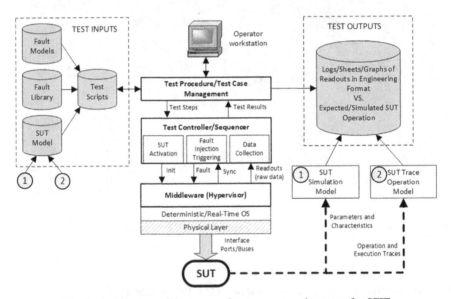

Fig. 6. Architecture of the proposed test system environment for SFIT.

In the proposed FIT-based test system, the cost of a fault injection campaign is influenced by the times to set up the environment, to run experiments, and to analyse the experimental data. The faults will be injected at runtime instead of the typical compile-time techniques in which the program is previously modified and then executed. To trigger a fault injection, the program in the SUT will have little or no modifications by means of timeouts and exceptions/traps that needs to be linked to the SUT interrupt handler vector, or applying the State Mutation and Trace Injection methods. The steps of the process from a fault injection is shown in Fig. 7, such as a stuck that affects a memory bit. A timer of the test system will be linked with the SUT interrupt handler vector, so a timeout will generate an interrupt to trigger the fault injection, providing emulation of transient faults and instantaneous hardware faults in the SUT. Unlike timeouts, the test system can also implement exception and traps to inject faults whenever events or conditions occurs. The State Mutation method can use SUT manufacturer frameworks and the Trace Injection method can register states and memory references.

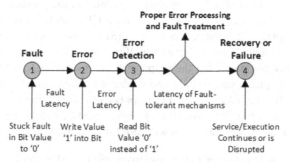

Fig. 7. Fault-error-failure process in a fault-tolerant system.

Various levels of abstraction may be identified for FIT implementation [10], such as the axiomatic models to provide a means to account for the behaviour of fault-tolerance mechanisms [44], which can be viewed as a shield intended to prevent errors. For experimental fault forecasting, the readouts R in an experiment can be used to determine the state S of the SUT, which can be defined in terms of the predicates P (fault activated, error signalled). A statistical characterisation of a test sequence as Bernouilli trials [3]:

$$P\{N(t) = k; n\} = \binom{n}{k}[C(t)]^k[1 - C(t)]^{n-k} \tag{1}$$

where n is the number of independent fault injection experiments, $N(t)$ the total number of assertions of P (coverages) observed in a time interval $[0, t]$ that follows a binomial distribution.

Accordingly, the mathematical expectation of the number of coverages is written as:

$$E[N(t)] = nC(t) \tag{2}$$

A rigorous approximation of the coverage requires that the effect of a second fault occurring during the processing of the first one be accounted in the analysis as:

$$C' \approx C(1 - \lambda' E[T_P']) \tag{3}$$

where C is the asymptotic coverage factor, λ' is the second fault occurrence rate, and T_P' is the random variable characterizing the coverage finite time.

The results of an experimental test sequence can be used for coverage estimation as:

$$\hat{C}(T) = \frac{N(T)}{n} \tag{4}$$

Although the proposed fault injection approaches are intended to be minimally invasive, they have some limitations, such as to inject faults into locations inaccessible to software, to measure latencies, and to appear low perturbations over SUT workload.

Onboard Data Processing based on COTS Devices. As commented, GPUs are currently one of the most promising devices for processing of AI algorithms, such as artificial neural networks (ANN). The use of COTS GPUs in low-cost space projects represents a great innovation instead of using rad-hard hardware, but the proposed FIT-based test system should consider the effects of a SEE on a GPU application, which can be a functional interrupt (FI), a silent data corruption (SDC), and a masked event. A FI is an error that causes an application to hang or malfunction, a SDC occurs when the application successfully completes but the output data is incorrect, and a in masked event the application completes successfully and no data errors are found in the output [14]. The test system shall be aimed to test FI errors because their mitigation is particularly important due to they often have strict timing requirements, and SDC errors because any data corruption in the processing chain will result in a permanent loss of data.

Relevant aspects to consider in the implementation of AI algorithms is to test their biased behaviour to reproduce and reinforce the bias that is present in the training data, and the verification of the dependency fairness in classification tasks as a special form of robustness. A neural network is fair if the output classification is not affected by different values of the chosen features, and in its verification several techniques can be used, such as generate discriminatory inputs, adaptive concentration inequalities for scalable sampling, or to repair the bias by introducing fairness properties in the code for runtime checking [43]. In several cases, such as deep neural networks, research has shown that they are not robust to small perturbations of their inputs and can even be easily fooled, so subtle imperceptible perturbations of inputs can change their prediction results [34] so their formal verification has mainly focused on safety properties [22]. Using the proposed test system, the testing of the ANNs implemented on a GPU could be performed by corrupting the input training data according to the fault model stored in the test system library, and calculating metrics such as F-score or the accuracy with the precision and the recall to assess the estimation. ANN training data could be received from satellite sensors, such as thermals or magnetometer, or controllers, such as the platform attitude control and objects detectors like asteroids [32]. The proper testing of the ANN model can provide a cost/time efficient implementation regarding a conventional controller, which could highlight the test system for the New Space era.

5 Conclusion and Future Work

In this article, a test system environment is proposed for FIT based on software, called SFIT, in order to integrate COTS devices in low-cost space projects, such as small scientific satellites. The proposed test system is composed of a test procedure/cases manager, a test controller, and a middleware to have independency from the physical layer and to obtain deterministic (quasi) real-time operations for communication with the SUT. The test system inputs are test sequences based on test scripts that can be developed using fault models, fault libraries, and a specific SUT model. The obtained outputs provide readouts from SUT that can be compared with the expected SUT operation according the initial test procedure. The proposed test system shows the advantage that it can inject faults on the messages exchanged with SUT through its communication interface ports in order to emulate controlled failures and errors in the SUT in real-time, avoiding modifications of application programs to inject faults in compile-time, thus improving the time of development and validation cycle. The real case study of the UPMSat-2 scientific microsatellite has been analysed, exposing the system overview, the model-based development process, and the validation approach using a HIL-based software facility.

The future aspects of this research work are related to apply the proposed SFIT environment to the development of new onboard applications for future versions of UPMSat-2 according to the time and cost needs of New Space trends, such as AI models built into COTS GPUs. Some applications can be processing of sensor-fusion data, such as magnetometers and reaction wheels, and the synthesis of auxiliary controllers, such as attitude control, by properly training machine learning models with payload data.

Acknowledgements. This work was supported by the Comunidad de Madrid under Convenio Plurianual with the Universidad Politécnica de Madrid in the actuation line of Programa de Excelencia para el Profesorado Universitario. We thank the support of the institute IDR and the department "Sistemas Informáticos" of ETSISI/UPM, Spain.

References

1. Airbus: New Space. https://www.airbus.com/public-affairs/brussels/our-topics/space/new-space.html. Accessed 06 Oct 2021
2. Airbus Oneweb Satellites: Building an Accessible Space for All: Revolutionizing the Economics of Space. https://airbusonewebsatellites.com/. Accessed 06 Oct 2021
3. Arlat, J., et al.: Fault injection for dependability validation: a methodology and some applications. IEEE Trans. Softw. Eng. **16**(2), 166–182 (1990). https://doi.org/10.1109/32.44380
4. Batista, C.L.G., et al.: Towards increasing nanosatellite subsystem robustness. Acta Astronaut. **156**, 187–196 (2019). https://doi.org/10.1016/J.ACTAASTRO.2018.11.011
5. Behrens, J.R., Lal, B.: Exploring trends in the global small satellite ecosystem. New Space **7**(3), 126–136 (2019). https://doi.org/10.1089/SPACE.2018.0017
6. Bernat, G., et al.: Probabilistic timing analysis: An approach using copulas. J. Embed. Comput. **1**(2), 179–194 (2005)
7. Bruhn, F.C., Tsog, N., Kunkel, F., Flordal, O., Troxel, I.: Enabling radiation tolerant heterogeneous GPU-based onboard data processing in space. CEAS Space Journal **12**(4), 551–564 (2020). https://doi.org/10.1007/s12567-020-00321-9
8. Buonaiuto, N., et al.: Satellite identification imaging for small satellites using NVIDIA. In: Small Satellite Conference (2017)
9. Burns, A., et al.: Guide for the use of the Ada ravenscar profile in high integrity systems. ACM SIGAda Ada Lett. **40**(2), 110–127 (2004). https://doi.org/10.1145/997119.997120
10. Carter, W.C., Abraham, J.: Design and evaluation tools for fault tolerant systems. In: AIAA Computers in Aerospace Conference, pp. 70–77 (1987)
11. Clark, J.A., Pradhan, D.K.: Fault injection a method for validating computer-system dependability. Comput. Long. Beach. Calif. **28**(6), 47–56 (1995). https://doi.org/10.1109/2.386985

12. Cotroneo, D., Natella, R.: Fault injection for software certification. IEEE Secur. Priv. **11**(4), 38–45 (2013). https://doi.org/10.1109/MSP.2013.54

13. Davidson, R.L., Bridges, C.P.: Adaptive multispectral GPU accelerated architecture for earth observation satellites. In: IST 2016 IEEE International Conference on Imaging Systems and Techniques, pp. 117–122. Institute of Electrical and Electronics Engineers Inc. (2016). https://doi.org/10.1109/IST.2016.7738208

14. Davidson, R.L., Bridges, C.P.: Error resilient gpu accelerated image processing for space applications. IEEE Trans. Parallel Distrib. Syst. **29**(9), 1990–2003 (2018). https://doi.org/10.1109/TPDS.2018.2812853

15. European Space Agency: ESA AI4EO Portal. AI4EO projects from ESA Φ-Lab, https://ai4eo.esa.int/. Accessed 22 Oct 2021

16. European Space Agency: GSTP Element 1 "Develop" compendium (2019)

17. Feiler, P., Gluch, D.: Model-Based Engineering with AADL: An Introduction to the SAE Architecture Analysis & Design Language. Addison-Wesley, Boston (2012)

18. González Harbour, M., et al.: MAST: modeling and analysis suite for real time applications. In: Proceedings - Euromicro Conference on Real-Time Systems, pp. 125–134 (2001). https://doi.org/10.1109/EMRTS.2001.934015

19. Haddad, N.F., et al.: Second generation (200MHz) RAD750 microprocessor radiation evaluation. In: Proceedings of the European Conference on Radiation and Its Effects on Components and Systems, RADECS, pp. 877–880 (2011). https://doi.org/10.1109/RADECS.2011.6131320

20. Harrison, L.H., et al.: Artificial intelligence and expert systems for avionics. In: Proceedings of the IEEE/AIAA 12th Digital Avionics Systems Conference, pp. 167–172. IEEE (1993). https://doi.org/10.1109/DASC.1993.283552

21. Hsueh, M.C., et al.: Fault injection techniques and tools. Comput. Long. Beach. Calif. **30**(4), 75–82 (1997). https://doi.org/10.1109/2.585157

22. Huang, X., Kwiatkowska, M., Wang, S., Wu, M.: Safety verification of deep neural networks. In: Majumdar, R., Kunčak, V. (eds.) CAV 2017. LNCS, vol. 10426, pp. 3–29. Springer, Cham (2017). https://doi.org/10.1007/978-3-319-63387-9_1

23. Iorgulescu, D.T., et al.: Artificial expertise in systems engineering (aerospace systems). Presented at the December 9 (2002). https://doi.org/10.1109/DASC.1991.177200

24. ITU: X.683 : Information technology - Abstract Syntax Notation One (ASN.1): Parameterization of ASN.1 specifications. https://www.itu.int/rec/T-REC-X.683. Accessed 22 Oct 2021

25. ITU: Z.100 : Lenguaje de especificación y descripción - Visión general de SDL-2010. https://www.itu.int/rec/T-REC-Z.100/es. Accessed 22 Oct 2021

26. Kaslow, D., et al.: A model-based systems engineering (MBSE) approach for defining the behaviors of CubeSats. IEEE Aerosp. Conf. Proc. (2017). https://doi.org/10.1109/AERO.2017.7943865

27. Kosmidis, L., et al.: GPU4S: embedded GPUs in space. In: Proceedings - Euromicro Conference on Digital System Design, DSD 2019, pp. 399–405. Institute of Electrical and Electronics Engineers Inc. (2019). https://doi.org/10.1109/DSD.2019.00064

28. de la Puente, J.A., et al.: The ASSERT virtual machine: a predictable platform for real-time systems. In: IFAC Proceedings, pp. 10680–10685. Elsevier (2008). https://doi.org/10.3182/20080706-5-KR-1001.01810

29. De La Puente, J.A., et al.: Model-driven design of real-time software for an experimental satellite. In: IFAC Proceedings, pp. 1592–1598. IFAC Secretariat (2014). https://doi.org/10.3182/20140824-6-ZA-1003.01967

30. Lentaris, G., et al.: High-performance embedded computing in space: evaluation of platforms for vision-based navigation. J. Aerosp. Inf. Syst. **15**(4), 178–192 (2018). https://doi.org/10.2514/1.I010555

31. Mehrparvar, A., et al.: Cubesat Design Specification Rev 13, the CubeSat Program. Cal Poly San Luis Obispo, US
32. Moore, C.: Technology development for NASA's asteroid redirect mission. In: International Astronautical Federal IAC-14,D2,8-A5.4,1,x22696 (2014)
33. Nanosats Database: World's largest database of nanosatellites, over 3200 nanosats and CubeSats. https://www.nanosats.eu/. Accessed 30 Sep 2021
34. Nguyen, A., et al.: Deep neural networks are easily fooled: high confidence predictions for unrecognizable images. In: IEEE Computer Society Conference on Computer Vision and Pattern Recognition, pp. 427–436. IEEE Computer Society (2015). https://doi.org/10.1109/CVPR.2015.7298640
35. Ottavi, M., Gizopoulos, D., Pontarelli, S. (eds.): Dependable Multicore Architectures at Nanoscale. Springer, Cham (2018). https://doi.org/10.1007/978-3-319-54422-9
36. Perrotin, M., Conquet, E., Delange, J., Schiele, A., Tsiodras, T.: TASTE: a real-time software engineering tool-chain overview, status, and future. In: Ober, I., Ober, I. (eds.) SDL 2011. LNCS, vol. 7083, pp. 26–37. Springer, Heidelberg (2011). https://doi.org/10.1007/978-3-642-25264-8_4
37. Poghosyan, A., Golkar, A.: CubeSat evolution: analyzing cubeSat capabilities for conducting science missions. Prog. Aerosp. Sci. **88**, 59–83 (2017). https://doi.org/10.1016/J.PAEROSCI.2016.11.002
38. Samudrala, P.K., et al.: Selective triple modular redundancy (STMR) based single-event upset (SEU) tolerant synthesis for FPGAs. IEEE Trans. Nucl. Sci. **51**(5), 2957–2969 (2004). https://doi.org/10.1109/TNS.2004.834955
39. Schmidt, A.G., et al.: Applying radiation hardening by software to fast lossless compression prediction on FPGAs. In: IEEE Aerospace Conference Proceedings (2012). https://doi.org/10.1109/AERO.2012.6187254
40. SpaceX: Astronomy Discussion with National Academy of Sciences. https://www.spacex.com/updates/starlink-update-04-28-2020/index.html. Accessed 06 Oct 2021
41. Takizawa, H., et al.: CheCUDA: a checkpoint/restart tool for CUDA applications. In: Parallel and Distributed Computing, Applications and Technologies, PDCAT Proceedings, pp. 408–413 (2009). https://doi.org/10.1109/PDCAT.2009.78
42. Troxel Aerospace: SEU Mitigation Middleware (SMM). https://troxelaerospace.com/products/software-solutions/. Accessed 06 Oct 2021
43. Urban, C., et al.: Perfectly parallel fairness certification of neural networks. In: ACM on Programming Languages. Association for Computing Machinery (2019)
44. Bouricius, W.G., Carter, W.C., Schneider, P.R.: Reliability modeling techniques for self-repairing computer systems. In: 24th ACM National Conference, pp. 295–309 (1969)

Author Index

Printed in the United States
by Baker & Taylor Publisher Services